The Animal Kingdom

VOLUME 2:
THE CLASS MAMMALIA 2

GEORGES CUVIER
EDITED AND TRANSLATED BY
EDWARD GRIFFITH

CAMBRIDGE
UNIVERSITY PRESS

CAMBRIDGE UNIVERSITY PRESS

Cambridge, New York, Melbourne, Madrid, Cape Town,
Singapore, São Paolo, Delhi, Mexico City

Published in the United States of America by Cambridge University Press, New York

www.cambridge.org
Information on this title: www.cambridge.org/9781108049559

© in this compilation Cambridge University Press 2012

This edition first published 1827
This digitally printed version 2012

ISBN 978-1-108-04955-9 Paperback

This book reproduces the text of the original edition. The content and language reflect
the beliefs, practices and terminology of their time, and have not been updated.

Cambridge University Press wishes to make clear that the book, unless originally published
by Cambridge, is not being republished by, in association or collaboration with, or
with the endorsement or approval of, the original publisher or its successors in title.

CAMBRIDGE LIBRARY COLLECTION

Books of enduring scholarly value

Life Sciences

Until the nineteenth century, the various subjects now known as the life sciences were regarded either as arcane studies which had little impact on ordinary daily life, or as a genteel hobby for the leisured classes. The increasing academic rigour and systematisation brought to the study of botany, zoology and other disciplines, and their adoption in university curricula, are reflected in the books reissued in this series.

The Animal Kingdom

Georges Cuvier (1769–1832), made a peer of France in 1819 in recognition of his work, was perhaps the most important European scientist of his day. His most famous work, Le Règne Animal, was published in French in 1817; Edward Griffith (1790–1858), a solicitor and amateur naturalist, embarked in 1824, with a team of colleagues, on an English version which resulted in this illustrated sixteen-volume edition with additional material, published between 1827 and 1835. Cuvier was the first biologist to compare the anatomy of fossil animals with living species, and he named the now familiar 'mastodon' and 'megatherium'. However, his studies convinced him that the evolutionary theories of Lamarck and St Hilaire were wrong, and his influence on the scientific world was such that the possibility of evolution was widely discounted by many scholars both before and after Darwin. Volume 2 is the second of four books on mammals.

Cambridge University Press has long been a pioneer in the reissuing of out-of-print titles from its own backlist, producing digital reprints of books that are still sought after by scholars and students but could not be reprinted economically using traditional technology. The Cambridge Library Collection extends this activity to a wider range of books which are still of importance to researchers and professionals, either for the source material they contain, or as landmarks in the history of their academic discipline.

Drawing from the world-renowned collections in the Cambridge University Library and other partner libraries, and guided by the advice of experts in each subject area, Cambridge University Press is using state-of-the-art scanning machines in its own Printing House to capture the content of each book selected for inclusion. The files are processed to give a consistently clear, crisp image, and the books finished to the high quality standard for which the Press is recognised around the world. The latest print-on-demand technology ensures that the books will remain available indefinitely, and that orders for single or multiple copies can quickly be supplied.

The Cambridge Library Collection brings back to life books of enduring scholarly value (including out-of-copyright works originally issued by other publishers) across a wide range of disciplines in the humanities and social sciences and in science and technology.

THE

ANIMAL KINGDOM

ARRANGED IN CONFORMITY WITH ITS
ORGANIZATION,

BY THE BARON CUVIER,

MEMBER OF THE INSTITUTE OF FRANCE, &c. &c. &c.

WITH

ADDITIONAL DESCRIPTIONS

OF

ALL THE SPECIES HITHERTO NAMED, AND OF
MANY NOT BEFORE NOTICED,

BY

EDWARD GRIFFITH, F.L.S., A.S., &c.

AND OTHERS.

VOLUME THE SECOND.

LONDON:

PRINTED FOR GEO. B. WHITTAKER,

AVE-MARIA-LANE.

MDCCCXXVII.

THE

CLASS MAMMALIA

ARRANGED BY THE

BARON CUVIER,

WITH

SPECIFIC DESCRIPTIONS

BY

EDWARD GRIFFITH, F.L.S., A.S., &c.

MAJOR CHARLES HAMILTON SMITH, F.R.S., L.S., &c.

AND

EDWARD PIDGEON, Esq.

_____

VOLUME THE SECOND.

_____

LONDON:

PRINTED FOR GEO. B. WHITTAKER,

AVE-MARIA-LANE.

MDCCCXXVII.

LONDON:
Printed by WILLIAM CLOWES,
Charing Cross.

THE

THIRD ORDER

OF THE

MAMMALIA.

THE CARNASSIERS

These form a considerable and very various assemblage of unguiculated quadrupeds, which, like man and the quadrumana, possess the three kinds of teeth. They all subsist on animal matter, and more exclusively so in proportion as their cheekteeth are more of a trenchant character. Such of them as have these teeth altogether or in part tuberculous, also take more or less of vegetable substances, while those who have them bristling with conical points, principally subsist on insects. The articulation of their lower jaw, directed cross-wise, and clasping like a hinge, does not allow of any horizontal motion ; it can only shut and open

Their brain, although tolerably furrowed, has no third lobe, nor does it form a second covering for the cerebellum in these animals any more than in the families which succeed them. The orbit is not separated from the fossa temporalis in their skeleton. Their cranium is narrow, and the zygomatic arcades are dispersed, and elevated, to give more volume and force to the muscles of their jaws. The predominant sense with these animals is that of smelling, and the pituitary membrane is usually extended over very numerous bony laminæ. Their fore-arm can turn, but with less facility than in the Quadrumana, and they have no thumbs on the fore-feet opposable to the other fingers. Their intestines are less voluminous, in consequence of the substantial nature of their aliments, and to avoid the putrefaction that flesh must undergo if it remained for any time in an elongated canal.

In other particulars their forms and the details of their organization vary considerably, and produce analogous varieties in their habits to such a degree as to render it impossible to range their genera upon a single scale. We are therefore obliged to divide them into several families, which are variously connected together by very numerous relations.

THE FIRST FAMILY OF THE CARNASSIERS.

The CHEIROPTERA

Have still some affinity with the Quadrumana by the pendulous penis, and the mammæ situated on the breast. Their distinctive character consists in a fold of skin extended between their fore-feet and their fingers, which sustains them in the air, and even enables them to fly, when the hands are sufficiently developed for that purpose. This arrange_ment demanded powerful clavicles and large shoulder-blades to give the shoulder the requisite degree of solidity; but it was incompatible with the rotation of the fore-arm, which would have weakened the force of the impetus necessary for flight. All of these animals have four large canine teeth, but the number of the incisors vary. They have long been divided into two genera according to the extent of their organs of flying; but the first of these two requires many subdivisions.

The BATS (*Vespertilio, Linn.*)

Have the arms, the fore-arms, and the fingers extremely elongated, and those with the membrane which fills up the intervals between them, form real wings, as much extended as those of birds. Accordingly the bats fly to a considerable height and with great rapidity. Their pectoral muscles possess a thickness proportioned to the movements which they are designed to execute, and in the middle of the sternum is a ridge like that of birds, to form a point of attachment for these muscles. The thumb is short and armed with a hooked nail, which these animals make use of to hang by and to creep. Their hind-feet are weak, divided into five toes of equal length and all of them are armed with nails. There is no cæcum to their intestines. Their eyes are extremely little, but their ears are often remarkably large; and, together with their wings, form an enormous extent of membranous surface. This is almost naked, and so sensible, that the bats can direct themselves into all the nooks of the labyrinth in which they nestle, even after their eyes have been taken out, probably by the diversity of impulsions from the external air. They are nocturnal animals, and in our climates pass the winter in a lethargic state. During the day they remain suspended in obscure retreats. They generally have two little ones at a birth, which they hold clinging to their breasts, and the size of which is very considerable in proportion to that of their mother.

This genus is very numerous, and contains several subdivisions.

At first they must be divided into

The Roussettes, *(Pteropus, Briss.)*

Which have sharp incisors in each jaw, and cheek-teeth with flat crowns, or more properly with two longitudinal and parallel projections, which are separated by a furrow, and wear away in course of time by attrition. From this conformation, as it might naturally be presupposed, they subsist principally upon fruits. They are, however, sufficiently dexterous in the pursuit of birds and the smaller quadrupeds. These are the largest of the bat-kind, and their flesh is used for food. Their habitat is the East Indies.

The membrane in this subdivision is sloped to a considerable depth between the legs. There is scarcely any tail. The fore-finger, about one half shorter than the middle, has a third phalanx and a little nail which is wanting in the other bats: but the other fingers have only two phalanges. **The** nose is simple, the ears small, and without parotides, and the tongue furnished with prickles, bent backwards. The stomach is a sack considerably elongated and unequally inflated.

1. ROUSSETTES *without tails, with four incisors in each jaw.*

The Black Roussette, (Pterop. edulis, Geoff.)

Of a brownish black, deeper in the under parts. Near four feet from the extremity of one wing to that of the other. An inhabitant of the Sunda and Molucca islands, where these animals conceal themselves in caverns. The flesh is remarkably delicate.

The Roussette of Edwards, (Pter. Edwardsii, Geoff.)

Fawn-colour, with the back of a deep brown. Habitat, Madagascar.

The Roussette of Buffon, (Pterop. vulgaris, Geoff.)
Buff. X. xiv.

Brown, the face and sides of the back fawn-coloured. From the Isles of France and Bourbon, where they inhabit the trees in the forests.

The Collared Roussette. Rougette of Buffon. (Pter. rubri-collis, Geoff.) Buff. X. xvii.

Grayish brown, the neck red. From the same isles, where it lives in the hollow trees.

ROUSSETTES *with a small tail: four incisors in each jaw.*

These comprehend all the species described for the first time by M. Geoffroy. One of them, woolly and grey, *(Pter. Ægyptiacus,)* lives in Egypt in the catacombs, vaults, &c. Another of a reddish hue, with a tail somewhat longer, and to a certain extent involved in the membrane, *(Pter. Amplexicandus,)* Geoff. Ann. Mus. t. XV. pl. IV., comes from the Archipelago of the East Indies, &c. To these may be added, the *Pteropus Griseus,* Geoff. Ann. Mus. tom. XV. pl. VI.; *Pteropus Stramineus,* Seb. I. LVII. 1, 2,; *Pteropus Marginatus,* Geoff. loc. cit. pl. v.; *Pteropus Minimus, id.*

3. Following the relations pointed out by M. Geoffroy, we shall moreover separate from the Roussettes the Cephalotes, which have cheek-teeth of the same character, but in which the index or fore-finger, though short and furnished like the preceding species with three phalanges, is, however, without a nail. The membranes of their wings, instead of being joined to the flanks, are both of them united together on the middle of the back, to which they adheie through the medium of a vertical and longitudinal partition. Very frequently they have but two incisors.

The CEPHALOTES *of Peron, (Cephalotes Peronii, Geoff.) Geoff. Ann. Mus. XV. pl.* IV.

Brown or red. Habitat, Timor.
When the Roussettes have been thus detached,

there will remain the genuine bats, which are all insectivorous, and all of which have cheek-teeth furnished with conical points. The index is never provided with a nail; and, with the exception of a single sub-genus, the membrane always extends between the two legs.

They must be divided into two principal tribes. The first has upon the middle finger of the wing three ossified phalanges, but the other fingers, and the index itself has but two.

To this tribe, which is altogether foreign, belong three sub-genera.

The MOLOSSI, (MOLOSSUS, *Geoff. Dysopes, Illiger.*)

Have a simple muzzle; ears large and short, originating from the angle of the lips, and uniting one to the other upon the muzzle; the parotis short, and not enveloped by the conch. We reckon but two incisors in each jaw. Their tail takes up the whole length of the interfemoral membrane, and often extends beyond it. All the species come from America, and are more or less brown. They were confounded by Gmelin under the common name of Vespertilio Molossus, but M. Geoffroy has already distinguished nine species, of which Buffon has only three, *viz.*, the *Molossus longicaudatus*, the *Molossus fusciventer*, and the *Molossus Guyanensis*. The description of the others will be found in the Ann. du Mus. VI. 150.

The NYCTINOMES *(Geoff.)*

Have four incisors below; the upper lip is high, and

considerably sloped. In other particulars, they resemble the Molossi. To these belong the *Nyctinome
of Egypt*, Geoff. Eg. mammif. 2, 2 ; *Vespertilio aceta-
bulosus*, Herm. Obs. Zool. p. 19 ; *Vespertilio plicatus*,
Buchannan.

The STENODERMES, *(Geoff.)*

The muzzle is simple, the interfemoral membrane
sloped as far as the coccyx. The tail is wanting,
and there are two incisors above and four below.

The NOCTILIONS, *(Noctilio, Linn. Ed. XII.)*

With a short muzzle, inflated, divided, and marked
by warts and curious furrows. The ears are separated. They have four incisors above and two below. The tail is short, and unconnected above the
interfemoral membrane.

There is but one species known, which belongs
to America, and is uniformly of a pale fawn-coloured
tint.

The PHYLLOSTOMES, *(Phyllostoma, Cuv. and Geoff.)*

The regular number of incisors is four in each jaw,
but from the lower some of these teeth frequently
drop out, being pushed aside by the growth of the
canines. These animals are also distinguished by
the membrane which, in the form of an upturned
leaf, crosses the termination of their noses. The
tragus of their ear resembles a little leaf, more or
less indented. The tongue, which is capable of great

elongation, is terminated by papillæ, which seem to be arranged so as to form an organ of suction, and their lips also are provided with tubercles symmetrically disposed. All this tribe is American; they run on the earth with more facility than the other bats, and are accustomed to suck the blood of other animals.

1. PHYLLOSTOMES *without a tail.*

The Vampire, (V. Spectrum, L.) Andira Guaçu of Brasil. Seb. LVIII. Geoff. Ann. Mus. XV. XII. 4.

The leaf is oval and hollowed in the form of a tunnel. This animal is reddish brown and about the size of a magpie. Habitat, South America. It has been accused of destroying men and animals by sucking their blood. But the truth appears to be, that it inflicts only small wounds, which may probably become inflammatory and gangrenous from the influence of the climate.

We may here add the Lunette, *(Vespertilio perspicillatus, L.)* spear-nosed bat. And three species given after Azzara, by M. Geoff. Ann. du Mus· XV. 181, 182.

2. PHYLLOSTOMES, *with a tail involved in the interfemoral membrane.*

Javelin Bat, (Vesp. Hastatus, L.) Fer de Lance, Buff. XIII. XXXIII.

The membrane of the nose very much resembles a javelin, or leaf of trefoil.

Here may be added *Vespertilio Soricinus*, (Anglicè leaf-nosed bat), Pall. Spic. Zool. fasc. III. pl. III. IV., cap. Schreb. XLVII.

3. PHYLLOSTOMES, *with a tail unconnected above the membrane.*

The indented Javelin Bat, (Ph. crenulatum, Geoff.)

The leaf of the nose is formed like a javelin, indented or furrowed at the side.

The second great tribe has but one ossified phalanx on the index, and the other fingers have each of them two.

This tribe is also divided into numerous subgenera.

The MEGADERMES, *(Geoffr. Ann. du Mus. XV.)*

These bats have on the nose a leaf more complicated than that of the Phyllostomes. The parotis is large, and most frequently forked or cloven; the shells or conchs of the ear are extremely ample and are united to each other on the summit of the head; the tongue and lips are smooth; the interfemoral membrane is entire, and there is no tail. They have four incisives below, but none have as yet been discovered above, and it would appear that their intermaxillary bone remains cartilaginous.

They all belong to the ancient continent, and are either African, as the *Feuille (Mega. Vons. Geoffr.)*, with an oval leaf to the nose, almost as large as the head: of Senegal, or of the Indian Archipelago, as

the *Spasm of Ternata (Vespert. Spasma, L. Seb. I.*
ʟⱽɪ.*)*.   The *Lyre, Geoffr. Ann. Mus. XV. pl.* xɪɪ.
The *Trefle of Java, Seb. ib. &c.*   The species are interdistinguished by the form of their leaves, like the
Phyllostomes.

The Rʜɪɴᴏʟᴘʜɪ (Rʜɪɴᴏʟᴘʜᴜꜱ, *Geoffr. et Cuv.)* vulgarly
Horse-shoe Bats.

They have the nose furnished with membranes and
crests exceedingly complicated, which are couched
upon the forehead, and present in the gross the
figure of a horse-shoe.   The tail is long and placed
in the interfemoral membrane.   They have four incisives below, and two very small ones above, situated in an intermaxillary cartilaginous bone.

There are two species, very common in France,
and discovered by Daubenton.

The *Great Horse-shoe Bat (Vesp. Ferrum equinum, L.) Buff.*
or *Rhinolphus bifer, Geoffr. Ann. Mus. XX. pl.* ⱽ. and
The Lesser (Vesp. hipposideros, Bechot.)   Buffon VIII.
xⱽɪɪ. 2, and xx. *Geoffr. loc. cit.*

They inhabit the quarries, remaining there isolated,
suspended by the feet, and enveloped in their wings,
so as to suffer no other part of their bodies to become
visible*

The Nʏᴄᴛᴇʀᴇꜱ (Nʏᴄᴛᴇʀɪꜱ, *Cuv. et Geoffr.)*

The forehead is hollowed by a small indenture

---

* Add the four other species represented. Geoff. Ann. Mus. XX.
pl. v., one of which is the *Vesp. Speoris,* Schn.

which is marked even upon the cranium, and the nostrils are surrounded by a circle of projecting laminæ. They have four incisors, without any interval above, and six below. Their ears are large, not joined, and their tail is comprised in the interfemoral membrane. These species belong to Africa. Daubenton has described one of them *(Vesp. hispidus, Linn.) Bearded Bat.* M. Geoffroy has ound others in Egypt*.

The RHYNOPOMES *(Geoffr.)*

Have an indenture not so strongly marked, the nostrils at the end of the muzzle, and a small lamina above: their ears are joined, and their tail extends considerably beyond the interfemoral membrane. One species of this subgenus is known: it belongs to Egypt, and chiefly inhabits the pyramids †.

The TAPHIENS (THAPHOZOUS, *Geoffr.)*

Have also a small foss or indenture on the forehead; but the nostrils are devoid of the elevated laminæ, and they have but two incisors above and four below. Their ears are separated, and their tail free above the membrane. M. Geoffroy has discovered a species in the catacombs of Egypt‡.

Nyctere of Thebaïs, 29, Mammif. I. 2, 2.

† *Rhinopome Microphyle,* Geoff. *Vespertilio Microphyllus,* Schr.

‡ The *Taphien Filet.* Eg. Mamif. I. 1. 1. *The perforated Taphien,* b. III. L. the *Vesp. lepturus.*

The COMMON BATS (VESPERTILIO, *Cuv. et Geoffr.*)

Have the muzzle without leaves or other distinctive marks, the ears separated, four incisors above, separated into couples, and six below sharp-edged, and triflingly notched. The tail is comprised in the membrane. This sub-genus is the most numerous of all. Its species are to be found in every quarter of the globe. Six or seven are enumerated in France. The first has been known for a long period.

The ordinary Bat (Vesp. murinus, Linn.) Buff. VIII. xvi.

> Gray, with oblong ears the length of the head. The other species have been discovered only by Daubenton.

The Serotine (Vesp. Serotinus, Linn.) Buff. VIII. xviii. 2.

> Fawn-colour, wings and ears blackish: the shell or conch of these is triangular and shorter than the head; the parotis pointed.
>
> They are frequently found beneath the roofs of churches, and other unfrequented buildings.

The Noctule (V. Noctula, L.) Buff. VIII. xviii. 1.

> Brown, triangular ears, shorter than the head, parotis rounded. A little smaller than the preceding. Found in the hollows of old trees, &c.

The Pipistrelle (V. Pipistrellus, Gm.) Buff. VIII. xix. 1.

> The smallest species found in this country: brown, with triangular ears and parotis.

M. Geoffroy further separates from the Vespertiliones, or Common Bats,

The OREILLARDS *or Great-eared Bats* (PLECOTUS, *Geoffr.)*

Whose ears, larger than the head, are united to
each other on the cranium, as is the case with
the Megadermes, the Rhinopomes, *&c.*

The common species *(Vesp. Auritus, L.) Buff.
VIII.* xvii. 1, is still more common here than
the ordinary Bat.   Its ears are nearly equal in
size to its whole body.   It inhabits the houses,
kitchens, *&c.*   We have another species discovered by Daubenton, called the *Barbastelle
(Vesp. Barbastellus, Gm.) Buffon VIII.* xix. 2.
Brown, with ears much smaller.

The GALEOPITHECI (GALEOPITHECUS, *Pall.) commonly
Flying Cats,*

Differ generically from the bats, because the fingers
of their hands, all furnished with trenchant nails, are
not more elongated than those of the feet; so that
the membrane which occupies their intervals, and
extends as far as the sides of the tail, can do little
else than perform the functions of a parachute.
Their canine teeth are indented and short, like the
molars.   Above are two incisors, also indented, and
considerably separated from each other.   Below are
six cut into narrow divisions, like combs, a structure
altogether peculiar to this genus.   These animals
live in trees in the Indian Archipelago, and pursue

PERUVIAN BAT OF PENN-ANT.   THE TAILED GLOSSOPHAG BAT.

NOCTILIO UNICOLOR.   *GLOSSOPHAGA CAUDIFER.*

London Published by G.&W.B.Whittaker. Sept.r 1824 .

J.t Basire sc.

THE TAILLESS GLOSSOPHAG BAT.   THE MEGADERMA LYRE BAT.

GLOSSOPHAGA ECAUDALA.   *MEGADERMA LYRA.*

London Published by G. & W. B. Whittaker Sept. 1824 .   J. Basire sc.

insects there, and probably birds. If we may judge
by the detrition which their teeth suffer in age, they
would appear to subsist also on fruits. They have
a large cæcum.

We know but one species distinctly, the fur of
which above is reddish gray, red below, varied and
radiated with different grays in youth. This is the
Lemur volans, Linn. Audeb. Galæop, pl. i. *et* ii. It inhabits the Moluccas, the Sunda Islands, *&c.*

All the other Carnassiers have the teats situated
under the belly.

THE INSECTIVORA,

Which form the second Family,

Have, like the Cheiroptera, cheek-teeth, with conical
points, and lead a nocturnal or subterraneous life.
They principally subsist on insects, and in cold
countries pass the winter in a lethargic state. They
do not possess lateral membranes like the bats, but
notwithstanding this they are never destitute of clavicles. Their feet are short, and their motions feeble.
The teats are situated under the belly, and the penis
in a case. They have no cæcum, and all of them
lean the entire sole of the foot on the ground in
walking.

There are two small tribes of these, distinguished
by the position and the relative proportion of their
incisors and canine teeth.

The first have two long incisors in front, followed

by other incisors and canine teeth, all shorter than the molar. This kind of dentition, of which the *Tarsiers*, among the quadrumana, have already furnished us with an example, approximates these animals in some degree to the rodentia.

The HEDGEHOGS (ERINACEUS, *Lin.*)

Have the body covered with prickles instead of hairs. The skin of their back, in bending the head and paws towards the belly, can close itself as if in a purse or bag, and present its prickly points on all sides to the adversary. The tail is very short, and all the feet have five toes. The two middle superior incisives are separated and cylindrical.

The common Hedgehog (Erinaceus Europæus) Buff. *VIII.* vi.

With short ears; sufficiently common in the woods and hedges; passes the winter in its burrow, and sallies forth from it in the spring, with the vesiculæ seminales in a state of the most incredible amplitude and complication. To the insects which form its ordinary regimen it adds the fruits which in time wear off the points of its teeth. Its skin was formerly made use of to hatchel hemp.

The Long-eared Hedgehog (Erinaceus Auritus.) Schreb *CLXIII.*

Smaller than the common hedgehog, with ears as

large as two-thirds of the head; in other respects, like ours in form and habits. It is found
from the north of the Caspian Sea as far as
Egypt*.

The SHREWS (SOREX, *Lin.*)

Are animals generally much smaller than the hedgehogs, and covered with simple hairs instead of
prickles. On each flank, under the ordinary skin, is
found a little band of stiff and close hairs from which
distils an odoriferous humour, produced by a peculiar gland.

The two middle incisors of the upper row are
crooked and indented at the base. They remain in
holes which they dig in the earth, seldom going out
till towards evening, and live on worms and insects.
But one species has for a long time been remarked
in France.

The common Shrew, or Musette (Sorex Araneus, Lin.) Buff.
VIII. x. i.

Gray, with a square tail as long as the body.
They are found in considerable numbers in the
country, in the meadows, &c. They have been
accused of causing a malady among horses by
their bite; but this imputation is false, and has
probably originated in the fact, that though cats

* Pallas has remarked, as an interesting fact, that hedgehogs
eat hundreds of cantharides with impunity, while a single one will
cause such horrible torments to cats and dogs.

will kill these animals readily, they refuse to eat them, in consequence of their powerful odour. Daubenton has made known another species.

The Water Shrew. (Sorex Fodiens, Gm.) Buff. VIII. xi.

Black above, white underneath ; squared tail, as long as the body. When it plunges in the water the ear is almost hermetically sealed by means of three small valves which correspond with the helix, the tragus, and the antitragus. The small stiff hairs which border the feet afford it the facility of swimming, and accordingly its favourite haunts are the banks of rivulets.

Herman, M. Gall, and M. Geoffroy, have added some other species.

The Desmans (Mygale, *Cuv.*)

Differ from the shrews by two very small teeth, placed between the two large incisors below, and also by having the two upper incisors of a triangular form, and somewhat flattened. The muzzle is lengthened into a very small and flexible snout, which is in a state of continual agitation. Their long tail, scaly and flattened at the sides, and their feet with five toes all united together by membranes, constitute them aquatic animals. Their eyes are very small and they have no external ears.

The Desman of Russia, vulg. *The Russian Musk Rat (Sorex Moschatus, Lin.) Buff. X.* i.

Almost as large as a hedgehog, of an ashy-gray,

very common along the rivers and lakes of Southern Russia. It subsists on worms, on the larvæ of insects, and more especially on leeches, which it easily draws out of the mud with its mobile snout. Its burrow, dug within the bank, commences under the water, and is raised in such a manner that the bottom remains above the level in the largest waters. This animal does not come voluntarily to dry land, but is often taken in nets. Its musky odour arises from a sort of pommade, secreted in small follicles under the tail. This odour is communicated even to the flesh of pikes, which prey upon these animals.

A small species of this genus has been discovered in the streams of the Pyrenees, and made known by M Geoffroy. Ann. du Mus. tom. XVII. pl. iv. f. i.

The Scalopes (SCALOPS, *Cuv.*)

Unite to the teeth of the Desmans, and to the simply-pointed muzzle of the Shrews, large hands, armed with strong nails, fitted for digging into the earth, and entirely similar to those of moles. Accordingly they lead the same sort of life.

The only species known is

The Scalope of Canada (Sorex aquaticus, Lin.). Schreb.
CLVIII.

Appears to inhabit a considerable portion of North America along the banks of rivers.

The CHRYSOCHLORES (CHRYSOCHLORIS, *Lacep.*)

Have, as well as the two preceding genera, two incisors above and four below ; but the muzzle is short,
large, and elevated, and their fore-feet have but
three nails, of which the outer one is very large and
the others diminish in proportion. The hind-feet
have five nails. These are also subterraneous
animals, and their fore-arm is supported, for the purposes of digging, by an additional bone placed under
the cubitus.

The Chrysochlore of the Cape, vulg. *Golden Mole.* *(Talpa
Asiatica, Lin.) Schreb. CLVII.,* and better, *Brown, III.
XLV.*

Somewhat less than our moles, without apparent tail. The only quadruped known which
presents any shade of those beautiful metallic
reflections, which glitter in such a variety of
birds, fishes, and insects. Its fur is green,
changing to the colour of brass or bronze. Its
ears have no conque, and its eyes are not perceptible *.

The second tribe of insectivora have four large

* The Red Mole of America, in Seba, I. pl. xxxii. f. 1, (*talpa
rubra*, L.) is probably of this genus; but the *tucan* of Fernandes,
ap. XXIV., which is confounded with it, appears rather a rat-mole,
by its long teeth in each jaw, and its vegetable regimen. It is probably to this first tribe of insectivora that *the long-tailed mole* belongs. Penn. arct. Zool. No. 68 ; but its dentition is not sufficiently
known to fix its place with accuracy.

canine teeth separated from each other, between which are small incisives, which is the most ordinary arrangement with the quadrumana and the carnassiers.

We find in this subdivision forms and habits analogous to those of the preceding tribe. Thus,

The TENRECS, *Cuv.* (CENTENES, *Illig.*)

Have the body covered with prickles like the hedgehogs; but, not to mention the great difference of their teeth, the tenrecs do not possess the faculty of rolling themselves up in a globular form. They have no tail, and the muzzle is remarkably pointed. Three species have been found in Madagascar, the first of which has been naturalized in the Isle of France. They are nocturnal animals, and pass three months of the year in a torpid state, although inhabitants of the torrid zone. Bruguière assures us that it is even during the greatest heats that they sleep in this manner.

The Tenrec (Erinaceus Ecaudatus, Lin.) Buff. XII. LVI

Covered with stiff prickles. Incisors sloping, and but four in number in the lower jaw. This is the largest of the three, and surpasses our hedgehog.

The Tendrac, (Erinaceus Setosus, Lin.) Buff. XII. LVII

With more flexible prickles, more resembling hairs. Six sloping incisors in each jaw.

The radiated Tenrec, (Erinaceus Semispinosus.)

Covered with hairs and prickles intermingled, radiated with yellow and black. Its incisors to the number of six, and the canines are slender and crooked. It is scarcely the size of a mole.

The MOLES, (TALPA, *Lin.*)

Are universally known by their subterraneous mode of existence, and their conformation eminently adapted to this kind of life.

An arm remarkably short, attached by a long shoulder-blade, supported by a vigorous clavicle and provided with enormous muscles, carries a hand extremely large, the palm of which is always turned upwards or behind. This hand is trenchant at its inferior edge : the fingers are distinguished with difficulty, but the nails which terminate them are long, strong, broad, and trenchant. Such is the instrument employed by the mole to tear the earth and push it back. The sternum, like that of birds and bats, has a ridge which gives to the pectoral muscles the magnitude necessary for their functions. To pierce the earth and throw it up, the mole uses its elongated and pointed head, the muzzle of which is armed at the end with a peculiar small bone. The cervical muscles are exceedingly vigorous, and the cervical ligament is even completely ossified. The hinder part of the mole is feeble, and the motions of the animal on the earth are as constrained and painful as they are rapid and vigorous below its surface. The sense of hearing in these animals is extremely fine, and the tympanum remarkably large, though the external ear is wanting. The eye, however, is very small, and so much concealed by the hair, that its very existence was for a long time denied. The jaws of the mole are feeble, and its nutriment consists of insects, worms, and some tender roots. This tribe have six incisives above and eight below.

Our Common Mole, (*Talpa Europæa, Lin.*) *Buff. VIII.* xii.

Pointed muzzle, hair fine and black. Some individuals are found white, fawn-coloured, and pied. This animal is extremely troublesome from the derangement it causes in cultivated grounds.

The radiaed Mole of Canada, (*Talpa cristata, Sorex cristatus, Lin*.*)

The two nostrils are surrounded with small points, cartilaginous and moveable, which when separated into radii, resemble a kind of star. It is less than our mole, blackish, and the tail, one half shorter than the body, is slightly covered with hair.

THE CARNIVORA
Will form a third family of the Carnassiers.

Though the epithet of Carnassiers is suitable to all unguiculated animals with three sorts of teeth, ex-

---

* We have satisfied ourselves by the inspection of its teeth that it is a true mole, and not a sorex. It is the *condyliura* of Illiger; but its characters, taken from the figure of La Faille and Buff. supp. VI. xxxvii., are false.

cepting the Quadrumana, inasmuch as they all subsist more or less on animal matter, still there are
many of them, and especially the two preceding families, which are reduced by their weakness and the
conic tubercles of their cheek-teeth, to live almost
entirely on insects. It is in the family now before
us that the sanguinary appetite is united with sufficient force to give it due effect. These animals
have always four large and long canine teeth separated, and between them six incisives in each jaw,
the second of which in the lower row has always its
root more deeply seated than the rest. The molars
are always either altogether trenchant or but partly
mingled with blunt tubercles, and never bristling
with conic points.

These animals are more exclusively carnivorous in
proportion as their teeth are more completely trenchant, and their regimen may be nearly calculated
from a comparison of the extent of the tuberculous
surface of their teeth with the part which is trenchant. The bears which can subsist entirely on
a vegetable diet have almost all their teeth tubercu-
lous.

The anterior molars are the most trenchant; then
comes a molar larger than the others, which is generally provided with a tuberculous heel of different
degrees of magnitude, and behind it are found one
or two small teeth entirely flat. It is with these
small teeth at the bottom of the mouth that dogs
chew the grass which they occasionally swallow.
This large molar above, and the corresponding one

below, we shall call, with M. Frederic Cuvier, *car-nivorous teeth (carnassières)*, the anterior pointed teeth we shall call *false molars*, and the posterior blunt ones, *tuberculous teeth*.

It is easily to be conceived, that the genera which have the fewer molars, and whose jaws are the shortest can bite with the greatest force.

It is on these differences that the genera may be most securely established.

We must however unite to them a consideration of the hind foot.

Many genera, like all those of the two preceding families, rest the entire sole of the foot on the ground in walking or standing upright; and this peculiarity is easily perceived from the absence of hairs under this whole part.

Others, much more numerous, never walk except on the end of their toes, elevating the tarsus altoge-ther. Their course is more rapid, and to this first difference they unite many others in their habits, and even in their internal conformation. Both have no clavicle except a bony rudiment suspended in the flesh.

THE PLANTIGRADES

Form the first tribe of the carnivora and walk on the en-tire foot, by which means they obtain a greater facility of raising themselves on their hinder legs. They par-ticipate in the slowness of motion and the nocturnal life of the insectivora, and, like them, are destitute of a cæcum. Most of the plantigrades which inhabit cold countries pass the winter in a lethargic state; they have all five toes on every foot.

The BEARS (URSUS, *Lin.*)

Have three large molars on each side in each jaw, entirely tuberculous; accordingly, notwithstanding their extreme strength, they seldom eat flesh except from necessity. The last but one in the upper row stands for the *carnivorous tooth (la carnassière)*. The last, which is the *tuberculous,* is the largest of all. In front of the three is another pointed molar, and in the interval between it and the canine, one or two very small and simple teeth, which often fall without inconvenience.

These animals are large, clumsy in the body, thick in the limbs, and have a remarkably short tail. The cartilage of their nose is elongated and mobile. They dig caves or construct huts for themselves, where they pass the winter in a state of somnolency more or less profound, and without taking any aliment. It is in this retreat that the female brings forth.

The species are not easily interdistinguished by sensible or obvious characters. We reckon

The Brown Bear of Europe, (Ursus Arctos, Lin.) Buff. VIII. xxxi.

With convex forehead, and brown fur, more or less woolly. Some are seen nearly yellow, others of a sleek and glossy brown, with almost a silvery reflection. The relative height of their limbs varies equally and without any constant analogy to the age or sex of the individual. The young bears are distinguished by a whitish collar. This animal inhabits the high mountains and large forests of Europe, and of a considerable part of Asia. The time of copulation is in June, and the birth takes place in January. Sometimes these bears lodge very high in trees. Their flesh, when young, is good for eating; but the paws at all ages are esteemed excellent.

Some think that it is possible to distinguish a *black Bear of Europe*. Those which had been exhibited as such, had a flat forehead, and the fur woolly and blackish, Also there has been mentioned a *bear of the Indies,* with blackish fur and a white spot on the breast, *&c.*

A species which we can with more certainty pronounce to be different, is,

The Black Bear of America, (Ursus Americanus, Lin.) Cuv. Menag. du Mus., in 8vo. II. p. 143.

Flat forehead, fur black and glossy, and fawn-coloured muzzle. We have always found in this species the small teeth behind the canine more numerous than in the bear of Europe. It has sometimes a fawn-coloured spot above each eye, and one of white or fawn-colour on the throat or chest. Individuals have been seen entirely fawn-coloured. It lives usually on wild fruits, often lays waste the fields, and repairs to the sea-coast for the purpose of catching fish when they are in abundance. It seldom attacks quadrupeds but in the want of every other alimentary supply. Its flesh is in estimation.

It is said that there is also in America a gray bear, larger than the black one, but it has not been described with sufficient accuracy.

The White Bear of the Icy Sea, (Ursus Maritimus, Lin.) Cuv. Menage. du Mus. in 8vo. p. 68.

Is another very distinct species, characterized by its elongated and flattened head, and by its white and glossy fur. It pursues the seals and other marine animals. Exaggerated accounts of its voracity have rendered it very celebrated.

The RACOONS (PROCYON, *Storr.*)

Have three hinder tuberculous molars, and three small pointed molars in front, forming a continued series as far as the canines. The tail is long; but all the rest of their exterior represents that of the bear on a minor scale. They rest the entire sole of the foot upon the ground only when standing; in walking they raise the heel.

The Racoon of the Anglo-Americans, Mapach of the Mexicans. (Ursus Lotor, Lin.) Buff. VIII. xliii.

Grayish-brown, with a white muzzle, a brown mark across the eyes; the tail ringed with brown and black. An animal about the size of

the badger, easily tamed, and never eating any
thing without having first plunged it in water.
It comes from North America; subsists on
eggs, insects, birds, &c.

The Racoon Crab-eater, (Ursus Cancrivorus.) Buff. Supp
VI. xxxii.

Of an ashy-brown colour, uniformly clear. The
rings of the tail less marked. Habitat, South
America.

The Coatis, (Nasua, *Storr.)*

Join to the teeth, the tail, the nocturnal life and
dragging walk of the racoons; the nose is singularly
elongated and mobile. The feet are demi-palmate,
and yet they climb trees. Their long nails serve
them for digging. They come from the warm regions of America, and subsist much in the same way
as our own martins.

The Red Coati, (Viverra Nasua, Lin.) Buff. VIII. xlviii.

Reddish fawn-colour. The muzzle and rings of
the tail brown.

The Brown Coati, (Viverra Narica, Lin.) Buff. VIII. xlviii.

Brown, white spots on the eye and muzzle.
We can scarcely introduce more fitly than here
the singular genus of the Kinkajous or Potto,
Cuv. *(Cercoleptes,* Illig.), which unites to the plantigrade motion a long and prehensile tail, like that of

the Sapajous, a short muzzle, a slender and extensible tongue; two cheek-teeth in front pointed, and
three tuberculous ones behind.

There is known but a single species *(Viverra cau-
divolvula,* Gm.), Buff. Supp. III. L., from the warm
parts of South America, and some of the great An-
tilles, where it is named poto. It is as large as a
pole-cat, with woolly hair, of a grayish or brownish
yellow. Nocturnal, of a disposition rather mild,
and capable of subsisting on fruits, honey, milk,
blood, *&c.*

The BADGERS, (MELES, *Storr.)*

Which Linnæus places like the racoons, in the genus
of bears, have a very small tooth behind the canine;
then two pointed molars followed above by one which
we begin to recognise as the true carnivorous tooth
(la carnassière), from the vestige of an edge discovered on its external side. Behind it is a squared
tuberculous tooth the largest of all. The *penultimus*
below begins also to exhibit some resemblance to
the inferior *carnassières;* but as it has on the interior side two tubercles as elevated as its edge, it
performs the part of the tuberculous tooth. The
last one is very small.

These are animals with a creeping walk and noc
turnal mode of life, like all the preceding. The tai.
is short, the toes deeply involved in the skin, and
they are eminently distinguished besides by a pouch
situated under the tail, whence oozes a fat and fœtid
humour. The nails of their fore-feet are consider-

ably elongated, and thus render them skilful in delving in the earth.

The Badger of Europe, (Ursus Meles, Lin.) Buff. VII. vii.

Grayish above, black underneath, a black band on each side of the head.

The GLUTTONS (GULO, *Storr.*)

Had also been placed in the genus of bears by Linnæus; but they approach more to the martins by their teeth, as well as by their entire constitution and character, resembling the bears only in their plantigrade motion. They have three false molars above and four below in front of the carnivorous tooth; and a small tuberculous one behind it, the upper of which is rather large than long. The upper carnivorous tooth has but a small tubercle. These animals have tails of moderate length, with a fold underneath instead of a pouch, and in other respects as to their gait, &c., they are sufficiently similar to the badgers.

The most celebrated species is the *glutton* of the north, *rossomak* of the Russians (*Ursus gulo*, Lin.) Buff. Sup. III. xlviii. As large as our badger, usually with a beautiful fur of a deep marron, with a disk somewhat browner on the back, but sometimes of paler tints. It inhabits the coldest countries of the north, is esteemed to be remarkably cruel, hunts by night, does not sleep during the winter, and contrives to master the largest animals

by leaping downwards on them from a tree. Its
voracity has been ridiculously exaggerated by some
writers.

The Wolverene of North America (Ursus Luscus, Lin.
Edw. CIII.

Does not appear to differ from the last by any
permanent characters.   Its tints are generally
somewhat paler.

Warm climates produce some species which cannot
well be ranged, except among the gluttons, not differing from them but by one false molar less in each
jaw, and by a long tail.   Such are those which the
Spanish Americans name ferrets (*hurons*), and which,
having in fact the teeth of our pole-cats and ferrets,
have also the same mode of life.   But they are distinguished from them by the plantigrade motion

The Grison (Viverra Vittata, Lin.) Buff. Sup. VIII.
XXIII. et XXV.

Black, the top of the head and neck gray; a
white band extending from the forehead to the
shoulders.

The Taïra, (Mustela Barbara, Lin.) Buff. Sup. VII. LX.

Brown, top of the head gray, a large white spot
under the neck.   These two animals are found
in all the warm regions of America, and diffuse
a musky odour.   Their feet are triflingly flat-

tened, and it would seem that they have sometimes been taken for otters.

It is probable that the *ratel (viverra mellivora, and
viv. capensis)*, an animal, about the size of a badger,
should be placed at the end of the gluttons and Grisons. It is gray above, black underneath, having
a white line between those two colours. It inhabits the Cape of Good Hope, and digs the earth
with its long front talons to discover the honey
there deposited by the wild bees. It is only known
by an incomplete description of Sparrman.

The DIGITIGRADES

Form the second tribe of Carnivora, that which
walks on the end of the toes.

There is a first subdivision of them which have
but one tuberculous tooth behind the upper carnivorous one. These animals have been named vermiform, in consequence of the length of their bodies
and shortness of their legs, which allows them to
pass through the smallest apertures. Like all
the preceding tribes they want a cæcum, but do
not fall into a state of lethargy during the winter.
Though small and feeble, they are exceedingly
cruel and live peculiarly upon blood. Linnæus
makes but one genus of them, that of

The MARTENS, (MUSTELA, *Lin.*)

Which we shall divide into four sub-genera.

The Polecats *(*Putorius, *Cuv.)*

Are the most sanguinary of all. Their carnivorous tooth below has no interior tubercle; the tuberculous tooth above is more broad than long. They have only two false molars above and three below. As to their exterior, they may be easily recognised by their muzzle being rather shorter and more thick than that of the martens. They all diffuse an infectious odour.

The common Polecat (Mustela putorius, L.) Buff. VII. xxiii.

Brown, the flanks yellow, with white spots upon the head. The terror of hen-roosts and warrens.

The Ferret. (Mustela furo, L.) Buff. VII. xxv. xxvi.

Yellowish, with red eyes, is perhaps only a variety of the polecat. In France we find it only in a domesticated state, and it is employed to pursue rabbits into their burrows. It comes to us from Spain and Barbary.

The Polecat of Poland, or Perouasca. (Mustela Sarmatica.) Pall. Spic. Zool. XIV. iv. 1, *Schreb. CXXXII.*

Brown, spotted all over with yellow and white. Its skin is much employed in the fur trade, on account of its beautiful variegation. It inhabits all the southern part of Russia, Asia Minor, and the coasts of the Caspian Sea.

To the polecats also must be referred two small species of our climates.

The Weasel. (Mustela Vulgaris, L.) Buff. VII. xxix. 1.

Altogether of an uniform red; and

The Ermine, (Mustela Erminea, L.) Buffon, VII. xxix. 2.
xxxi. 1.

Which is red in summer, white in winter, with the tip of the tail always black. Its winter coat forms one of those furs most universally known.

It is probable that we must still refer to this race

The Polecat of Siberia. (Mustela Sibirica, Pall.) Spic. Zool.
XIV iv. 2.

Altogether of a clear uniform fawn-colour; and

The Mink, Norek, Noerz, or Polecat of the Northern Rivers.
(Mustela Lutreola, Pall.) Spic. Zool. XIV. iii. 1. *Les*
Mem. de Stockh. 1739, *pl* .xi

Which frequents the edges of the water in the north and east of Europe, from the Icy as far as the Black Sea, and subsists on frogs and crab-fish. The feet are triflingly flattened between the bases of the toes, but the teeth and the round tail approach it more to the polecats than to the otters. It is brown with a whitish jaw. Its odour is that of musk, and its fur is extremely fine.

The Polecat of the Cape. (Zorille of Buff. Viverra Zorilla,
Gm.) Buff. XIII. xli.

Radiated irregularly with white and black. It had been so long confounded with the mouffettes that the name Zorillo (fox's cub), which the Spaniards applied to those fetid animals of America was given to it. It has, however, nothing in common with them except the nails adapted for delving. This circumstance indicates a subterraneous mode of life, which might induce us to distinguish this species from the other polecats.

The MARTENS, properly so called, (MUSTELA, *Cuv.*)

Differ from the polecats by an additional false molar above and below, and by a small interior tubercle in their carnivorous tooth below. Two characters which somewhat diminish the cruelty of their nature.

Europe has two species very nearly approaching to each other,

The common Marten (Mustela Martes, L.) Buff. VII. xxiii.

Brown, with a yellow spot under the neck; inhabits the woods.

The Fouine. (Mustela Foina, L.) Buff. VII. xviii.

Brown, with all the upper part of the throat and neck whitish. Frequents houses. Both these species occasion much mischief.

One species is known in Siberia.

The Sable (Mustela Zibellina) Pall. Spic. Zool. XIV. iii. 2. *Schreb. CXXXVI.*

So much celebrated for its rich fur. It is brown, with some spots of white on the head,

and is distinguished from the foregoing by having
hair even under the toes. It likewise inhabits
the most frozen mountains. The hunting of this
animal in the midst of winter, through eternal
snow, is one of the most painful of human labours. The pursuit of sables first gave occasion to the discovery of the eastern regions of
Siberia.

Northern America produces also many of the
marten tribe, which travellers and naturalists have
pointed out under the ill-defined appellations of
Pekan, Vison, Mink, Foutereau, &c.

The species to which we apply the name of *Vison*
(*Mustela Vison*), is altogether brown, with the little
point of the chin white. The fur is remarkably
brilliant. It is found in Canada and the United
States.

That which we shall name, *Pekan,* and which
comes from the same countries, has the head, the
neck, the shoulders, and the top of the back, mixed
of gray and brown. The nose, the crupper, the
tail, and the limbs, are blackish.

Both have hair even under the toes.

The Mouffettes (Mephitis) *Cuv.*

Have, like the polecats, two false molars above and
three below, but their upper tuberculous tooth is
very large and as long as broad, and their inferior
carnivorous tooth has two tubercles on its internal
side, which approximates them to the badgers, as
the polecats approach the grisons and gluttons The

mouffettes have, besides, like the badgers, the front
nails long and proper for digging.   The resemblance
holds good even in the distribution of colours.   In
this family, so remarkable for its unpleasant odour,
the mouffettes are distinguished for a stinking pre-
minence above all the other species.

The mouffettes are usually radiated with white
upon a black ground.   But they appear to vary in
the same species by the number of streaks, and they
have not been interdistinguished with any sufficient
accuracy. All those which come from America have
a long and tufted tail, but M. Lechenaud has lately
reported the existence of one species in the island
of Java altogether destitute of this appendage.

The OTTERS (LUTRA, Storr.)

Have three false molars above and below, a strong
talon on the upper carnivorous tooth, a tubercle on
the internal side of the lower, and a large tuberculous tooth, almost as long as broad above. The head
is compressed and the tongue partly rough.   They
are moreover distinguished from all the preceding
sub-genera by their palmate feet, and their tail horizontally flattened; two characters which constitute
them aquatic animals.   They are supported on fish.

The common Otter. (Mustela lutra, L.) Buff. VII. xi.

Brown above, whitish beneath.   Habitat, the
rivers of Europe

The Otter of America, (Mustela lutra Brasiliensis, Gm.)

Altogether brown or fawn-coloured, with a
white or yellow neck, and somewhat larger than

our common otter. Of the rivers of the two Americas.

The Sea Otter, (Mustela lutris, L.) Schreb. CXXVIII.

Twice as large as ours, with a body considerably elongated, tail three times less than the body, hind feet extremely short. Its blackish covering as smooth and brilliant as velvet, is the most valuable of all furs. It frequently is whitish on the head. The English and Russians pursue this animal in all the northern parts of the Pacific Ocean, for the purpose of selling its skin to the Japanese and Chinese.

The second subdivision of the digitigrades has two flat tuberculous teeth behind the upper carnivorous tooth, which itself has a heel or protuberance tolerably broad. They are carnassiers, but without showing much courage in proportion to their strength. They frequently live on carrion. They have all a small cæcum.

The Dogs (Canis, *L.*)

Have three false molars above, four below, and two tuberculous teeth behind each carnivorous one. The first of these tuberculous teeth in the upper row is very large. The upper carnivorous tooth has but a single small tubercle within, but the lower one has its posterior point altogether tuberculous. Their tongue is soft. The fore-feet have five toes and the hinder four.

The domestic Dog. (Canis Familiaris, L.)

Is distinguished by a curved tail, and varies besides *ad infinitum* as to size, form, colour, and the quality of the hair. The dog is the most complete, the most remarkable, and the most useful conquest ever made by man. Every species has become our property ; each individual is altogether devoted to his master, assumes his manners, knows and defends his property, and remains attached to him until death; and all this proceeds neither from want nor constraint, but solely from true gratitude and real friendship. The swiftness, the strength, and the scent of the dog, have created for man a powerful ally against other animals, and were perhaps necessary to the establishment of society. He is the only animal which has followed man through every region of the earth.

Some naturalists think that the dog is a wolf, others a domesticated jackal. The dogs, however, which have become wild again in desert islands do not resemble either of these species. The wild dogs and those belonging to barbarous people, such as the inhabitants of New Holland, have straight ears, which would lead us to the belief that the European races, which approximate the most to the original type, are our shepherd's dog and our wolf-dog. But the com parison of crania points to a closer approximation in the *mastiff* and *Danish dog :* after which

come the *hound,* the *pointer,* and the *terrier,*
which do not differ between themselves except in size and the proportions of the limbs.
The *greyhound* is more lank, and its frontal sinuses are small and its scent more feeble. The
shepherd's-dog and the *wolf-dog* resume the
straight ears of the wild dogs, but with a
greater development of the brain, which proceeds increasing, with a proportionate degree
of intelligence in the *barbet* and the *spaniel.*
The *bull-dog,* on the other hand, is remarkable
for the shortness and vigour of its jaws. The
small chamber-dogs, the *pugs, spaniels, shock-
dogs,* &c., are the most degenerate productions,
and constitute the most striking marks of that
power which man exercises over nature.

The dog is born with the eyes closed. He
opens them the tenth or twelfth day. His
teeth begin to change in the fourth month, and
his growth terminates at two years of age. The
female goes with young sixty-three days, and
brings forth from six to a dozen young ones.
The dog is old at five years, and seldom lives
more than twenty. The vigilance, the bark, the
singular mode of copulation of this animal, and
his striking susceptibility of a varied education,
are universally known.

The Wolf. (Canis Lupus, L.) Buff. VII. I.

A large species with a straight tail; fur of a
grayish fawn colour, with a black streak on the
fore-limbs of the adults. It is the most mis-

chievous of all the carnassiers known in our
countries. It is found from Egypt as far as Lap-
land, and appears to have passed into America.
In northern regions its fur becomes white in the
winter season. It attacks all our animals, but
yet by no means displays courage in proportion to its strength. It often preys on carrion.
Its habits and physical development have many
close relations with those of the dog.

The Black Wolf. (Canis Lycaon, L.) Buff. IX. xli.

Inhabits also in Europe, and is found even in
France, but very rarely*. Its fur is of a deep
and uniform black. It is said to be more ferocious than the common wolf.

*The Red Wolf. (Canis Mexicanus, Lin.) Agoura-Gouazou
of Azzara.*

Of a fine cinnamon red, with a short black mane
along the entire spine. Found in the marshes
of all the hot and temperate regions of America.

*The Jackal, or Golden Wolf. (Canis Aureus, Lin.) Schreb.
XCIV.*

Somewhat less than the three preceding;
grayish brown, the thighs and legs of a clear
fawn-colour; some red upon the ear. It inhabits in troops a great part of Asia and Africa,
from India and the environs of the Caspian Sea

---

* I have seen four individuals taken or killed in France. It
must not be confounded with the Black Fox, with whose synonymes
Gmelin has mixed it up.

as far as Guinea. It is a voracious animal, which hunts after the manner of a dog, and seems to resemble him more nearly than any other wild species in conformation and facility of being tamed.

The FOXES may be distinguished from the wolves and dogs by a longer and more tufted tail, by a more pointed muzzle, by pupils calculated for nocturnal vision *, and by upper incisives less sloping. They diffuse a fetid odour, dig themselves burrows, and only attack weak animals. This sub-genus is more numerous than the preceding.

The common Fox. (Canis Vulpes, L.) Buff. VII. vi

More or less red, the end of the tail white. Is spread over most climates from Sweden even to Egypt. Those of the north are distinguished only by a more brilliant fur. We observe no constant difference between those of the Old Continent and those of North America. *The Coal Fox (Canis Alopex) Schreb. XCI.*, which has the end of the tail black, and is found in the same countries as the common, and *The Cross Fox (id. XCI. A.)*, which is distinguished only by a streak of black along the spine and over the shoulders, are probably but varieties of the common fox. But the following species are very distinct.

* The Baron seems to conclude that elongated pupils are adapted for nocturnal habits, a conclusion which we have elsewhere ventured to think unfounded.—ED.

The Corsac, or Small Yellow Fox, (Canis Corsac, Gm.) Buff.
Supp. III. xvi., under the name *Adive.*

Of a pale yellowish gray, some blackish waves
on the base of the tail. The end of the tail black,
and the jaw white. Common in the vast heaths
of Central Asia, from the Volga to the East
Indies; possesses the habits of the common
fox, and never drinks.

The Tri-coloured Fox of America, (Canis Cinereo-argenteus.)
Schreb. XCII. A.

Ash-coloured above, white beneath, a band of
cinnamon red along the flanks. Habitat, all
the hot and temperate climates of the two Americas.

The Silvery or Black Fox*.

Black, but the ends of the hairs are white, except on the ears, shoulders, and tail, where they
are purely black. The tip of the tail is altogether white. From North America. Its fur is
one of the finest and most highly prized.

The Blue Fox, or Isatis, (Canis Lagopus), Schreb. XCIII.

Deep ashen colour; the under part of the toes
furnished with hairs. It is often white in winter. From the north of Siberia. Likewise
very much esteemed for the fur.

---

* Gmelin has confounded this with the black wolf, under the
name of *Canis Lycaon.*

The Cape Fox, (Canis Mesomelas,) * *Schreb. XCV.*

Yellow on the flanks, the middle of the back black, mixed with white, and finishing in a point at the end †.

The CIVETS (VIVERRA),

Have three false molars above and four below, the anterior of which occasionally drop out; two tuberculous teeth, tolerably large above, one only below, and two projecting tubercles on the internal side of the lower carnivorous tooth in front, the rest of this tooth being more or less tuberculous. The tongue is covered with sharp and rough papillæ. Their claws are partly straightened as they walk, and near the anus is a pouch more or less deep, where an unctuous and odoriferous matter exudes from peculiar glands.

They are divided into four sub-genera:

The CIVETS, properly so called, (VIVERRA, *Cuv.*)

In which the deep pouch situated between the anus and the organ of generation, and divided into two

---

* Gmelin has confounded it with the Adive of Buffon, a factitious species not differing from the Sachal.

† The Fennec of Bruce, which Gmel. names *Canis Cerdo*, and Ilig. *Megalotis*, is too little known to be classified. It is a small animal of Africa, whose ears almost equal the whole body in size and which climbs trees. Neither the teeth nor toes have been described.

bags, is filled with an abundant unction, of a
strongly-musked odour.

The Civet (Viverra Civetta, Lin.)   Buff. IX. xxxiv.

> Gray with brown or blackish spots, the tail
> brown, less than the body.   Along the entire
> back and tail is a mane capable of elevation.
> This animal comes from the hottest parts of
> Africa.

The Zibeth, (Viverra Zibetha, Lin.)   Buff. IX. xxxi.

> Gray, shaded with brown, long tail tinged with
> black.

The GENETS, (GENETTA, *Cuv.*)

In which the pouch is reduced to a slight hollow
formed by the projection of the glands, and almost
without any sensible excretion although there is a
most manifest odour.

The Common Genet, (Viverra Genetta, Lin.) Buff. IX. xxxvi.

> Gray, with small round black spots, and a tail
> tinged with black.   As large as a marten, and
> still more slender.   Seems to inhabit from
> Southern France as far as the Cape of Good
> Hope*.

* *The Civet of Malacca* of Sonnerat, the *Genet of the Cape,*
Buff., the *Cape-Cat,* Forster, the *Bisaam-Cat* of Vosmaer, of which
Gmelin has made so many species, appear only common Genets.
To this subdivision must be referred the *radiated Pole-cat* of India,
Buff. Sup. VII. lvii. (*Viv. fasciata, Gm.*)

The Fossane of Madagascar, (Viv. Fossa,) Buff. XIII. xx.

Has those parts of a fawn colour which are black in the genet, and scarcely any rings to its tail.

The Mangoustes, *Cuv.* (Herpestes, *Illiger.*)

In which the pouch is voluminous, simple, and has the anus bored in its depth.

The Mangouste of Egypt, so celebrated under the name *Ichneumon, (Viverra Ichneumon, L.) Buff. Sup. III.* xxvi.

Gray, with a long tail terminated by a black tuft; larger than our cats, as slender as our martens. It searches peculiarly for the eggs of crocodiles, but also subsists on all kinds of small animals. Domesticated in houses, it hunts mice, reptiles, *&c.* The Europeans at Cairo call it *Pharaoh's rat*, the people of the country *Nems*. What the ancients related of its jumping down the throat of the crocodile to put it to death, is fabulous.

The Mangouste of the Indies, (Viverra Mungos, L.) Buff., XIII. xix.; *and that of the Cape, (Viv. Cafra, Gm.) Schreb. CXVI. B.*

Both have the tail pointed, and the fur gray or brown, but uniform in the latter, streaked crosswise with black in the former, which also has the jaws streaked with fawn-colour.

The mangouste of the Indies is celebrated for his combats with the most dangerous serpents, and by the fame of having made known the virtue of the *ophiorhiza mongos* against their bite.

The SURIKATES, (RYZÆNA, *Iliger*)

Which have a strong resemblance to the mangoustes, even to the very tints and transverse streaks of the fur, but which are yet distinguished from them and from all the Carnivora hitherto treated of, by having only four toes on all the feet. Their pouches extend into the anus like those of the preceding.

But one species is known, a native of Africa, (*Viverra tetradactyla, Gm.*) Buff., XIII. VIII., somewhat smaller than the mangouste of India *.

The last subdivision of the digitigrades has no small teeth whatever behind the large molar below. It contains the most cruel animals, and the most de- idedly carnivorous of the whole class. There are two genera:

The HYÆNAS, (HYÆNA, *Storr.*)

Which have three false molars above and four below all conical, blunt, and singularly thick and clumsy The upper carnivorous tooth has a small tubercle within, and in front. But the lower one has none and presents only two strong trenchant points. This powerful apparatus enables the hyænas to break the

* The *Zenik* of Sonnerat, deuxieme voy. pl. 92, does not seem to differ from the Surikate but by being ill-drawn.

bones of the strongest animals which become their prey. Their tongue is rough. All their feet have four toes like the Surikates, and under the anus is a deep and glandulous pouch. They are nocturnal animals, voracious, living particularly on carcasses, and seeking them even in the tombs. There are many superstitious traditions relative to these animals. Two species are known:

The Striped Hyæna (Canis Hyæna, Lin.) Buff. Sup. III. XLVI.

Gray, irregularly radiated across with brown or black. A mane along the nape of the neck and the back, which it elevates in the moments of anger. It inhabits from the Indies as far as Abyssinia and Senegal.

The Spotted Hyæna, (Canis Crocuta, Lin.) Schreb. XCVI. B.

Gray, spotted with black, from the South of Africa. It is the tiger-wolf of the Cape.

The CATS, (FELIS, *Lin.)*

Are of all the Carnassiers the most powerfully armed. Their short and round muzzle, their short jaws, and above all their retractile claws, which, drawn upwards and concealed within the toes by the effect of elastic ligaments when in a state of rest, thus never losing their point or edge, render them, especially the larger species, very formidable animals. They have two false molars above and two below. Their upper carnivorous tooth has three lobes, and a blunt heel within, the lower two have pointed and trenchant lobes without any heel. Finally, they have but a very small upper tuberculous tooth, without any thing to correspond below. The species of this genus are very numerous, and very various in size and colour, although all similar in form. They cannot be subdivided but according to the somewhat unimportant characters of size and the magnitude of the fur.

At the head of this genus stands,

The Lion, (Felis Leo. Lin.) Buff. VIII. i. ii.

Distinguished by its uniform fawn-colour, the tuft of hair at the tip of its tail, and the mane which clothes the head, neck, and shoulders of the male. This is the strongest and most courageous of all animals of prey. Formerly the species was spread through the three divisions of the old world, but at present it seems almost confined to Africa, and some neighbouring parts of Asia. The head of the lion is more squared than that of the following species

The tigers are large species with smooth skin, very frequently marked with bright spots.

The Royal Tiger, (Felis Tigris,) Buff. VIII. ix.

As large as the lion, but with a more elongated body and rounder head. Of a bright fawn-colour above and pure white underneath, radiated irregularly and cross-wise with black. The most cruel of quadrupeds and the most terrible scourge of the East Indies. So great are its force and swiftness that, during the march of armies, it has been known to snatch a horseman from his horse, and carry him off into the recesses of the woods without the possibility of rescue.

The Jaguar or American Tiger. The great Panther of Furriers. *(Felis Onça, Lin.) d'Azzara, Voy. pl.* ix.

Almost as large as the oriental tiger, and almost as dangerous. A bright fawn-colour above, marked along the flanks with four ranges of black spots in the form of eyes, that is, with rings more or less complete, with a black point in the middle. White underneath, radiated cross-wise with black.

There are some individuals black, on which the spots, of a still deeper hue, are visible only at certain points of exposition.

The Panther, (Felis Pardus, Lin.) The *Pardalis* of the Ancients. *Cuv. Menag. du Mus. 8vo. I. p.* 212.

Fawn-coloured above, white underneath, with six or seven ranges of black spots in the form of roses, that is to say, formed by an assemblage of five or six simple spots on each flank.

E 2

The Leopard, (Felis Leopardus, Lin.)

Like the panther, but with ten ranges of smaller spots,

These two species are of Africa, and smaller than the Jaguar. Travellers and Furriers designate them indistinctly under the names of leopard, panther, African tiger, &c.*

The Guepard, or Hunting Tiger of India, (Felis jubata, L.) Schreb. CV.

Of a clear fawn-colour, with small black simple spots equally distributed. This animal is smaller than the panther, has greater height in the limbs, and longer hair on the nape of the neck. In India they are trained like dogs for the chase, for which purpose the panther is also employed in some countries.

The Couguar, Puma, or pretended American Lion. (Felis discolor, L.) Buff. VIII. xix.

Red, with small spots of the same colour a little deeper, and which are not easily distin guished. Of the whole of America.

---

* Buffon has mistaken the jaguar for the panther of the Old World, and has not well distinguished the panther and leopard. Thererore we forbear to cite him.

The Melas or Black Panther, (Felis Melas, Peron.)

Black, with simple spots of a deeper hue
Of the East Indies.

The Ocelot, (Felis pardalis, Lin.) Buff. XIII,
pl. xxxv. xxxvi.

With lower limbs than the preceding. Gray,
with large fawn-coloured spots bordered with
black, and forming oblique bands on the flanks.
Habitat, the whole of America.

Among the inferior species, the lynxes should be
distinguished, which are remarkable for the brushes
of hair with which their ears are adorned.

The Common Lynx (Felis Lynx. L.) Buff. VIII. xxi.

Reddish fawn-colour, most frequently spotted
with blackish; tail very short. Of all the old
continent. It was formerly found in France,
and it is not long since the last of the race have
disappeared from Germany.

The Lynx of Canada, (Felis Canadensis, Geoff.) Buff.
Supp. III. xliv.

Whitish gray with some spots of a pale brown
It appears to form a distinct species.

The American Lynx, (Felis rufa. Güld.) Screb. CIX. B.

Reddish fawn-colour, patched with brownish.
Brown waves on the thighs. Somewhat smaller
than the lynx. Of the United States.

The Lynx of the Marshes, Booted Lynx, &c. (Felis Chaus. Güld.) Schreb. CX. Bruce's Travels, pl. xxv.

A yellowish gray brown; hinder part of the legs blackish. Inhabits the marshes of the Caucasus, of Persia, Egypt, and Abyssinia. Hunts water-fowl, *&c.*

The Caracal (Felis Caracal, L.) Buff. IX. xxiv. and Supp. III. xlv.

Of a vinous red nearly uniform. Inhabits Persia, Turkey, *&c.* Is the true lynx of the ancients.

The inferior species, the ears of which have no pencils of hairs, resemble more or less our domestic cat. Such are,

The Serval, (Felis Serval, L.) Buff. XIII. xxxv.

As large as a lynx. Yellowish, with irregular black spots.

The Jaguarondi, (Felis jaguarondi,) Azzara, Voy. pl. x.

Elongated, and altogether of a brownish black. Both these last live in the forests of South America.

The Common Cat, (Felis catus, L.) Buff. VI. i. &c.

Originally a native of our European forests. In its wild state it is grayish brown, with trans-

verse waves of a deeper colour. The under
part is pale ; the inside of the thighs and four
paws yellowish. Three bands on the tail, and
the lower third blackish. In a domestic state,
it varies, as every one knows, in colour, in the
length and fineness of the fur, but infinitely
less so than the dog. It is also much less submissive and attached.

The AMPHIBIOUS ANIMALS

Will form the third and last of the small tribes into
which we shall divide the Carnivora. Their feet are so
short and so much enveloped in the skin, that they can
use them on the earth for no purpose but creeping.
But as the intervals of the toes are filled by membranes, they form excellent oars. These animals accordingly pass the most considerable portion of their
life in the sea, and do not come to land except to bask
in the sun, and suckle their little ones. Their elongated bodies, the great mobility of their spine, and
the strength of its flexor-muscles, their narrow pelvis, their hair smooth and tightened, as it were,
against the skin, are properties, which combined
together, are well calculated to make them excellent swimmers ; and all the details of their anatomy
confirm this opinion, and correspond with the result
of our first superficial observations.

But two genera have hitherto been distinguished,
the *Seals* and the *Morses.*

The SEALS (PHOCA, L.)

Have four or six incisors above, four below, pointed
canines and cheek teeth to the number of twenty,
twenty-two, or twenty-four, all trenchant or conical,
without any tuberculous part. Five toes on all the
feet, those of the fore-feet decreasing gradually
from the thumb or great toe to the little one ; while,
on the contrary, in the hind feet, the great and little
toe are the longest, and the intermediate ones decrease in size. The fore-limbs are enveloped in the
skin of the body as far as the wrist, the hinder nearly
as far as the heel. Between them is a short tail.
The head of the seals resembles that of a dog,
and they likewise possess the kind and intelligent
expression of countenance peculiar to that animal.
They are easily tamed, and soon become attached
to those who feed them. Their tongue is smooth,
and sloping towards the end. The stomach is
simple, the cæcum short, the canal long, and tolerably equal. These animals live on fish ; they eat
always in the water, and can close their nostrils
when they dive by means of a kind of valve. As
they dive for a long time, it has been supposed that
the botale foramen remained open in them as in the
fœtus. But this is not the case; there is, however,
a large venous sinus in the liver which assists them
in diving, and renders respiration less necessary to
the circulation of the blood. Their blood is very
abundant and extremely black.

The SEALS, *properly so called, or without external ears,*

Have pointed incisors, the external parts of which above are longer than the other teeth. They have trenchant molars with many points. All their toes possess a power of motion, and are terminated by pointed nails placed on the edge of the membrane which unites them.

The common Seal. (Phoca Vitulina, L.) Buff. XIII. XLV. *and Supp. VI.* XLVI.

From three to five feet in length: of a yellowish gray: more or less waved or spotted with brown according to the age. It becomes white in old age. It is common on our coasts and found to a considerable distance in the north. We are even assured that it is this species which inhabits the Caspian Sea, and the large lakes of fresh water in Russia and Siberia, but this assertion does not appear to be founded on a very exact comparison.

The Crescent Seal, or Swartside. (Phoca Groënlandica.) Egede Groënl. fig. A. pag. 62.

Yellowish gray, spotted with brown when young, marked afterwards with an oblique brown semilunar mark five feet long. Of the Icy Sea.

The White-bellied Seal. (Phoca monachus, Gm.) Buff. Supp. VI. pl. xiii.

From ten to twelve feet long. Blackish brown, with white belly. Of the Mediterranean, and more especially of the Adriatic Sea.

The Bottle-nosed Seal. (Phoca Leonina, L.) Sea-Lion of Anson ; *Sea-Wolf* of Pernetty ; *Sea-Elephant* of the English and of Peron. *Peron, Voy. L.* xxxii.

From twenty to twenty-five feet long ; brown; the muzzle of the male is terminated by a wrinkled sort of horn or snout, which swells up when the animal is angry. It is common in the southern latitudes of the Pacific Ocean, in Terra-del-Fuego, New Zealand, and Chili. It is hunted on account of the abundant oil which it furnishes.

The Hooded Seal. (Phoca Cristata, Gm., Phoca Leonina, Fabricius.) Egede, Groënl. pl. vi.

Eight feet long. A sort of moveable cowl or crest attaches to the summit of the head, with which the animal covers its eyes and muzzle when threatened. Inhabits the Icy Sea.

The SEALS, *with external ears.* (OTARIES, *Peron.*)

Might deserve to form a genus apart, since, beside the external projecting ears, they have the four middle incisives in the upper row with double edges, a form hitherto unremarked in any other animal. The external ones are simple, and smaller, and the four lower ones are forked. All the molars are simply conic; the toes of the anterior web-feet are almost immoveable; the membrane of the hinder feet is prolonged in a small strap; beyond each toe all the nails are flat and slender. Their hair is less smooth than that of the preceding tribe.

The Maned Seal. (Phoca jubata, Gm.) Sea Lion of Steller, Pernetty, *&c. Buff. Supp. VII.* XLVIII.

From fifteen to twenty feet long and more. Yellow; the neck of the male covered with hairs thicker and more frizzled than those on the rest of the body. It may be found in the entire Pacific Ocean, if, as seems probable, those of the Straits of Magellan do not differ from those of the Aleutian Isles.

The Sea Bear. (Phoca Ursina, Gm.) Buff. Supp. VII. XLVII.

Eight feet long, without a mane, varying from brown to whitish. Of the North Pacific Ocean. Some other seals are found in this sea, not differing much from the sea-bear except in size and colour. Such is the *small Black Seal* of Buffon. *(Phoca pusilla) Buff. XIII.* LIII. *The Yellow Seal* of Shaw, *&c.*

The Morses, (Trichecus.)

Resemble the seals in the limbs and general form of

the body, but differ much in the head and teeth. Their lower jaw wants incisors and canine teeth, and takes in front a compressed form that it may be placed between two enormous canines, or rather tusks, which grow from the upper jaw, and are directed towards the lower, being sometimes nearly two feet in length and of a proportionable thickness. The enormous size of the alveolars necessary to form a lodgment for canines of this description elevates the top of the upper jaw in the form of a large inflated muzzle, and the nostrils are nearly directed upwards, and do not terminate the snout. The molars have all the figure of short and obliquely truncated cylinders. Four are reckoned on each side, above and below; but at a certain age two of the upper ones fall out. Between the two canines are, moreover, two incisors, like the molars, and which the generality of writers have not recognised to be incisors, though they are inserted in the intermaxillary bone. Between them are also two small and pointed ones in young individuals.

The stomach and intestines of the morses are pretty much the same with those of the seals. It would seem that they are sustained as well on fuci as on animal substances.

But one species has been as yet distinguished, which is called* *The Sea-Cow, Sea-Horse, Great Toothed Beast,* &c.

* Shaw suspects that there may be two distinguished by tusks more or less thick and more or less convergent.

(Trichecus rosmarus, &c., *Lin.) Buff. XIII.* LIV. and better Cook, *III. Voy.*

It inhabits all parts of the Icy Sea, surpasses the strongest bulls in breadth, attains even to twenty feet in length. It is covered with a smooth and yellowish hair. It is in great request for its oil and tusks, the ivory of which, though grained, can yet be employed in the arts. Excellent main-braces for carriages are also made of the skin*.

The MARSUPIATA, *or* POUCHED ANIMALS,

Which we range at the end of the Carnassiers, as a fourth family of this great order, but which might almost form a separate order, as they present so many peculiarities in their economy.

The first of these peculiarities is the premature production of their young, which are born in a state scarcely comparable to the development at which ordinary fœtuses arrive within a few days after conception. Incapable of motion, scarcely shewing the germs of limbs and other external organs, these little ones remain attached to the teats of the mother until they are developed as far as the young of other animals are at their birth. The skin of the abdomen is most usually disposed in the form of a pouch round the teats, and the imperfectly developed young

---

* Previous naturalists have very inconveniently joined to the Morses the Lamantins and Dugongs, animals which are much more closely allied to the Cetacea.

are there preserved as in a second matrix, and even after they have learned to walk, they constantly return hither when they apprehend any danger. Two peculiar bones, attached to the pubis and interposed within the muscles of the abdomen, afford a support to the pouch, and, strange to say, are found in the males as well as females, and in those species where the fold of skin which constitutes the pouch is scarcely visible.

The matrix of the animals of this family is not opened by a single orifice in the bottom of the vagina; but it communicates with this canal by two lateral tubes in the form of a handle. It would seem that the premature birth of the young is connected with this very singular organization. The males have the scrotum situated in front of the penis, unlike all other quadrupeds. Another peculiarity of the marsupiata is, that notwithstanding a general resemblance between the species, so striking, that for a long period they were formed but into a single genus, yet they differ so strongly in the teeth, feet, and organs of digestion, that were these characters rigorously attended to, these animals might be divided among various orders. We find in them insensible shades from the carnassiers to the rodentia, and without attending to the peculiar bones of the pouch, and regarding all which possess them as marsupiata, many might be found which it would be proper to insert among the Edentata; and, in fact, we shall have them there under the name of *Monotremes*.

We might say, in a word, that the Marsupiata

form a distinct class, parallel to that of the ordinary quadrupeds, and divisible into similar orders. This holds good so far, that if we were to place the two orders in two opposite columns, the Sarigues, Dasyuri, and Perameles, might stand against the insectivorous Carnassiers with long canines, such as the Tenrecs and the Moles. The Phalangers and Kangaroos, might pair with the Hedgehogs and Shrews. The Kangaroos, indeed, properly so called, cannot with propriety be compared to any other animals, but the Phascolomys might rank with the Rodentia.

Linnæus arranged all the species which he knew under his genus *Didelphis,* a word which signifies double matrix. The pouch may doubtless, in some respects, be considered as a second. The first subdivision of the Marsupiata. is distinguished by long canines and small incisives in the two jaws, back molars bristling with points, and, in general, by all the characters belonging to the teeth of the insectivorous Carnassiers. Accordingly we find that the regimen of these two families is almost entirely similar. The thumb of the hinder feet is also opposable, which has originated for these animals the name of *Pedimana.* It wants a nail. The two first sub-genera have the four other toes distinct.

The Sarigues* (Didelphis, *L.*)

Have ten incisors above, the middle ones a little

---

* *Carigueia* is their Brasilian name, according to Margrave, whence have been formed *Sariguoi, Cerigon, Sarigue.* They are named *Micouré,* in Paraguay; *Manicou,* in the islands; *Opossum,* in the United States; and *Thlaquatrin,* in Mexico.

longer than the rest, and eight below. Three anterior cheek-teeth compressed, and four posterior ones bristling with points, the upper of which are triangular and the lower oblong. In all they have fifty teeth, the greatest number yet observed among quadrupeds. The tongue is rough and bristly, the tail prehensile and partly naked. The thumb of the hinder foot is long and considerably separated from the other toes. Their mouth being very deeply divided, and their large naked ears, give them a very peculiar physiognomy. They are fetid and nocturnal animals, very slow in their motions. They lodge in trees, and there pursue birds, insects, &c., without, however, rejecting fruits. Their stomach is small and simple, the cæcum of moderate size, and not turgescent.

In certain species the females have a deep pouch, within which are their teats and where they can enclose their little ones.

The Sarigue, with bi-coloured ears, Opossum of the Anglo-Americans. (Didelphis Virginiana.) Penn. Hist. Quad. 302.

Almost as large as a cat, with fur mingled with black and white ; the ears equally divided between white and black, and the head almost entirely white. Inhabits the whole of America; comes by night into the frequented places to attack hens, eat their eggs, &c. Their young ones, sometimes sixteen in number, do not weigh above a single grain at their birth. Although blind and almost unformed, they find the teat by instinct, and adhere to it until they have grown to the size of a mouse, which does not take place till after the fiftieth day, when they open their eyes. They do not discontinue their visits to the pouch until they are as large as rats. The period of gestation in the uterus is only twenty-six days.

The Crab-eater, or Great Sarigue of Cayenne, Brazil, &c. (Did. Marsupialis et Did. Cancrivora, L.) Buff. Supp. III. LIV.

Of the size of the preceding ; yellowish, mingled with brown; a brown line on the forehead ; lives in marshy places near the sea-shore, and is sustained principally on crabs.

The four-eyed, or middle-sized Sarigue of Cayenne. (Didelphis Opossum, L.) Buff. X. XLV. XLVI.

Chestnut or fawn-coloured above, whitish below. A spot of pale yellowish above each eye. Larger than a large rat.

Some other species have no pouches, but merely a fold on each side of the belly, which is but the vestige of a pouch. Their custom is to carry their little ones on their backs, with the tails twisted round that of the mother.

The Cayopollin. (Did. Cayopollin, Did. Philander, Did. Dorsigera, L.) Buff. X. LV.

Yellowish gray; the circumference of the eyes

and a band upon the nose brown, the tail spotted with black. It is about as large as the Surmulot or Brown Rat *(mus decumanus)*.

Marinose, or Marine Opossum. (Did. Marina.) Buff. X.
LII. LIII.

Yellowish gray, a brown spot in the midst of which stands the eye. Tail not spotted. Less than a rat.

The Touan. (Did. brachyura.) Buff. Supp. VII. LXI.

The back blackish, the flanks of a lively red, the belly white, the tail shorter than the body. Less than a rat.

These three species belong to South America.

Finally, one species is known with the feet webbed, and must be aquatic. It is not known whether it has a pouch or not. It is

CHIRONECTES, *Illig. (Didelp. Palmata, Geoff. The small Otter of Guiana, Buff. Supp. III.* XXII. *Lutra memina, Bodd.)*

It is brown above, with three transverse gray bands broken in the middle, and white underneath. Larger than a Surmulot.

The DASYURI, (DASYURUS, *Geoff.)*

Have two incisors and four cheek-teeth less in each jaw than the Sarigues. Thus there remain to them only forty-two teeth, and their tail covered all over with long hairs (from which their name is derived δασὺς and οὔρος) is not prehensile. Their hinder thumb is also much shorter, and like a tubercle. They inhabit New Holland, and live on insects, carcasses, &c., sometimes penetrating even into the houses, where their voracity renders them very unseasonable guests. Their mouth is less divided, their muzzle less pointed, their ears are hairy, and shorter than in the sarigues. They do not climb trees.

The Dog-headed Dasyurus, (Did. Cynocephala,) Harris, Soc. Lin. IX. xix.

As large as a dog (three feet and a half long without the tail, which is nearly two,) tail compressed ; gray fur.

The Rough Dasyurus, (Did. Ursina, id. ib.)

With long, coarse, black hairs, and some white spots irregularly distributed. It inhabits with the preceding the north of Van Dieman's Land. Mr. Harris gives it eight incisors above and ten below ; makes the tail slightly prehensile, and naked on the under side. It may probably form a new sub-genus when better known.

The Long-tailed Dasyurus, (Das. Macrouros), Geoff. Peron. Voy. pl. xxxiii.

As large as a marten, with a tail as long as the body. Brown fur spotted with white covering the body and the tail.

The Dasyurus of Maugé.

Olive-coloured, spotted with white, but without spots on the tail. Somewhat less than the preceding.

The Dasyurus of White, (Did. Viverrina, Shaw,) Gen. Zool.
CXI. White Bot. b. App. 285.

Black, spotted with white, without spots on the tail. One-third less than the preceding.

The Tapod-Tafa, White, Bot. b. App. 281.

Uniformly grayish.

The Pencilled or Brush-tailed Dasyurus, (Did. penicillata, Shaw) Gen. Zool. I. II. pl. cxiii.

Gray, the tail covered with black and rough hairs.

The Dwarf Dasyurus.

Less than a rat, of an ashy-red. The thumb longer, and the teeth more equal and more contiguous than in the preceding species. Inhabits the South of Van Diemen's Land.

The PERAMELES, (PERAMELES, *Geoff*) *Phylacis, Illig.*

Have the hinder thumb equally short with the Dasyuri, and the two succeeding toes are united together by a skin as far as the nails. The great and small toe of the fore-feet have the form of simple tubercles. The upper incisors are ten in number, the external ones of which are pointed and separated. The lower incisors are six only. Their molars are the same as those of the Sarigues. They have forty-eight teeth in all. The tail is hairy and not prehensile. They also inhabit Australasia. Their large nails, almost straight, denote that they dig into the ground, and their long hind-feet that their course must be rapid.

The Perameles with pointed muzzle, (Perameles Nasutus, G.) Ann. du Mus. IV.

With muzzle exceedingly elongated, pointed ears, and fur of a grayish-brown. It resembles the tensee at the first glance.

The second subdivision of the Marsupiata possesses in the lower jaw, two long and large incisors, pointed, and with trenchant edges, inclining forwards, and having six corresponding ones in the upper jaw. Their upper canines are also long and pointed. But the lower canines are such excessively small teeth, that they are often concealed by the gums. The last sub-genus is even sometimes found to want these lower teeth altogether.

Their regimen is in a great measure frugivorous. Their intestines, and above all, their cæcum, is accordingly longer than those of the Sarigues. They have all a large thumb, so much separated from the other toes as to have the appearance of being directed backwards almost like that of birds. It is destitute of nail, and the two toes which follow it are joined together by the skin, even as far as the last phalanx. This arrangement has caused these animals to be termed,

PHALANGERS, (*Phalangista, Cuv.*)

The PHALANGERS PROPERLY SO CALLED, (*Balantia, Illig.*)*

Have not the skin of the flanks extended. They have in each jaw on each side four back grinders, each presenting four points on two ranks, in front a large conical one compressed, and between this and the upper canine two small and pointed ones; to which correspond the three very small ones below, of which we have spoken. The tail is always prehensile.

Some have the tail in a great measure scaly. They live in the Moluccas on trees, where they seek for insects and fruits. When they see a man they suspend themselves by the tail, and it is possible to make them fall through lassitude, by continuing to

* The name *phalanger* was given by Buffon to a single species known in his time, on account of the union of the two toes of the foot. That of *philander* is not, as might be supposed derived from the Greek, but from the word *pelandor*, which in Malay means a rabbit, and which the inhabitants of Amboyna give to a species of the kangaroo. Seba and Brisson have applied it indistinctly to all pouched animals. In the Moluccas the phalangers are called *couscous* or *coussous*. The first travellers, not having sufficiently distinguished them from the *sarigues*, gave occasion to believe that this last genus was common to the two continents. *Balantia* comes from Βαλαντιον, a purse or pouch.

stare at them for some time. They diffuse an unpleasant odour, yet their flesh is eatable.

As to their colours, some are whitish, some gray spotted with black, some red with a brown streak along the spine (which are most common), and some brown with white crupper. But the limits of their species have not yet been precisely determined. Linné's denomination of *Didelphis Orientalis* embraces them all. (Buff. XIII. x. xi.)

The Fox-like Phalanger, (Did. Lemurina et Vulpina, Shaw.)
Bruno de Viq. d'Az. White, Voy. 278.

As large as a rat, or even as a cat, grayish-brown, more pale underneath, with a tail chiefly black.

The Phalanger of Cook, (Cook, Last Voy., pl. viii.)

Less than a cat, reddish-gray, white under neath, red in the flanks, a white interval towards the end of the tail.

The FLYING PHALANGERS, (PETAURUS, *Shaw.) Phalangista, Illig.*

Have the skin of the flanks more or less extended between the legs, like the polatouches (flying squirrels) among the rodentia, which permits them to sustain themselves for some instants in the air and to make more extended leaps. They are also found no where but in New Holland.

In some of this species are found some lower canine teeth, but extremely small. Their upper canines, and the three first molars, both above and below, are remarkably pointed. The back molars have each four points.

The Flying Dwarf Phalanger, (Did. Pymæa, Shaw,) Gen. Zool. pl. cxiv.

Of the colour and almost the size of a mouse. The hairs of the tail are very regularly disposed on each side like the barbs of a pen.

Some other species want the lower canines, and the upper are very small. Their four back molars also present four points, but somewhat curved in the form of a crescent, which is pretty nearly the form of those of the ruminantia. In front there are two above and one below, less complicated. This structure renders them still more frugivorous than the preceding species.

The Great Flying Phalanger, (Did. Petaurus, Shaw,) Gen. Zool. pl. cxii. *White, Voy.* 288.

Resembles the taguan and the galeopithecus in size. Its fur is soft and copious, its tail long and compressed or flattened. They are of different shades of brown. Some are variegated and others whitish.

The Long-tailed Flying Phalanger, (Did. Macroura, ib.)

Deep brown above, white underneath. As large as a Surmulot, with a slender tail, about one and a half as long again as the body.

Our third subdivision has the incisives, and the
upper canines like the second. The two toes of the
hind feet are also united in a similar manner. But
they want the posterior thumb, and the lower canines. It comprehends but a single genus.

The KANGUROO-RATS, (HYPSYPRYMNUS, *Illig.**)

The last animals of this family which preserve any
thing of the general characters of the Carnassiers.
Their teeth are pretty nearly the same as those of
the phalangers, and they have an additional pointed
canine above. The two middle incisors in the upper
row are longer than the others and pointed. Below
they have but two, inclining forward. They have a
long molar in front, trenchant and indented, followed
by four others with four blunt tubercles. What,
however, most eminently distinguishes these animals is their hind legs, being much longer in proportion than the fore. The feet of the latter want the
thumb or great toe, and have the two first toes
joined together as far as the nail; so that it might be
imagined, on a slight glance, that there were but
three toes, the internal one of which had two nails.
They often walk on two feet, and employ their
long and strong tail to support themselves. They
have the form and habits of the kanguroos, from
which they differ only by having their canine tooth
in the upper jaw. Their regimen is frugivorous,
and their stomach large, and provided with many

* Ὑψιπρυμνος, raised from the hinder part.

convolutions; but the cæcum is rounded and of moderate size.

There is but one species known, of the size of a small rabbit, and of a mouse-coloured gray, which has been called *Kanguroo-rat,* (*Macropus minor,* Shaw.) It comes from New Holland, where the inhabitants call it *potoroo.* White, Bot. B. 286.

The fourth subdivision does not differ from the third, but by the total absence of canine teeth. It is that of

The KANGUROOS, (MACROPUS, *Shaw. Halmaturus, Illig.**)

Which exhibit all the characters which we have just assigned to the preceding genus, except that the upper canine is wanting, and that the middle incisors surpass the others. The irregularity of their limbs is still more remarkable. This is so great, that they walk on all-fours with difficulty and slowness, but bound with prodigious vigour on their hindfeet, the large nail in the middle of which, almost in the shape of a wooden shoe, serves them also as a weapon of defence; for, resting on one leg and their enormous tail, they can give very violent blows with the foot which is at liberty. In other respects they are animals extremely mild, and their regimen is herbivorous. Accordingly, we find that their cheek-teeth exhibit only transverse cones. They have but five, the anterior of which fall out in age, so that the old ones have no more than three. Their

---

* Fitted for leaping, ἄλλομαι.

stomach is formed of two long pouches, divided into cavities like the colon. Their cæcum is also large and cavernose. The radius allows a complete rotation to their fore-arm.

The penis of these two genera is not divided, but the female organs are the same as in the other pouched animals.

The Gigantic Kanguroo (Macropus Major, Shaw. Didelphis Gigantia, Gm.) Schreb. CLIII.

Sometimes is six feet high; the largest animal in New Holland. It was discovered by Cook, in 1779, and at the present day it propagates in Europe. They say that its flesh resembles that of the stag. The little ones, which have but one thumb when born, betake themselves to the pouch of the mother, even when they are old enough to graze, which they do by thrusting their muzzles out of the mother's pouch when she is herself grazing. They live in troops conducted by the old males. They make enormous leaps.

It would appear that under this name many species of New Holland and the adjoining countries have been hitherto confounded, the fur of which, more or less gray, varies only by very trifling shades*.

* M. Geoffroy distinguishes the *Smoked Kanguroo*, the gray of which is somewhat deeper. The *Kanguroo with Moustaches*, which has some white on the front of the upper lip. The *Red-necked Kanguroo*, rather less than the others, and having the nape of the neck tinted with red.

Very lately has been discovered,

The Elegant Kanguroo, (Macr. elegans,) Peron. Voy. I. xxvii.

Of the size of a large hare, gray-white, radiated cross-wise with brown. Of the island of St. Pierre.

Another species has been known much more anciently :

The Kanguroo of Aroé, (Didelphis brunii, Gm.) Schreb. CXLIII. Named Pelandor Aroé, or the Rabbit of Aroe, by the Malays of Amboyna.

European naturalists did not pay sufficient attention to the description given by Valentine and Le Bruyn of this animal. It is larger than a hare, brown above and fawn-coloured underneath, and is found in the islands of Aroé, near Banda, and in that of Solor.

The fifth subdivision has in the lower jaw two long incisors without canines, and the upper two long incisors situated in the middle, some small ones on the sides, and two small canines. It comprehends but one genus.

The Koalas,

With clumsy body, short legs, and without a tail. Its front toes, to the number of five, are divided into two groups for the purpose of grasping : the thumb and index are on one side, the three other toes on the opposite. The thumb is wanting in the hind-foot, the two first toes of which are joined together as in the preceding genera.

But one species is known, with ash-coloured hair, which passes a part of its life in trees and another part in burrows dug with its feet. The mother carries her young one for a long time on her back.

Finally, our sixth division of the Marsupiata will be

The PHASCOLOMES. (PHASCOLOMYS, *Geoff**.)

These are true rodentia by the teeth and the intestines. The only affinities they preserve to the Carnassiers consist in the articulation of the lower jaw, and in a rigorous system it would be necessary to class them with the rodentia. We should, indeed, have so classed them, had we not been conducted to them by an unbroken series from the didelphes to the phalangers ; from those to the kanguroos ; and from the kanguroos to the phascolomes ; and also if the organs of generation had not been precisely similar in all the pouched animals.

They are heavy animals, with a large flat head, short legs, and shapeless body. They are without a tail, and have five nails on the front-toes, and four, with a small tubercle instead of thumb, on the hinder feet, all long and proper for digging. Their gait is excessively slow. They have in each jaw two long

---

* Pouched Rat, from φάσκωλον and μῦς.

incisives almost like those of the rodentia, and their cheek-teeth have each two transverse cones.

They live on grass, have a pyriform stomach, and a broad and short cæcum, provided, like that of man and the ourang-outang, with a vermiform appendage. The penis is forked as in the sarigues.

But one species is known, about the size of a badger, with a copious fur more or less yellowish. It lives in King Island, south of New Holland, in burrows. It would propagate with facility in our climates, and its flesh is said to be excellent. It is

The Didelphis Ursina of Shaw.   The natives call it *Wombat**. (*Peron, Voy. pl.* xxviii.)

* M. Bass has described an animal externally the same as the Phascolomes, and to which he also gives the name of *Wombat,* but which would appear to have six incisors, two canines, and sixteen molars in each jaw.   If there be not some erroneous combination of two different descriptions, this would be an additional sub-genus to place among the perameles.   M. Illiger has already established it under the name *Amblotis.*—See Mém. de Petersb. 1803 to 1806, p. 444, and the Bulletin des Sciences, No. 72, An. XI.

PL. 1. OF THE REGNE ANIMAL.

1. THE GREAT GALAGO. 1.P.230. 3. THE AMERICAN OTTER. II.P.38.

2. THE VISON. II.P.37 4. THE RED WOLF. II.P.42. 5. THE KOALA. II.P.76.

kar. Sep. 1825.

SUPPLEMENT

TO

THE CARNASSIERS.

THE name given by our author to his second order of mammiferous animals, and which is adopted here without alteration, may demand some explanation. It certainly has an
unscientific sound to scientific ears ; nor is it really unex-
ceptionable ; but as difficulties attend its rejection, and the
consequent substitution of a better term, greater than
those which accompany its use, prudence dictates the choice
of the lesser evil.

Carnassiers (or flesh-eaters simply) is applied by our
author to eaters of flesh, either partially or exclusively. In
that comprehensive sense it would indeed include many
animals not arranged by him in this order, belonging both
to this and even other classes ; but it is restricted again to
such flesh-eaters as have the three kinds of teeth, together
with the jaws articulated exclusively for a vertical opening
and the other minor particulars pointed out by the Baron in
his introductory observations on the order.

The word carnivora is more familiar to the English
reader, which, as it conveys a signification of a voracious
appetite for flesh, rather than simply the means or inclination to eat, it is aptly enough applied exclusively to those
which feed altogether, or almost altogether, on animal matter.   These may be also conveniently termed beasts of prey.

The Carnassiers, then, according to Cuvier, include, as we
have seen, five families :—the cheiroptera, or bats ; insectivora ; carnivora, or beasts of prey ; the amphibia ; and the
marsupiata, or pouched animals.

The flying membrane of the bats, the subterraneous habits of the insectivora, the sanguinary voracity of the beasts of prey, the aquatic location of the amphibia, and the double matrix of the marsupiata, would offer grounds for separating each of these families into distinct orders with sufficient precision and exactness ; but the same influential character of dentition pervades them all—the analogy is general—incisive teeth, canine and cutting cheek-teeth are found throughout, varying in aptitude for carnivorous regimen, in accordance with the degree of carnivorous impulse of the several species.

In zoological arrangement, as has been observed, the most influential characters must be exhausted before those of minor consequence are resorted to, and none is more important in its consequences than that of dentition*; whatever other discrepancies, therefore, may be observed between the different families of this order, principle requires that apparent convenience should be sacrificed to real propriety, and that all the animals with flesh-eating teeth should be included in one order, for which the word carnassiers, as applicable to the character of the teeth, is equally serviceable with any classico-barbarous appellation that might be coined for the purpose.

The animal kingdom, when viewed abstractedly in regard to the carnivorous propensities of so very large a portion of it, presents a painful and distressing picture, a picture the eye of humanity could never bear to contemplate but for a simultaneous consciousness and self-conviction of the mind, of a propensity and practice in ourselves, which we revolt from in inferior beings. In this instance it seems indeed necessary to suppress and stifle our better feelings, rather than

---

* This remark is subject to some qualification. In the cheiroptera, as we shall see, dentition is not always the most important character.—P.

by refinement and sensitiveness, vainly to endeavour at counteracting the impulse of nature, or rashly to oppose the imperious dictates of necessity. Nor can we very satisfactorily account in reasoning and theory, for that which disgusts us so much in practice. Why the life of one should necessitate the death of another ; why many indeed should die that one m ay li ve, is very difficult to answer.

The Mosaic history affords negative evidence, that the carnivorous necessity was not coeval with animal creation. " And to every beast of the earth, and to every fowl of the air, and to every thing that creepeth upon the earth, wherein there is life, I have given every green herb for meat." Man, according to the same sacred authority, was not authorized to eat flesh till after the flood. It seems, therefore, that the change generally took place after the pristine state of animal life, and must have been accompanied with great changes, both in physical character and mental impulse. We are not able to surmise any rational cause for this, unless we refer it to the fall of man, which " brought death and all our woes into the world." Should the suffering of innocence for the crimes of guilt be deemed unjust, and therefore improbable, we may observe analogous dispensations of Providence around us ; the fate of many species is more or less hard now, according to the direction in which man can turn them to his use. In society also no man is so isolated as not to confer credit or shame, misery or happiness, on many connected with him. Even the mysterious doctrine of atonement may be said to be in some degree analogous to the subject in question.

Fossil osteology, it is true, as far as speculative theory has been hitherto deduced, from the facts offered by that branch of science to our notice, offers no confirmation of an age ex-cluding animal food. The oldest fossil remains, presumed by their location in remote strata, to be of incalculable antiquity, are of carnivorous animals now known, of others of the same regimen, apparently still more formidable than any that now exist, as well as of species of the herbivorous races, both known and unknown, and if the antiquity of these bones be established, it seems to refer the era of original creation to a more remote period than is ordinarily understood.

The improvements in modern science lead us to the conclusion that new creation and annihilation never now take place—varied modifications, and consequent unceasing activity, is the business of a large portion of matter, of all of it, indeed, which enters into the composition of organization. Organized bodies, vegetable as well as animal, are minute when first endowed with vitality; they accumulate, by degrees, till maturity, perform the important office of germination, and almost immediately proceed in the same gradual way to nonentity again, not indeed by decreasing in size, but in vital energy, till life becomes extinct, and the disorganized particles separate.

These particles sooner or later again appear to enter into the composition of some other living body, modified in shape indeed, but engaged in the same office of vitality, till again dismissed, and taken up by a third; and thus does all matter capable of organization proceed successively in its destined office.

The modes by which this is performed, in animal and vegetable bodies, are very analogous; the plant sucks up its nutrition by its roots, not from the steril sand or rock, but from the disorganized particles in the earth and water, which yield nutrition in proportion as they are saturated with such particles. The roots of animals, that is, the stomach, in like manner, derive support from disorganized matter; in some species, from vegetables only, in others, from flesh alone, and in many, from both. The animal, as well as vegetable, having played its part, dies, but still is essential, by administering to the necessities of its survivors, more or less, directly of either kind.

It may be observed, that flesh decomposed is much more favourable to the sustentation of the living plant, even than vegetable compost. Abstract the immaterial *anima* therefore from the flesh, and in its elements, at least, it is nearly allied to the mere vegetables.

The conclusion, from these premises, therefore, if correct, appears to be, that all aliment, which must be either flesh or vegetable, however dissimilar in present character, is essentially the same.

The food of animals is, to a certain extent, a subject of volition and choice, but they are restricted in their election by physical adaptations. Many of the order before us, have not the power either of masticating or digesting vegetable substances, others can do it partially—the former possess strength, agility, and the power of enduring abstinence to a very great degree, the latter sort possess these qualities, proportioned to the necessity for carnivorous regimen ; and all are endowed according to such necessities, with an appetite and mental impulse for destruction, by a power over which they can exercise no control.

It may, perhaps, be questioned whether some of the feelings to which we might wish to give another origin, as a taste for sport, as it is called, by the destruction, in various ways, of less powerful beings, do not owe its existence to some such principle inherent even in the mind of man.

The several physical peculiarities, above alluded to, which are displayed most obviously in the teeth, may be observed to form the principal grounds of the subdivisions of this order, combined with other characters more obvious, perhaps, to the senses, but less comprehensive in extent.

Supplementary Essay on the Cheiroptera.

ACCORDING to the opinion of some clever writers the quadrupedal form is the natural and primitive construction of the mammiferous tribes. It appears, say they, to be best suited for the reciprocal relations which subsist between those animals and the various localities by which they are environed. Accordingly, they consider the deviations from this original plan in the light of anomalies ; and the progressive locomotion of man, for example, on one pair of extremities, while the other is adapted to different uses, and fits him for new destinies, they believe an exception to the general rule .

If relatively to man this consideration be curious, how much more reason have we for surprise on the contemplation of other exceptions to this common law which also seem to depart from all rules of proportion and all adaptation of means to ends. The original plan seems utterly overturned, and the combinations which result from its subversion, can scarcely be deemed less than monstrous.

Such are the sentiments which a general view of the bats is likely to produce in the mind of an observer. We are prejudiced against their deformity, and revolted by their disgusting ugliness. Such ideas, indeed, have been carried so far that nations have not hesitated to pronounce these animals impure, and not only to abstain from touching them but even to avoid all knowledge of them whatsoever.

The writings of naturalists sufficiently attest the early ignorance respecting the peculiarities of this singular genus. Aristotle defines them to be birds with membranous wings. He hesitates, however, to class them with the winged tribes, on account of their feet, and perceiving them unprovided with four distinct feet, he cannot prevail upon himself to rank them exactly among quadrupeds. His theoretical deductions, from their want of tail and crupper, are by no means founded on any positive observations.

Pliny notices them only to remark that there are birds which are viviparous and suckle their offspring by means of teats.

On the revival of letters in Europe, men were contented in this respect, as in every other, to copy the ancients. Aldrovandus was the first who expanded a little on the subject of the bats ; but in conformity with the prejudices of his age, he made one family of them and the ostrich, because these two species of *birds* partake equally of the nature of quadrupeds. Scaliger makes quite a sort of miracle of the bat: he discovers that it has two and four legs : that it walks without paws, and flies without wings : that it sees in darkness and is blind in light : in short, that it is the most remarkable of all birds, for it has teeth while it is destitute of a beak.

Even at a later period the organization of the cheiroptera was but little studied. They were considered only as far as it was necessary to comprehend them in certain methodical distributions, and those points of their conformation alone attended to which corresponded with the bases which had been arbitrarily established for zoological systems. However, a pretty correct idea was soon attained of the affinities of the bats, for, fortunately, such external characters were chosen as grounds of distinction, as corresponded with anatomical characters more general and more profound. The bats were no longer separated from viviparous quadrupeds : a profounder study of their organization confirmed those indications which their teeth had furnished.

The bats, like the viviparous quadrupeds in general, have the bilocular heart, the cellular lungs, suspended and enclosed in the pleura ; the muscular diaphragm interposed between the cavity of the thorax and that of the abdomen :

an ample and compacted brain ; and the cranium composed of a similar number of pieces equally convoluted. Their sensitive, digestive, and secretive systems are similar. The teeth of three sorts, the body equally covered with hair, and they are viviparous and mammiferous. Their bones, their muscles, their vessels, are all upon the same construction as in the quadrupeds in general, and the resemblance is so great that the minutest details of their structure would suffice solely and separately to shew that they are true mammalia, and must be comprehended in that class.

Far, however, beyond a result like this were the bold views of Linnæus, who ranged them unhesitatingly in the same order with man and monkeys, giving to both a similar name : sometimes that of *Anthropomorphæ* (beings with a human visage) ; sometimes that of *Primates* (animals of the foremost rank in creation). Extraordinary as this classification may appear, it has been consecrated by the illustrious name of its inventor.

A new school followed, which admitted among all living beings certain successive and graduated relations, and a progressive advance from the simple to the composite. The case of the bats, constituted like the mammalia, but hovering in the air like birds, furnished an example of transition from one class to another too tempting not to be eagerly adopted. But the faculty of flight in birds and in bats results from a very different mode of organization, and to establish a relation between them from the circumstance of both being sustained in the air, is to confound the effect with the cause. This is produced in each by very different instruments, and the anomaly which this faculty presents in the cheiroptera is clearly derived from the type of the mammalia.

The parts which correspond to fingers are in birds very nearly effaced. They exist only in a rudimentary state, attenuated and sown as it were together. The hand of a bird is consequently a mere rudiment. The wing exists beyond it, resting and adjusted on this extremity of the limb, and consisting in its long terminal pennæ ; so that in fact, in the last analysis we find that the greater part of this organ is composed of branches or elements belonging to the epidermic system.

In the bat, on the contrary, it is the limb itself and principally the hand, which are so wonderfully aggrandized. Let us imagine the hand of an ape, the solid parts of which should have passed almost to a filament, and separated from the carpus, like the radii of the segment of a circle, and we shall have a precise idea of the hand of a bat.

The thumb alone does not experience the same modifications. It remains short, free, and susceptible of the most varied movements.

Such is also the thumb of apes. As it is not employed as an organ of flight by the cheiroptera, that it may preserve its ordinary function, it remains in its full entirety, provided with its last phalanx and with its nail.

On the other side, the four fingers, which their immeasurable length changes into instruments of volitation, are no longer susceptible of their original destination ; so that it is with much pain and difficulty that the bats occasionally employ them to move their bodies upon a horizontal plain, or to hold their little ones.

Another anomaly renders these four fingers worthy of attention. They are no longer entire. They are destitute of nails, and it is remarkable that the phalanx which terminates them, and which, in other instances, always appears with an impression beneath the nail, finishes exactly where the nail should begin.

The long phalanges of the bats, act as supports to their wings, and seem destined to retain in a state of tension the membranous substance which resists the air. This last is produced in the bats by a prolongation of the skin of the flanks. The back and the belly furnish each a leaf as we may be assured of by separating the membrane of the wings. Notwithstanding, however, that this membrane is formed of two skins, pasted as it were together, it appears to us but a sort of slender, light, and transparent net-work. Thus, as the bones of the hand are only elongated by diminishing in thickness, in like manner the tegumentary system is not extended but by growing slender in a similar proportion. Here it is worthy of remark, that a general law of all organization is made to contribute in a wonderful manner to facilitate the aërial movements of the cheiroptera, for bones more compact, and a membrane of greater density, especially at such a distance from the *vis motrix*, would have added a momentum to the bodies of these animals, that all their efforts to surmount would not have overcome.

This analysis of the wing of a bat, by presenting to us the arm and hand of a mammiferous animal, the metacarpi and phalanges of which are united by membranes, is not only sufficient to convince us that there is no analogy between it and that of a bird, but must also lead to the consideration of such extremities in the mammalia as are the best calculated for seizing objects, and the most profoundly divided. This will be necessary before we can perfectly understand all the anomalies of this singular organ. Now we find that the mammalia which possess the deepest digitations are the quadrumana; and as we find that in this particular the bats are next, we are led to conclude that Linnæus judged correctly of their natural affinities. But we are still more completely led to this conclusion by considering other traits of their conformation.

The farther we remove from the quadrumanous groups, who have the mammillary glands situated on the thorax, the more we observe those glands descend from the breast to the abdomen. All the bats, with the exception of one subdivision (the Rhinolphi) have teats exactly resembling those of the quadrumana, both in number and position. The organs of generation are similar; but it is the teeth in particular which leads us to the belief of identity of type in the quadrumana and cheiroptera. If it were otherwise, how could we account for the exact repetition of form in parts so complicated and so little essential to life as the incisive teeth? The roussettes have precisely the same sort of incisors as the monkeys, and the vespertiliones as the lemurs. The molars stand in similar relations, being formed in the latter by a crown, bristling with conic points, and in the former by a simple edge. The cheek pouches likewise, which most of the monkeys of the ancient continent possess, and which are so perfectly in conformity with their gluttony and restless character, are found also in the bats. They fill them with insects when they hunt, and reserve them as a banquet to gormandize upon in their retreats.

Such numerous relations between the quadrumana and bats prove that when Linnæus placed the genus Vespertilio after the Lemurs, he presented those animals in the natural order of affinity: but he certainly went too far in placing man, monkeys, lemurs, and bats, in one family, under the title of *Primates*. It might have been sufficient to observe that these families were derived from each other; but when one organ, like the hand, becomes from its disproportionate extension the predominant organ in the bats and the most actively influential on the entire mode of their existence, it is certainly incumbent on the naturalist to distinguish them by a more general term of classification. The arm of the bat not only prescribes the destiny of its life among created beings, but also demands a correlative adaptation of all the other parts of its organization.

One of the points of their organization most worthy of remark is the disposition of the cutaneous system to extend beyond the outline of the animal itself, and to communicate to the organs of sense more compass and activity. Sufficient attention has not perhaps been bestowed on the manner in which this extension takes place. The skin of the flanks is not only carried over the arms, and distributed between the phalanges of the metacarpi and fingers, but it also embraces the hinder extremities: it is prolonged, more or less, in the different species, between the legs, and spread to the length of the tail, so as to form a surface round the animal utterly disproportioned to the smallness of its body. In truth, a surface of this extent could alone impart such exquisite tact to the organs of touch and hearing. Spallanzani, who has observed their phenomena, attributes them to a sixth sense.

The external ears participate in this tendency to extension in the cutaneous system ; insomuch that sometimes, as in the *vespertilio auritus*, there is a portion of the ears prolonged over the forehead, and partly joined together, equalling in length the animal itself. They participate in this tendency too, in a manner not a little singular, being double in the majority of the bats. Independently of the external ala of the ear, which differs from that of other animals only by a greater degree of extension, there is a second which borders on the meatus auditorius.

Although this little ear, or auricula*, is not positively to be found except in the bats, it is not an organ of which there are no traces to be discovered elsewhere. Nature operates with a certain number of materials, which vary only in dimension. This auricula is derived from the tragus, or rather it is the tragus itself, which we might almost be tempted to consider as a separate part, in consequence of its extent and peculiar uses.

This susceptibility of increase in the tegumentary system is manifested at the entrance of other cavities of the organs of sense. In many of the cheiroptera we find the nose bordered with crests, or foliaceous appendages, formed by a duplicature of the skin. These membranes are disposed in the form of a tunnel, the end of which serves for an entrance

* The French term is " *oreillon*," an excellent word for which we have no equivalent.—P.

into the nasal cavities. Thus it is the same with the organ of smell as with that of hearing ; both are provided with an external ala or cornet.

Parts thus extended and multiplied cannot fail to exercise a very powerful influence, and hence the sphere of sensation and perception of the bats is very considerably enlarged. They acquire the notion of many minute corpuscles, to which no other animal can be sensible. From the observations of Spallanzani we learn, that immediate contact is not necessary to advertise the bats of the presence of external objects by the sense of touch ; it is sufficient if they feel the air interposed between them and such objects, and appreciate the mode of its re-action on the membranes of their wings. We find also that those large tunnels, placed in front of the organs of hearing and of smell, render these animals sensible to the feeblest emanations of sound, and the slightest odoriferous exhalations.

With these means of increasing the power of sensibility, and perception of external objects, the bats are also gifted with the faculty of excluding such perceptions at pleasure. This, indeed, is a most wise and necessary provision, as otherwise they would be completely overwhelmed by the effects of such sensitive acuteness. The perfection of their organs would be converted into a source of destruction. The little ear, or auricula, is made to border in such a manner on the meatus auditorius, that it serves as a valve to close that passage : a trifling inflection of the ear is sufficient for the performance of this operation, and in some species a mere contraction of the cartilages.

The excessive extension of the hand of bats exercises a predominating influence not only over the organs which set it in motion, but also on those of a higher order, and, in fact, over the entire frame, submitting as it were, to itself all the materials of organization. The organs of sense, confined in other animals within such narrow limits, present in the bats the most astonishing complications, and even the heart is in some measure displaced, and situated higher up in the cheiroptera. The pectoral muscles strongly experience this influence ; they are more voluminous, and have their seat and points of attachment on a sternum, composed of pieces remarkable for their size and perfect ossification. The sternum of the quadrumana, on the other hand, is weak, small, and almost entirely cartilaginous.

In the last mentioned animals the bones of the fore-arm are susceptible of pronation and supination, an immense advantage to animals designed to live in trees ; but this faculty would prove a serious inconvenience to the bats, as at every flapping of their wings, the resistance of the air might occasion a rotation of their hands. They are, there-fore, deprived of it. This is accomplished by the sacrifice of the cubitus, which, however, does not entirely disappear. The tertius humeralis remains, and this portion connected to the radius, contributes to give it sufficient force and solidity to sustain the carpus and the entire hand.

When we compare the anterior and hinder extremities of the bats, we may calculate in some measure the wonderful augmentation of the former. The latter remain within or-dinary dimensions, and are but partially involved in the membranes of the flanks. The foot is free. The interfe-moral membrane has its final points of attachment on the tarsus, and is sustained when in a state of developement by one of the small bones of that part, projecting outwards in a spiral form.

The hinder toes are small, compressed and equal, and always five in number. The thumb is indistinguishable. All are terminated by claws or little horny laminæ, formed like the quarter of a circle, very sharp at the point, and re-markable from their equality and parallelism.

This conformation of the toes enters of necessity into the constituent plan of these animals, and never experiences any change or modification. The functions, delegated elsewhere to the anterior extremities, are in the bats concentrated in the hinder ones, where alone any genuine toes are found. In the others, as we have observed, but one remains, the other four being nothing but solid stalks fitted to extend or fold up the membrane.

Such are the resources of the bat for terraqueous locomotion. On a superficial consideration they would appear not easily manageable for such a purpose, yet the animal can draw some advantage from them in case of necessity. The wings, when furled, are converted into fore-feet. The bat then becomes a quadruped, and moves itself along with a tolerable degree of velocity. But there is much pain, effort, and diversity of action necessary for this operation. At first we behold the bat push forward the end of its wing and cling to the soil, by burying in it the nail of its thumb : then from this " point d'appui" it draws its hinder limbs under the belly, rises from this squat position, and makes a kind of tumble which propels its body forward. But as it fixes itself to the ground only by the thumb of one wing, its motions are necessarily performed in a diagonal line, and it is thrown by the leap it makes on that side on which it was fixed. For the next step it employs the thumb of the opposite wing, and tumbling forward in an opposite direction, it proceeds by these alternate deviations on its destined course. This exercise is not a little fatiguing, and the bat seldom employs it except in the perfect security of its cavern, or when a series of accidents have occasioned it to fall on a horizontal plane. When a bat finds itself in this predicament, it endeavours to escape from it as soon as possible. For in such a situation it cannot rise from the ground and resume its flight. Its wings have too much extent, and the efforts which it makes to rise, only causes them to dash against the ground, and procure it a new fall. But if it can gain an elevated place, a tree, or even a hillock, it easily resumes the only position which is completely suitable to its nature.

It is in the air alone that the bats really enjoy themselves. It is there alone that they feel possessed of perfect liberty ; can avail themselves of their resources ; and experience the most boundless confidence. Sometimes, indeed, this confidence carries so far as to make them brave real dangers.

But these aerial courses cannot be perpetual : repose must follow. For this critical moment the bats reserve all their prudence. The perception of the dangers to which they are then exposed leads them to seek the most profound and inaccessible retreats, and to take the precaution of suspending themselves to the vaulted roof of caverns, with the head hanging downwards. Simply fastened in this manner, by the nails of their hinder feet, they have only to let go their hold to escape by flight every unforeseen attack.

We may now see the reasons for this inverse position, to which it is remarkable that none but the bats are restrained. From no other situation could they so conveniently resume the modes of operation to which they are familiar—from no other could they derive so many facilities of escape from their enemies, and of losing themselves in the vast immensity of air. When the bats are ready to shoot forth and have to unfold the embarrassing mantle formed by the membrane of their wings, it is necessary that they should have a space at its sides proportioned to the amplitude of its extent. That they should fall, therefore, from an elevated position, is obviously necessary to the complete fulfilment of their intended flight.

The hinder feet of the bats, which are intended to affix these animals to the ceilings of their retreats, must have a form appropriated to this destination. It is easy, on this principle, to account for the parallelism and equality of their toes, as also for the curved form and steely point of their talons. That system, which gives to the different parts of the organs of locomotion corresponding uses, and uses determined by necessary relations, is thus completed in the bats by extremities of this description, which are ever invariable in their forms.

It is impossible to enter the subterraneous abodes of the bats without being at first considerably affected by the odour of their dung. It is generally gathered in considerable heaps under the soil towards the centre of the spaces which they occupy, and it is perfectly obvious from what quarter the excrements proceed, namely, from the vault of the cavern.

The mode in which the bats void their excrements is somewhat singular. Their ordinary position, fastened as they are by their hinder feet to the roof of their habitation, is by no means favourable for such an operation. Therefore it must be altered. The bat, accordingly, first sets one of its paws at liberty, and strikes the vault with it repeatedly. Its body put in motion by this means oscillates, and is balanced on the five nails of the other foot, which form, by their equality and parallelism a right line, like the axis of a pair of hinges. When the bat has arrived at the highest point of the curve which it is describing, it extends its arm and seeks a resting point on which to fasten the nail which terminates it, namely, that of the thumb of the anterior extremity. It is sometimes the body of a neighbouring bat which serves for this purpose, a side wall, or some other solid object. Having thus fixed the nail, the animal has attained its end. It is placed in a horizontal position, which is the one most suitable for its purpose.

With respect to the organs of digestion in the cheiropterous tribes, we here discover, in a remarkable manner, the ascendant exercised by that type, from which the bats are an obvious deduction. All the traits of the quadrumana are reproduced, and what is more singular, with certain slight modifications, all of which have a close relation with certain trifling changes in the termination of the wing.

This is sufficient to prove the prodigious dominating influ ence of this organ over the general structure of the animal.

The majority of the bats live on insects. Their stomach is simple without thread or complication. The intestinal canal is short and of a diameter pretty equal throughout, and the cæcum is entirely wanting. The teeth correspond with this arrangement. The incisives are lobular, the canines long and sharp, and the molars bristling with points. Some bats (the Roussettes), which live principally on fruits, vary a little in the conformation of the teeth and intestines, and are also characterized by a lesser prolongation of the dermis : accordingly, it is with some difficulty that we concede to them the name of bats at all.

The sharp teeth of most of them are the only weapons with which nature has provided them to attack, seize, and lacerate those insects which form their subsistence. For, catching them in their flight, they possess one facility which has not been very generally remarked. This consists in the largeness of their mouths.

The opening of the lips in the mammiferous animals in general does not extend beyond the canine teeth. We would be almost inclined to assert that the upper lip followed the lot of the intermaxillaries ; that it was subordinate to them, and formed as it were their natural covering. In fact, the mouth is never large, and deeply cut, except in those animals which possess very long intermaxillaries; and on the other hand, is always extremely narrow where these bones are small. To this rule, however, the bats, at least those of them which are insectivorous, form a remarkable, and, we believe, a solitary exception. The commissure of their lips is prolonged considerably behind and corresponds with the last molar but one. Their cheek-pouches may be regarded as the cause of this anomaly. The cheeks, which are rendered flabby by these appendages, unfold and extend with the lips, and the lower jaw is thus capable of being so far separated from the upper as to form with it a right angle.

The bats resemble the smaller insectivorous mammalia in their gloomy habits, their nocturnal life, the susceptibility of the sensitive organs, which forces them to avoid light and noise, and the very small degree of their specific heat. They pass the winter, or more properly speaking, a considerable portion of the year in a state of lethargy. Exquisitely sensible to the slightest impressions of cold and humidity, they rarely sally from their retreats excepting in the fine evenings of summer. Then they enjoy a full portion of activity, and excited to a very high degree, they are attentive to nothing. Occupied by the chase with immeasurable ardour, they themselves in turn become an easy prey to the rapacious birds of night, or fall into the snares which have been laid for them. They are taken with nets, or with a line, for they strike with avidity against every object which hovers in the air around them.

The bats being thus derived from the quadrumanous type, and presenting besides numerous relations with the small insectivorous carnassiers, might perhaps be considered as an order possessing fixed limits and altogether distinct within itself.

It may not prove uninteresting, previously to noticing the most important of the cheiropterous sub-genera, to give a brief sketch of the sentiments of the principal systematic writers on these animals. We shall thus see how far they were able to execute a just classification, by means of the zoological characters then in use.

Belon was the first who gave a figure of a bat, namely, the great-eared bat, or oreillard. Aldrovandus reproduced this figure, and added a second of the largest European species. Belon had moreover marked a third species which he saw in Egypt. In the course of a little time it was ascertained by the descriptions of travellers and the iconographes of naturalists, that every country possessed in some sort, bats peculiar to itself. Although this, perhaps, at first, was not distinctly asserted, it was certainly the result of the publications of Clusius, Pison, Bontius, Flaccourt, Seba, and Edwards.

These materials were possessed from the year 1748, though still it was believed that there were but five species of bats. The catalogue of Linnæus does not include a greater number. But, at all events, there was no dispúte respecting the bats being a distinct family. This was a point indeed which might have been considered as almost instinctively established previously to the invention of all zoological systems.

Brisson, in 1756, adopted some new views on this subject. He had ranged winged quadrupeds according to the numerical order of the incisive teeth. He perceived that according to this principle of arrangement, the bats branched out into two series, and therefore considered himself obliged to divide them into two genera, to which he gave the names of *Pteropus* and *Vespertilio.* So little regard was paid at that time to the natural affinities of animals that no one was surprised to see these two groups separated from each other, and the interval filled by beings that had no relation to the bats whatever.

While things were in this state, Daubenton began his researches upon animals for his comparative anatomy. He soon found in France four of the bat family, which had not been before observed. This discovery occasioned him to review the labours of his predecessors on the mammalia, and to put forth a monography on the subject. This work, especially valuable at the epoch of its publication, was printed in the collection of the Academy of Sciences, in 1759. The monography of this celebrated naturalist was also enriched by many foreign species found at Paris in the public collections, and by others, then lately brought by Adanson from Senegal.

From this period, the family of the bats was established on solid foundations. A guide was procured, which was appreciated and followed. Linnæus gave the first example of its good effects in withdrawing from his genus *Vespertilio*, the hare-lipped bat, and making it, in the twelfth edition of the *Systema Naturæ*, the genus *Noctilio* of his *glires*.

The employment of the incisive teeth had hitherto done so well for the establishment of genera, that it was natural to set some value on the character. Great then was the astonishment of naturalists to learn, first from Brisson, then from Daubenton, that the bats differed among themselves in this very important respect.

The number of these animals, as yet known, was not considerable, nor had sufficient attention been bestowed on the affinities of the animal world. Naturalists continued, after the example of Daubenton, to comprise all the bats, with which they were acquainted, within a single genus; and, by way of apology for so doing, they affected to insist on the discordance of their generic characters, alleging that the anomalies of these animals were utterly inexplicable and irreconcilable with all the principles of systematic arrangement.

Erxleben alone renewed the division of Brisson, of *Pteropus* and *Vespertilio*, and in his mode of doing it exhibited but little judgment—for he destroyed the essential character of the genus *Vespertilio* by defining it like Brisson, and yet placing in it the new species of bats discovered by Daubenton, to which the definition of Brisson could by no means apply.

Subsequent naturalists did nothing but copy their predecessors and each other. They adhered to the plan of a single genus, and seem to have imagined that they fully satisfied the increasing demands of science by an enumeration of the incisive teeth of each species.

The naturalist to whom we are most indebted for the

H 2

most precise and scientific notions concerning these animals, is unquestionably M. Geoffroy St. Hilaire. The character of the incisive teeth, and its modifications, had originated the modes of division to which we have alluded. This acute observer soon perceived that one circumstance connected with these teeth, the fact of their being most frequently indented or crenulated, had proved a source of error even to the most expert zoologists. Pallas, for instance, reckoned eight incisors in the lower jaw of the *vesp. pictus*, instead of six, the real number ; and Daubenton had not remarked as many in the upper jaw of the *ferrum equinum*.

Another fact respecting these was calculated to mislead observers. This is, that being smaller than their alveoli or beds, they are easily detached from thence, and are found wanting in some individuals.

These teeth are also considerably dependant on the organs in their vicinity. In other animals, there is usually but one modification or condition for those organs, which are situated near the incisive teeth. They are generally contained within fixed limits, and do not interfere with the development of the intermaxillary bone, which itself supplies the incisors with a suitable basis and degree of solidity. This arrangement being unaffected by any thing immediate, the incisors grow in their proper beds, according to the action exercised upon them by the constituent elements of the whole being. Accordingly, as they are influenced by causes, disseminated as it were through the entire system of organization, they may be employed to indicate these causes in a general way, and consequently will form a most excellent generic character.

But the contrary to all this takes place in the Cheiroptera. Their organs of sense are complicated in consequence of the tendency of the dermis to acquire such considerable augmentation.

The organ of smelling, among others, is very frequently

obstructed and filled by certain kinds of pipes and tunnels. But as an extraordinary degree of development rarely occurs in one place, without becoming an impediment elsewhere, so we find that this development of the fossæ nasales is influential on the intermaxillary bone. The latter becomes smaller in proportion to the extension and prolongation of the former. It is sometimes so considerably diminished as to become little more than a bony point which goes off and is lost in the dermis: sometimes, indeed, it is altogether wanting.

The incisors which necessarily follow all the conditions of this organ, which become small as it lessens, and disappear when it is gone, are, in the bats, thus crossed in their development by a specific influence. They are no longer obedient to the controlling power of the general organization, and are no longer to be received as a test of its constitution. They vary, on the contrary, according to the intensity of the local action which presses them, and become a character of less value than where the natural progress of their growth is unimpeded.

But though they yield in importance to the organs of sense in their vicinity, they become, under a new aspect, an object worthy of consideration. They serve as a medium for appreciating the different modifications of these organs, and, in conjunction with them, to establish the characters for some particular groups and subgenera;—and, as these different modifications are also simultaneous with others, which affect either the organs of digestion, the wings, the tail, or the interfemoral membrane, it follows that we have a certain portion of characters sufficiently elevated to arrange the bats in divisions strongly marked, and to dispose them in small natural families or groups.

M. Geoffroy has divided the bats into fifteen of these groups, other naturalists have added more, and the number seems likely to be augmented, as observation is extended in different parts of the world in regard to these very curious animals. We scarcely need add, that these several groups are termed by Geoffroy, and their other inventors, genera, though in the present work they are more simply treated as subdivisions or subgenera of the single genus cheiroptera or bat. For the names and characters of these, as well as for the specific descriptions, we refer to the table.

We shall first speak of the ROUSSETTES (*Pteropus.*)

This division of the Cheiroptera, until the time in which M. Geoffroy made his communications to the Ann. du Mus. (1810), was not considered as composed of many species. It was singular enough that the various observations of Seba, of Clusius, of Brisson, of Edwards, and of Buffon, respecting the Roussettes, should have been all attributed to the same animal, which was distinguished by a single specific name, *viz.*, the *Vespertilio Vampyrus.*

The researches of naturalists in Egypt, in Bengal, at Timor, and Java, very considerably augmented this little family; and by affording the means of comparison with such as were already known, and had been taken but for simple varieties of age and sex, proved that a number of these animals existed, sufficiently resembling to constitute species of one genus, and sufficiently distinct to constitute separate species.

The roussettes are easily distinguished by their gait, their long and conical head, their slender and pointed muzzle, their small and simple ears, and, lastly, by the smallness of the interfemoral membrane. They have little or no tail; the posterior extremities are simply bordered, but not united by the interfemoral membrane, and the membrane of the wings extended on the upper part of the legs, and passing the metatarsus above, touches on the origin of the fourth toe. They are the only bats which have the second

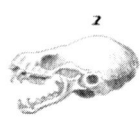

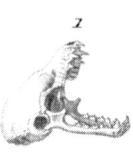

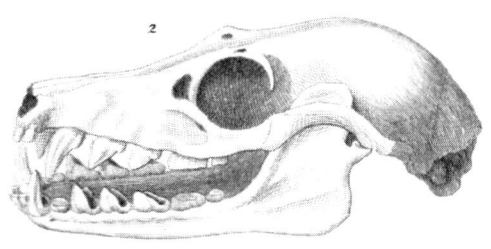

2

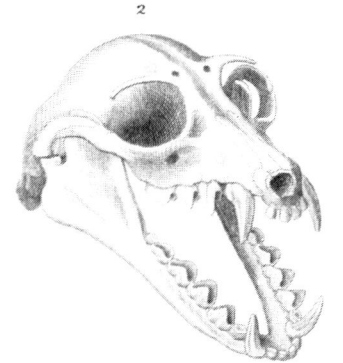

2

J.^s Basire sc:

SECTION OF CHEIROPTERA WITH   SECTION WITH RIDGED OR FURROWED.

POINTED CHEEK TEETH.   *CHEEK TEETH.*

Species Vespertilio auritus.   *Species Pteropus vulgaris.*

Natural size.

London Published by G.& W.B.Whittaker. Septr 1824.

finger of the hand provided with a nail, and with the phalanx belonging to it, and the only ones which are deprived
of the second external ear, or at least of that part of the ear
which is formed by the replication and excessive development of the tragus. Their tongue is rough and papillous,
and their teeth, in form and number, resemble those of the
simiæ. One peculiarity in these teeth must not be over-
looked; the smallness of the first and last molar, prevents
their being of any great utility in mastication, but the other
molars supply this defect by being considerably larger They
have, upon the whole, a form which is not found in any
other animal; their coronals are not bristling with tubercles; they present a long and straight surface, the plane of
which is oblique, and detrition exercises its action more on
the centre than on the edges, which project in sensible
ridges. The inspection of these molars would be sufficient,
(had not observation already established the fact,) to prove
that the habits and dietetic regimen of the roussettes are
different from those of the bats of our climates.

Their osteological characters are much the same as those
of our bats, except that their shoulder-blade is more triangular than square; the cubitus, almost effaced in the other
bats, is more apparent and more disengaged from the radius, which it accompanies for about two-thirds of its
length. The sternum forms a very strong projection, and
the first sternal piece, more large, more robust, and more
deeply separated in front, reminds us of the form and uses
of the furca in birds.

The second finger of the wing is half-turned from within
outwards, an effect perhaps of the development of the membrane during flight. It is a little less so, in the metacarpian
phalanx, more in the penultima, and still more in the unguiculated phalanx. The consequence is, that the nail, attached to the extremity of this last phalanx, is in a situation
exactly opposed to the plane or direction of the wing. This

second finger, too, though it wants no phalanx, is yet shorter than the corresponding finger in the other bats, though they want the little bone to which the nail is affixed. Finally, it is necessary to remark that the roussettes are destitute of the leaves or membranes which surround the nose in some other families of this order.

It would not be easy to find a group of animals more completely circumscribed, and more perfectly isolated among its congeners than the roussettes. But the advantage resulting from this is counterbalanced by some inconveniences. The difficulty of studying the species becomes more enhanced, as the characters which we are reduced to employ for that purpose, must of necessity be of an inferior order, and somewhat arbitrary.

These bats, however, possess one very distinctive mark, of which the eminent naturalist above-mentioned has availed himself for their classification. The larger roussettes have no tail, and the others have a small one. Dividing them on this principle, M. Geoffroy has made eleven species of the roussettes, five larger, with tails, six smaller, without. We proceed to notice whatever is interesting in these, without troubling the reader in this place by their enumeration.

The *Pteropus Edulis,* or *eatable roussette,* or *Kalou,* is a species discovered by M. M. Peron and Lesueur, in their voyage to Australasia, at the island of Timor. It received its name from these learned travellers, because its flesh, which is white, delicate, and remarkably tender, is regarded by these islanders as no small delicacy. The inhabitants of Timor confound it with all the other species of Cheiroptera, under the name of *Malanou Bourou,* (bird of night). The Malays call it *Kalou.* It measures more than five feet from the extremity of one wing to that of another, and is about a foot in length, from the point of the muzzle to the end of the crupper. The iris is very brown, and the nails of the toes are long and remarkably sharp. The muzzle resembles that of a dog, with the end of its nose cut in two. The skin is rough. From the occiput to the shoulders it is reddish, and in all the other parts black, mixed with some white hairs.

" The Pteropus javanicus," says Dr. Horsfield, in his elegant zoological researches, " is the largest species of the genus hitherto discovered : in adult subjects, the extent of the expanded wings is full five feet, and the length of the body one foot. In the specimen which I have placed before me in this description, the extent of the wings was five feet and two inches. The length of the arm and forearm together, from the union with the body to the origin of the phalanges, is fourteen inches ; the latter are distributed as in other species of pteropus. The naked thumb projecting beyond the membrane, measures two inches ; and the claw, which is strong and sharp, has an extent of nearly one inch along its curvature. On the index the claw is minute, and by the particular inflexion of the phalanges, which was first pointed out by M. Geoffroy, it obtains a direction opposed to the plane of the membrane. The length of the posterior extremities is eight inches and an half. The toes, which are slender, compressed, and distinct, agree in size, with the exception of the exterior toe, which is almost imperceptibly smaller ; they are disposed on the same plane. The claws have nearly the same size and extent of curvature as the claw of the thumb. The interfemoral membrane is regularly cut out in a circular manner, and forms a border along the inner side of the posterior extremities, about an inch and a half in breadth.

" The pteropus javanicus is extremely abundant in the lower parts of Java, and uniformly lives in society. The more elevated districts are not visited by it. Numerous individuals select a large tree for their resort, and suspending themselves with the claws of their posterior extremities to the naked branches, often in companies of several hundreds, afford to a stranger a very singular spectacle. A species of ficus, in habit resembling the ficus religiosa of India, which is often found near the villages of the natives, affords them a very favourite retreat, and the extended branches of one of these are sometimes covered by them. They pass the greater portion of the day in sleep, hanging motionless : ranged in succession, with the head downwards, the membrane contracted about the body, and often in close contact, they have little resemblance to living beings, and by a person not accustomed to their economy, are readily mistaken for a part of the tree, or for a fruit of uncommon size suspended from its branches. In general these societies preserve a prefect silence during the day ; but if they are disturbed, or if a contention arises among them, they emit sharp piercing shrieks, and their awkward attempts to extricate themselves, when oppressed by the light of the sun, exhibit a ludicrous spectacle. In consequence of the sharpness of their claws, their attachment is so strong, that they cannot readily leave their hold, without the assistance of the expanded membrane ; and if suddenly killed in the natural attitude during the day, they continue suspended after death. It is necessary therefore to oblige them to take wing by alarming them, if it be desired to obtain them during the day. Soon after sunset they gradually quit their hold, and pursue their nocturnal flights in quest of food. They direct their course, by an unerring instinct, to the forests, villages, and plantations, occasioning incalculable mischief, attacking and devouring indiscriminately every kind of fruit, from the abundant and useful cacao-nut, which surrounds every dwelling of the meanest peasantry, to the rare and most delicate productions, which are cultivated with care by princes and chiefs of distinction. By the latter, as well as by the European colonists, various methods are employed to protect the orchards and gardens. Delicate fruits, such as mangos, jambus, lansas, &c., as they approach to maturity, are ingeniously secured by means of a loose net or basket, skilfully constructed of split bamboo. Without this precaution, little valuable fruit would escape the ravages of the kalong.

" There are few situations in the lower parts of Java, in which this night wanderer is not constantly observed; as soon as the light of the sun has retired, one animal is seen to follow the other at a small but irregular distance, and this succession continues uninterrupted till darkness obstructs the view. The flight of the kalong is slow and steady, pursued in a straight line, and capable of long continuance. The chase of the kalong forms occasionally an amusement to the colonists and inhabitants, during the moonlight nights, which in the latitude of Java are uncommonly serene. He is watched in his descent to the fruit trees, and a discharge of small shot readily brings him to the ground. By this means I frequently obtained four or five individuals in the course of an hour; and by my observations I am led to believe, that there are two varieties which belong to one species, as they appear all to live in one society, and are obtained promiscuously."

One observation we shall make here is, that all the species of one genus are found to inhabit the same region, to the exclusion of another, and particularly the torrid zone of a Continent. This is particularly applicable to the subgenus *Pteropus*, none of the species of which have yet been discovered out of the warm climates of the old world.

The *Roussette* of Edwards, or Madagascar bat, the *common roussette*, *Vespertilio ingens* of Clusius,) and the *red-necked roussette*, (*Rougette* of Buffon,) and confounded by Linnæus and Gmelin under the synonyme of *Vespertilio*

Vampyrus. Brisson composed his genus *Pteropus* of the two last, and a third, the *Vespertilio Spectrum,* which, in truth, does not belong to this genus, but to the *Phyllostomes.*

The *Pteropus griseus gray roussette* is another of the species discovered by Messrs. Peron and Lesueur. It is remarkable for nothing so much as the shortness of its ears. The membrane of the wings does not grow precisely from the flanks, but takes its origin almost from the central line of the back.

All the roussettes of Timor inhabit the trunks of old trees, or the hollows of rocks. The large species alone inhabits caverns, and usually the deepest and most obscure. In speaking of the *Pteropus stramineus,* (lesser *Ternate bat* of Pennant,) M. Geoffroy deems it possible that the circumstance of the roussettes of Timor living so much in trees may impede the perfect growth of their hair.

The roussette of Egypt was discovered by M. Geoffroy himself, who detached many of them with his own hand from the ceiling of the great Pyramid. There is nothing very remarkable about them, but that the head is shorter in proportion, and larger than in the other species. Its hair is thick and soft, of a grayish-brown, and its incisors are remarkably small, slender, and symmetrically arranged.

The *Pteropus amplexicandatus* is distinguished above the others by the dimensions of its tail, which yet does not exceed the thigh in length. The interfemoral membrane is not so much sloped as in the preceding species, but extends from one side to the other, so as to pass above the tail, and cover one-half of it—whence the name.

The *Kiodote Pteropus minimus* of Geoffroy, has been formed into a separate genus by F. Cuvier, under the name of MACROGLOSSE. Besides the form of its head, which gives it a strong mark of distinction above the other frugivorous

Cheiroptera, it is separated from them still farther by very important modifications in its system of dentition. The kiodote, like the other roussettes, has four incisors and two canines in each jaw; but it has none of the anomalous false molars, neither are its hinder molars small and rudimentary, but as large and as fully developed as the others. These molars are ten in number, in the upper jaw, five on each side, and the two first are pointed. There are twelve in the lower jaw, six on each side. The first, almost at the basis of the canine, is 'separated by rather a wide interval from the second, and the three anterior ones are pointed. The others, in the two jaws, are plain and even, and very long, in comparison with their bulk. It may also be remarked, that the teeth of the kiodote are the smallest that we are acquainted with, as yet, in the class mammalia, though there are many smaller species to be found in that class than the kiodote.

The Roussettes, and still more the Cephalotes, have a large and thick muzzle, which indicates powerful jaws, and the faculty of biting with force. The kiodote, on the contrary, by its large head, and very slender muzzle, which narrows off suddenly from the eyes announces, what is really the fact, very weak jaws, incapable of acting with any degree of force. The resemblance, in this respect, subsisting between the kiodote and some of the edentata, with long muzzles, which live on ants, &c., would lead us to suppose that these small teeth must be of little use, unfit to encounter any substance that offered much resistance, and incapable of any great effort at mastication.

This animal, according to M. Leschenault de la Tour, has a tongue two inches (French measure) in length, which is double the length of the head, and which it can thrust completely out of its mouth, and draw back again, like the pangolin (*Manis pentadactylæ*, L.). This organ is not covered with rough papillæ, like the tongue of the roussettes, and is by no means calculated to act on bodies by friction. This learned traveller, to be sure, adds, that the kiodote lives on fruits: It is very certain, however, that its habits and mode of subsistence are very little known.

Its length is above two inches from the occiput to the posteriors ; its head about one inch, and it is ten inches from the extremity of one wing to that of another. It is the smallest species known of the roussette family, if indeed it belongs to them.

The details of its organization are but brief. Its limbs are like those of the roussettes ; the finger, corresponding with the index, has a nail, and the tail is but a mere rudiment. Its eyes are large; projecting, and have round pupils. The muzzle, at its termination, is divided by a sort of furrow, and the nostrils, circular and projecting, open on its sides. The external ear is simple, and is marked transversely by wrinkles, which result from the manner in which it is folded up, when closed. The membrane of the wings is entirely naked, except that part which forms a posterior border for the hinder limbs. All the rest of the body is covered with a fur, extremely fine and soft; tolerably thick, not so long upon the head as elsewhere, and apparently of a woolly character. On the head, neck, shoulders, arms, back, crupper, interfemoral membrane, and thighs, it is of a beautiful, uniform, fawn-colour ; elsewhere, it has a slight brown tint ; the iris is yellow. The cry is very sharp.

According to M. Leschenault, *kiodote* is the Javanese name of this bat, but Horsefield says, that the natives call it *Lorvo-assee*, dog-bat. He has given it the scientific appellation of *rostratus*. It is not very common in Java, and is very destructive to the fruit.

The most remarkable of all these species seems to be the mantled roussette (*pteropus paliatus*). The membrane of the wings grows from the central line of the back. The head is large and rounded, approaching to the form of an ellipsis. The muzzle short and thick. The individual noticed by M. Geoffroy was but young, and of course not completely developed. Its teeth were not entirely formed : the canines had but just appeared, and scarcely exceeded the molars. Its incisive teeth were very distinct, four in number in each jaw. The upper ones equal, and at a small distance from each other, the lower smaller and closer. The intermediate incisors were finer than the lateral.

But to pass over minor peculiarities, there were two very remarkable characters in this species which occasioned its describer to conjecture that it might one day be withdrawn from the roussettes, and elevated to the rank of a genus. These were,—1. The absence of a nail on the index finger, which is, however, as short as that of the other roussettes, and equally provided with all its phalanges. 2. The insertion of the wings, the membranes of which adhere together

Nothing can be more singular than such an organization In the other bats, the membrane extended between the fingers of the hand grows from the sides, being formed by a prolongation of the skin, which grows more slender in proportion to its extension. But in this roussette, the same membrane actually grows from the central line of the back, where the skin forms a slight projection before it is extended horizontally and to the extremities. It has exactly the appearance of a mantle thrown over the shoulders of the animal, and gave occasion to M. Geoffroy to give it the appellation of *paliatus*.

An arrangement of this kind must have for its final cause the advantage and comfort of the animal in which it occurs. The Deity does nothing in vain, nothing without a benevolent purpose. Accordingly, we find that this disposition of the skin, first, by increase of surface, renders the body of this bat specifically lighter, and assists it in its flight ; and secondly, that when the wing is folded, it forms an ample and profound pouch, which constitutes a convenient envelope for the young bats which are yet suckling, and a place of shelter where they find all the protection and all the heat necessary for their security and development.

We shall next speak of the Cephalotes, two species, one described by Pallas, and another discovered by M. Peron, which have a close affinity with the roussettes, but yet differ sufficiently to constitute a separate sub-genus. They have the conical head, the sharp muzzle, the ears without tragers, the short index provided with all its phalanges, the shortness and particular position of the interfemoral membrane, the small tail, the papillous tongue, and the remarkable form of the molar teeth peculiar to the roussettes. But they differ in other points. Their head is shorter and larger than that of the others, and the face still more than the cranium. The latter retreats more and is narrower in front. The teeth are but twenty-eight, four incisors, four canines, and twenty molars, eight of which are in the upper jaw and twelve in the lower. Thus the incisives are but half as many as in the roussettes, nor is this occasioned by too close an approximation or too excessive a development of the canine teeth. The upper incisors are at a certain distance between them, and perfectly isolated, which, however is not the case with the lower.

Such an anomaly in a character of this importance cannot exist alone. Those who are most superficially acquainted with the laws of Zoology, are aware that such a modification produces others. This is the consequence of what our author has termed the *subordination of characters*. Every thing in the organization of an animal is connected, and without knowing why or wherefore, we always find a correlation to exist, even where it is impossible to perceive any necessary connexion.

But in truth this want of two incisors can be considered as no anomaly, but when we class the two species in question with the roussettes. For, in pursuing the examination of their characters, we shall soon establish in all their principal organs other analogous differences, which must oblige us to refer them to a distinct and particular type. We shall thus avoid deranging a genus so accurately limited as that of the roussettes.

The molar teeth of the cephalotes, though closely resembling those of the roussettes, are not, however, of an identical form with them. The upper jaw has two less, *viz.*, the small anterior molars before mentioned. The last but one is also proportionally longer. Finally, those of the lower jaw are straiter, and the first is so small that it is covered by the gum and scarcely perceptible. The effect of detrition on their coronals is also remarkable in these teeth. In the roussettes the bony substance is more worn than the enamel, but in the cephalotes both are equally so. The surface of these teeth, and especially that of the back molars is altogether plane. This is characteristic of herbivorous animals alone. Must we conclude from this that the cephalotes do not use exactly the same regimen as the roussettes ; that they do not even eat the sweeter kind of fruits but content themselves with a simpler kind of vegetable nutriment ?

The organs of motion in the cephalotes exhibit a proportional difference, as in the parts just described. The wings are conformed as in the mantled roussette. The common teguments, as in that very singular species, are elevated on the central line of the back, and from those a lamina which becomes the point of departure for the membranes, which are elongated over the arms, and extended between the fingers.

Pallas gives us nothing similar in his description of the *vespertilio cephalotes*. But an arrangement so strange and novel might well have escaped him. At least we may be led to suppose so, from a certain tendency to systematic prejudices that exists in the minds of the greatest philosophers. It is by no means uncommon for the corporeal eye to see nothing but what has been perceived in anticipation by the intellectual organ.

Such are the differences upon which the distinction between the cephalotes and the roussettes has been founded. Had it not been for the *mantled roussette*, the limits of the two genera would have been still more strongly marked, and the interval between them still wider. The mantled roussette may be considered as a link uniting and attaching together these two little tribes.

We have remarked that these animals have four incisive teeth in each jaw. M. Geoffroy asks, whether it be possible that the renewal or growth of certain teeth might occasion the disappearance of others? The falling out of the incisive teeth, in the bats in general, is a case of common occurrence. But this is a matter easy of observation, and the progressive steps of which are obvious. These teeth, fixed in a bed of no great depth, are but feebly retained by the gums. As ossification advances, the alveolar cavity is filled with more promptitude than in other animals, and it is not astonishing that the incisives, under such circumstances, should be shaken and speedily disappear. But this produces no influence or re-action on the canine teeth. These being more deeply lodged in the maxillary bone, preserve their position, and experience no other variation than a slight degree of wear, in consequence of their mutual friction.

This being understood, it is easy to decide the question as it regards the cephalotes. If the incisives fell from age or accident, their places would be easily found. But in the cephalotes there is no place which the incisors could possibly have occupied beyond the two which we have mentioned. The canine teeth, and the lower ones still more than the upper, are infinitely closer than in the roussettes. It is manifest, therefore, that the smaller number of teeth in the cephalotes is the result of natural condition, and not the effect of age.

The characters which we use in the classification of animals do not all possess an equal value, or rather the same characters do not possess the same value in the different subdivisions of the animal world. They acquire importance when they are observed to be permanent in certain natural groups of living beings; but on the other hand, they lose it, and can only be employed in a secondary manner, when in other genera they are found to vary from one species to another.

The application of these principles to the consideration of the skin of the bats, undoubtedly bestows a certain preeminence on the characters derived from the various modifications of the cutaneous system. This, however, constitutes little more than the knowledge attainable by the commonest observers. A bat is recognised by the dimensions of its arms, the membrane of which proceeding from the sides forms one of the principal attributes of the animal. No attention is paid in this case to the conformation of the other parts, to the state of the viscera, to the number and structure of the teeth, and finally to the habits of the being. The consideration of all these appeared superfluous, after the predominant character of the family had been ascertained.

The bats, however, were found to differ very materially from each other. As an example of this difference, we may mention two genera, in one of which the development of the cutaneous system is the least considerable, and the greatest in the other. The first are the frugivorous bats, which we have just noticed under the names of roussettes and cephalotes; the second are those sanguinary cheiroptera,

I 2

which are easily recognised by the surrounding membranes of the nose.

The first may be considered as bats, upon the very slenderest claims. The general tendency to augmentation in the cutaneous system is carried on by them only in the wings. No other part is similarly developed. There is no volume in the tragus forming a second ear, no tunnels about the nostrils, no interfemoral membrane. There are but some trifling vestiges of the latter, extending along the internal edges of the legs.

It is far otherwise, however, with those bats which destroyed the first establishments of the Europeans in the New World. They are buried as it were, and lost amid the multiplied foldings of their integuments. Their ears are simple and double, and their nostrils surrounded with leaves and bordered with semicircular crests. Their interfemoral membrane occupies the entire space comprehended between their legs, which are themselves of very remarkable dimensions. This augmentation is visible also in the membrane of the wings, the size of which is considerably increased by an additional phalanx on the third finger. The body of the animal itself is scarcely to be distinguished amidst all this integumentary profusion. Their appearance is thus rendered more gloomy, and their physiognomy more ferocious. There is something vague and indefinite about their forms, which aggravates the horror which the remembrance of their devastations has inspired.

What is most remarkable in these two examples, and altogether conformable to the physiological views here taken generally, is the curious correlation of all the tegumentary parts with one another, their conduciveness to the same result, their great influence, and above all, the permanence of their forms in the generic groups where they are observed. The teeth are by no means so fixed a character. They vary for instance, in the Roussettes and Cephalotes, where the integumentary distinctions are comparatively trivial.

The bats which have the nose surrounded by membranes, are divided into three sub-genera, the Rhinolphi, the Phyllostomata, and the Megadermes.

The Rhinolphi have the nasal leaf exceedingly complicated, the tail long, the intermaxillary bone small, and provided with two teeth only, and (a remarkable character which they share with the Roussettes and Cephalotes) the ears simple and without any extra-devlopement of the tragus. They are the only insectivorous bats which are without this appendage. Some naturalists are of opinion that there is no such thing as any certain distinctions of genus, and that frequently nothing more is wanting than one or two species to bind by an indissoluble link, groups, between which the largest intervals have been supposed to exist. The genera of the bats are, however, by no means favourable to this opinion. In truth, it is remarkable that in every part of the globe where these animals are found, however distant from each other, their organization accurately corresponds with that of some one of the families already known to us.

This limitation of genera, the consideration of the Rhinolphi alone will suffice to illustrate. As a sub-genus it is most strictly circumscribed, and its species are distinguished with unusual accuracy.

The number of the mammæ is one of the most remarkable characteristics of this sub-genus. Beside the two pectoral in common with the other bats, the Rhinolphi have two others situated close to each other above the os pubis. This is a singular anomaly and holds throughout all the species.

In consequence of the disposition of the ear above alluded to, the Rhinolphi betake themselves to the most profound excavations, and bury themselves to a considerable extent

under ground. Deprived of the faculty of rendering themselves deaf at pleasure, they fly to retreats where the noise
and cries of diurnal animals cannot reach them.

In compensation for this simplicity of form in the ear,
the organ of smell, as we before observed, is singularly complicated. We find the entrance of the nasal cavities as
favourably disposed in the Rhinolphi for the conveyance of
odours as the ears are for that of sounds. They are formed
by a sort of conch, as if the odorous emanations, in a mode
analogous to the communication of sounds, were collected
and directed into the olfactory chambers.

We cannot consider this peculiar organization, common
to the Rhinolphi, the Phyllostomata, and the Megadermes,
as accidental. An arrangement so scrupulously accurate in
all its details, must involve a fixed design, and mark a very
distinct type.

The nasal chambers do not extend in the Rhinolphi beyond
the first molars. But they are inflated and globulous, and
the entrance to the nostrils is both in front and underneath.
It is a large opening, terminated by the intermaxillary bone,
which is reduced to a simple lamina, and is obedient to the
motion of the lips.

These last, which are elevated by their swelling almost to
the height of the forehead, leave a vacuum between themselves and the nasal chambers, at the bottom of which, and
as it were in a tunnel, are the two openings of the nostrils.
A fold of skin protects and furnishes the circuit of this
tunnel, and thus forms the conch of which we have spoken.
This fold is extended in front of the nostrils like a horse-
shoe (from which one species derives its name), and it is
detached and elevated behind, like a leaf, differing in form
according to the species.

The thickness of the lips results from an aggregate of
muscular fibres, which are glued, as it were, one over the
other, and opposite in their direction. The occasional con-

traction of this fleshy mass, draws along with it the inter-maxillary bone.

The index finger has no phalanx ; the others have two, or three, if we reckon the small bone of the metacarpus. The tail is long and almost entirely comprehended in the inter-femoral membrane.

One of the most remarkable species is the Purse Rhinolo-phus, (*Rhinoliphus Speoros.*) A most curious character is a sort of purse on the back of the leaf, situated on the fore-head, the internal sides of which are naked, and its entrance is distinguished by a sort of cushion, and opens by a sphincter. It is tolerably spacious, and we have no reason to believe that it leads any where. It is in general pretty exactly closed, and when the eyelids are down, it looks like the eye of a cyclops. This cavity has been found entirely empty, and it is not easy to conjecture its use. This bat is a native of Timor.

The Phyllostomata do not belong to the same countries as the Rhinolphi, but, on the contrary, are exclusively confined to the warm regions of the New World. They differ from the Rhinolphi in all the preceding characters. We shall first notice their organs of sense, beginning with that of touch. Their wings have proportionally greater length, and owe it in part to an additional phalanx on the middle finger. This is the unguical phalanx, but instead of being terminated by a nail, it is terminated by a cartilage, which the tension of the membrane draws along, and causes to bend on the interior side. Similar cartilages are seen on the fourth and fifth fingers. The membrane which unites all the parts of the wing extends to the posterior extremities, sideways, and without passing the tarsus. The feet are, therefore, less engaged in it than among the roussettes and other bats.

The interfemoral membrane furnishes no generic cha-racter. It differs in the different species ; so does the tail, which is found in some and wanting in others.

All the phyllostomata have an indented tragus, or second ear, interior, and placed on the border of the auricular foramen. Another lobe is observed within the ear, not far from the opening.

The nasal leaf, though not so complicated as that of the Rhinolphi, does not less merit the attention of the naturalist. Its seat is circumscribed by thick swellings, so that the nasal openings appear, as it were, at the bottom of a tunnel. The edges of this cavity are detached in a thin lamina, the semi-curve of which, is a good representation of an horse-shoe. It is from the middle of this curve that the leaf, properly so called, arises, consisting of a thick and elongated sort of cushion, the edges of which, are accompanied by membranes. It is terminated in a point at its extremity, from which it has been compared, in some species, to a javelin.

The movements of this apparatus are regulated by the muscles of the nostrils, and of the lips. The nostrils are hermetically sealed, when the leaf is lowered, and descends into the tunnel, and when the horse-shoe is raised.

The tongue is remarkable ; its breadth to its length is as one to six ; it is flattish above, and rounded underneath ; in length and narrowness, like the tongue of the ant-eaters ; it also resembles them in the faculty of being completely thrust out ; its surface is slightly and regularly shagreened ; close to its extremity may be observed a kind of organ of suction ; it is a cavity, the centre of which is filled by a point in relievo, the circuit of which is marked by eight warts.

The eyelids open and close sideways.

In speaking of the teeth of the phyllostomata, we may observe, that though the teeth usually correspond with the digestive organs, as trenchant teeth agree best with such animals as have a simple stomach, and short intestinal canal, and large and flat teeth, with such as have large and

ample intestines, yet this agreement does by no means exist invariably. What is best, is not always found to be
observed ; were it so, we should have only either animals
altogether carnivorous, or altogether herbivorous. We
know, on the contrary, that all the degrees comprised within these limits, are pretty nearly filled. An attentive observer will soon perceive that the abdominal viscera may undergo some variation without any corresponding change in
the structure of the teeth, and these last, in their turn, may
be modified, without any reciprocal alteration in the digestive organs.

Besides, it is certain, that many different structures of
teeth may produce the same effects ; and in this case,
cæteris paribus, this diversity of form cannot, of itself,
prove a sufficient ground of generic distinction. What is
individual, should be transferred to specific characters.

Of this the sub-genus, we are upon, is an excellent example. It is composed of species which perfectly resemble each
other, except in the structure, arrangement, and number of
the cheek teeth. These differences, (if we divest ourselves
of deference for an imaginary theory, and stick to observation,) we shall find to contain nothing essential. They by
no means depend on causes inherent in the nature of the
teeth, and simply relate to a change of proportion in the
maxillary bones. We have phyllostomata with short muzzles, and others with muzzles more elongated, but all make
the same use of their teeth, whatever differences may exist
between them.

In comparing the crania of the Javelin bat, and the
Vampire, we are struck by the difference of their proportions. That of the vampire is narrower and longer ; this
contraction is peculiarly observable in the lower jaw,
which does not, however, prevent the canines, which terminate that jaw, from being very thick at their roots. The
incisives, whose growth is somewhat impeded by this

development, do, however, exist, though very small, and crammed close together in front of the canines. Another peculiarity in the lower jaw is, that it exceeds the upper. There are four incisives in each. The molars of the vampire ten above and twelve below, are of the carnivorous character ; the first are very short, and almost plain ; the others are trenchant, and terminate in three or four points. Those below are compressed, and remarkable for one of the points which extends considerably beyond the rest ; the upper molars differ from each other in form and dimensions ; the second are triangular, the last large, but have no great depth.

The Javelin-bat resembles the Vampire, as to the teeth, merely in the disposition of the upper incisors ; the branches of the lower jaw being more separated, keep the canines at a certain distance, and thus leaving more space for the incisors, allow them to be ranged on a single line.

The Javelin-bat has four molars less than the Vampire, that is $\frac{8}{10}$. The construction of its teeth altogether, more closely resembles that of the insectivora, while the teeth of the Vampire exhibits relations with those of the animals which feed on flesh. The occipital cavity is also stronger in the latter than in the Javelin-bat.

In the smallest of the phyllostomata, (the *Soricinum*,) there are but three molars in each rank : twelve in all.

All observers have agreed in attributing to the phyllostomata the faculty of sucking the blood both of men and animals. Pison has given us some very circumstantial details on this subject. Similar accounts are also to be found in the narratives of Peter Martyr, of Father Jumilla, of the brothers Ulloa, and of M. de La Condamine, all which accounts are to be found in a French work, entitled *Histoire Naturelle*, (vol. xiii. p. 58,) where they are transcribed from the original text of these authors. M. Raume de St. Laurent, in the same work, confirms the veracity of these writers by his own testimony, and it has received further confirmation from the judicious remarks of an observer, equally distinguished by his accuracy and discrimination, Don Felix d'Azzara, from whom we shall translate the following passage :

" The species, with a leaf upon the nose, differ from the other bats in being able to turn, when on the ground, nearly as fast as a rat, and in their fondness for sucking the blood of animals. Sometimes they will bite the crests and beards of the fowls while asleep, and suck the blood. The fowls generally die in consequence of this, as a gangrene is engendered in the wounds. They bite also horses, mules, asses, and horned-cattle, usually on the buttocks, shoulders, or neck, as they are better enabled to arrive at these parts from the facilities afforded them by the mane or tail. Nor is man himself secure from their attacks. On this point, indeed, I am enabled to give a very faithful testimony, since I have had the ends of my toes bitten by them, four times, while I was sleeping in cottages in the open country. The wounds which they inflicted, without my feeling them at the time, were circular, and rather elliptical ; their diameter was trifling, and their depth so superficial, as scarcely to penetrate the cutis. It was easy, also, on examination, to perceive, that these wounds were made by suction, and not by puncture, as might be supposed. The blood that is drawn, in cases of this description, does not come from the veins, or from the arteries, because the wound does not extend so far, but from the capillary vessels of the skin extracted thence, without doubt, by these bats, by the action of sucking or licking." *Hist. Nat. du Paraguay*, tome 2. p. 273.

Buffon, in investigating the possibility of the Vampires sucking blood, without causing, at the same time, a pain of sufficient acuteness to awaken a sleeping person, concludes that the operation must be performed with the tongue ; and he adds, that we may form an idea of the *modus operandi*, by examining the tongue of a roussette, of which, with its hand, slender and sharp papillæ, directed backwards, he gives a figure. The tongue of the phyllostomata is not formed in a similar model, as we have already remarked ; the conjecture of Buffon, is not, however, the less well-founded. It is most certain, that a man who was sleeping ever so profoundly, and that animals, whose sleep is much more light than ours, must, unquestionably, be awakened, and that abruptly enough, by the pain of a bite inflicted with teeth. It is the tongue alone which can make apertures sufficiently subtile to open the extremities of the veins, without causing an acute sensation of pain. This conjecture becomes certainty, when we discover a portion of the tongue, such as we have above described, exactly constituted as an organ of suction, and designed, in fact, for the performance of that identical function.

It must not, however, be imagined that the phyllostomata are absolutely and exclusively nourished by the blood of animals. They have attained sufficient of a terrible celebrity, by destroying, altogether, at Borja, and other places, the cattle which the missionaries had introduced, without adding to their evil reputation by any marvellous exaggerations. They all live on insects, in the manner of the other bats ; this fact has been proved, by opening the stomachs of several of them ; d'Azzara declares that they would not venture to attack the cattle during the night, except when prompted by hunger, arising from the deficiency of other alimentary matter.

All of these bats, whether the jaws be short or elongated, suck the blood of animals.   Peter Martyr relates it of the phyllostomata, of the isthmus of Darien ; the two Ulloas, of those of Carthagena ; Raume, of the vampire of the isle of Trinity ; and Don Felix d'Azzara, of all the species which he discovered in Paraguay. Pison, previous to the time of these travellers, reported that this thirst of blood, was a necessity with this genus of the bats, and he was acquainted with two species.

It is not true, however, that the wounds which they inflict on men are so dangerous as Father Jumilla relates. It is impossible, indeed, to believe that the feeble effort which they make to draw a few drops of blood, could be attended with such pernicious consequences; but the testimony of d'Azzara is positive on this head, and must set the question at rest. " No one," says he, " in our neighbourhood, fears these animals, or gives himself any trouble about them ; notwithstanding a prevalent and most absurd report, that previously to sucking the blood of their victim, they flap their wings upon the part intended for banquet, for the purpose of lulling and deadening its sensibility."

Pison was the first who presented us with any researches on the phyllostomata, of which he gave a notice rather than a detailed description of two species, under the Brazilian names of *andira* and *andira-guacu*. He has spoken, however, at sufficient length to prove that the figure placed in front of his description belonged to no animal brought from Brazil. This figure represents a roussette, which the editors of Pison's book procured in some of the cabinets of Europe, and took for a vampire, in consequence of its size.

Sloane appears to have found the *andira* again at Jamaica, or the smallest of the two above-mentioned species.

But these two species were not truly known until Seba gave figures of them, and Linnæus described them, together with a third species, in his catalogue of animals, under the names of *Vesp. spectrum, V. perspicillatus, and V. spasma.* The only deficiency of these figures, which are all of the natural size, is that the interfemoral membrane has been represented of a square cut, and is destitute of the long osselets which support it.

Edwards has since produced another figure in his history of birds of one of these phyllostomata, that of Jamaica. But it is much more incorrect than Seba's.

Buffon, established the species of the javelin-bat, named since by Linnæus, *V. hastatus.* First he had given this new phyllostoma as the *V. perspicillatus,* or *V. Americanus* of Seba, but afterwards he reproduced it as a new species, under the title of the great javelin-bat, *(grand-fer-de-lance),* Hist. Nat., Supp., tom. 7, tab. 74.

Pretty nearly about the same time, Pallas gave a complete history of the smallest species of this genus, which he compared in size to the shrew, and for this reason called it *V. soricinus.*

These are all the phyllostomata of which mention has been made in systematic writers. Shaw's General Zoology, which appeared in 1800, contains no more species than that of Gmelin.

Among the ten species of bats discovered by d'Azzara in Paraguay, four belong to this genus. His *brown bat,* however, and *reddish-brown bat* are not the vampire and javelin-bat, as he believed, but ought rather to be considered as entirely new ; and also his *brown-striped bat.* M. Geoffroy makes nine species of the phyllostomata, for which we refer the reader to our table.

The name vampyrus, which Linnæus has given to the roussettes known in his time, was appropriated by Buffon to the *phyllostoma spectrum,* as he was assured the habits which authorized this appellation belonged exclusively to this species.

Every thing leads us to believe that it is the same of which Pison has spoken under the name of *andira-guacu.* He has described it to be about the magnitude of a pigeon.

Seba has given a figure not correct as to the interfemoral membrane. Schreiber afterwards gave a better figure, and added to its incorrectness. The slight stroke in Seba's figure meant to represent a tendon, has become under the graver of the German artist, a real tail. Schreiber afterwards gave a better and original figure of the vampire.

The length of muzzle in the vampire, the size of its ears, and the smallness of the nasal leaf, compose altogether a very singular physiognomy. The membrane of the wings is prolonged along the entire edge of the metatarsus, and terminates at the origin of the first toes. The interfemoral fills all the space comprised between the legs. Its terminal edge forms a salient angle, which is the product of three lines equal between themselves, the two extremes corresponding with the osselets of the tarsus and the third with that part of the membrane which is deprived of support.

The fur is soft to the touch, marron-colour above, and of a reddish yellow beneath.

The *phyllostoma perspicillatum*, or *lunette*, *(spectacle bat,)* was called by Buffon *grand fer-de-lance*, for no other reason that we can discover but that it is invariably found to be smaller than his other *fer-de-lance*, *(common javelin bat)*. He gives it as wanting the second ears or development of the tragus, and the figure in front of his description exhibits them very distinctly. In fine, he discovers that there are no incisive teeth in the upper jaw, and M. Geoffroy has reckoned four in all the subjects submitted to his examination. Be it remembered, however, in reference to this last remark, with what facility all the bats lose their incisive teeth; and more particularly the upper ones.

In this species, the lips are bordered by a series of warts, and very strong nodosities are conspicuous on the articulations of the third and fourth finger.

The interfemoral membrane forms a re-entering angle; it is almost without support, in consequence of the smallness of its osselets. The colour of the back is blackishbrown, and of the belly clear brown.

Under this species, as a variety, is ranged the first bat of d'Azzara, or the *obscure and striped bat.* It is one third larger than the lunette, and of an obscure colour bordering on reddish. It resembles that species by having two white bands on the head. This, however, cannot be considered as a sufficient proof of identity of species, as the same character is found in another phyllostoma of Paraguay, very different from this.

The *musette,* or *phyllostoma soricinum,* inhabits Surinam, and the adjacent isles: Pallas has published a very complete description of parts of its viscera, and skeleton. It is the smallest of all the known phyllostomata. The muzzle is tolerably long, but less narrow than that of the vampire. The canine teeth are, consequently, at a greater distance from each other, and the incisors are less incumbered, and are arranged upon a single line. The leaf is small, altogether at the extremity of the muzzle, in the form of a heart, (whence its English name, *heart-nosed bat,*) larger at the base, in the males, and terminated by rather a sharp point. The ears are small and oblong, and the interfemoral membrane forms a re-entering angle, and is supported by very short osselets.

The tongue is very large, especially long, and channeled towards its extremity. The borders of the furrow, are furnished with papillæ divided into two branches. This organ of suction, in the Javelin-bat, is circular. The disposition in this species, is the same in its result, and there is no doubt that all the phyllostomata equally make use of these depressions of the tongue to open the extremity of the veins, and determine thither the flowing of the blood.

Five other species have been added by M. Geoffroy, in the details of which, there is nothing very interesting for this part of our work : we shall, therefore, proceed now to the MEGADERMES.

The phyllostomata possess relatives in the old continent much nearer of kin than the Rhinolphi. Such is the

family to which has been given the generic name of ME-
GADERMES. They are thus named, because in them the
cutaneous system is carried to its full extent. When we
devote ourselves to researches on the natural affinities of
beings, we sometimes meet with continued series in organization. The Megadermes form a true intermediate link,
which unites the phyllostomata to the rhinolphi : but this
link remains perfectly circumscribed. It is a group, on
each side of which are intervals or hiatuses very strongly
marked, and which is as completely separated from the
phyllostomata as from the rhinolphi.

The character which forms the common bond between
these three families, is the singular apparatus in the form
of conchs at the entrance of the nasal cavities : still it
is very different in the three genera. In the phyllostomata
it is simple; rather complicated in the megadermes ; and so
much so in the rhinolphi that it is not very easy to form a
precise notion of it from description.

The Megadermes, which are provided with the additional
ear, and are destitute of a tail, must not be confounded with
the rhinolphi ; in this respect, they approximate more to the
phyllostomata, though differing from them in other very
essential points. Their tongue is short, and without any
furrow, at least, at the extremity. It is without warts
or papillæ of any kind, and, consequently, not organized
for the purposes of suction. Their lips, also, which are
hairy and tuberculous, are not more adapted for the same
operation.

In none of the bat tribe are the organs of sense more
powerfully seconded by the cutaneous system. The wings
are of a very great extent, though they do not possess the
unguical phalanx, which it is so surprising to find in the
phyllostomata, on the third finger. They have, proportionally, as much breadth as length, and reach to the
hinder-feet, between the fourth and fifth toe.

The ears are of such an excessive amplitude, that they meet, and unite on the top of the head.

Even to the leaf itself, there is a sort of supplement at its base, in the form of a lamina, which forms a second covering for the basis of the cone, and which is disposed on the sides into auricles for the nasal apertures.

It sometimes happens that certain organs augment at the expense of the neighbouring parts. It is a question if the development of the leaf may not exercise some influence of this kind on the intermaxillary bone. It is most certain that this bone is reduced to so mere a rudiment, that M. Geoffroy could discover no traces of it in the two megadermes of India, nor Daubenton in the Senegal species. To deny its existence altogether, seems, however, to be a proposition inconsistent with zoological analogy, and indeed with common sense. It is more natural and reasonable to suppose that there is an intermaxillary bone in the megadermes as in the rhinolphi, but that it is small, suspended in the cartilages, and that it may frequently disappear. Its character of fragility, and very accessible situation, render it difficult for it to resist the least efforts.

We shall not be surprised to find that the upper incisives do not exist, when the piece in which they should be inserted is wanting. But though they have not been found in such of the megadermes as have been submitted to the inspection of naturalists, we are not to suppose they may not exist in other individuals. It is most likely that, as with the rhinolphi, they share the fate of the intermaxillary bone, and that they exist with it, probably, to the number of two. This conjecture acquires additional force from the perfect resemblance of the upper maxillaries in the megadermes and the rhinolphi.

It may also be observed that this resemblance in the disposition of the maxillaries, serves to remove the megadermes a degree further still from the phyllostomata.

This distinction holds good, also, in the teeth. The lower incisors of the megadermes are four, well-ranged, and slightly furrowed on the edges. The upper canines are remarkable by three facettes, and the lower ones, by their curvature backwards: a direction which is very rare, and which may also contribute to prevent the development of the intermaxillary bone.

The cheek-teeth are eighteen, eight in the upper jaw, and ten in the lower. The first upper one is trenchant, compressed, and terminated by a long and fine point. The two teeth which follow, present the figure of two M's placed side by side, and the extreme points of which, are marked by sharp tubercles. The last cheek-tooth, from its smallness, might be taken for a moiety of the preceding one.

The lower cheek-teeth are compressed ; the two first are simple, triangular, and with but a single point ; the three others somewhat longer, bristling with four points, formed on a kind of double plan, the most projecting side of which, is in front.

It is easy, from this description, to perceive that these teeth approximate more to the cheek-teeth of flesh-eating animals than of the insectivora. The megadermes too, may be probably distinguished from the other bats, with nasal leaves, by a more decided taste for flesh.

There are two characters in which the megadermes resemble the vampire. They manifest no appearance of tail, and the interfemoral membrane, which comprises the entire space between the under extremities, is cut in a square form, from that point where it ceases to be supported by its osselets.

Such are the grounds upon which M. Geoffroy was led to consider the *Vesp. Spasma*, and its congeners, as forming an isolated group, separated from the other cheiroptera by

K 2

characters sufficiently marked, to be entitled to the rank of generic distinctions.

We shall not trouble our readers with any details concerning the species of the megadermata, which, with their characters, will be found in the table.

The genus or sub-genus nycteris, was established by M. Desmanets and M. Geoffroy St. Hilaire, and afterwards admitted by Illiger.

These bats have teeth similar in number and in form to those of the vespertilio.

A single, but a very essential difference distinguishes these teeth from the teeth of the vespertilio: this is the disposition of the incisors.

They are smaller, especially the lower ones, which are hardly to be distinguished by a single view; above, they are not (as in the Lemurs) separated in pairs, but, on the contrary, they are arranged on a continuous line, commensurate with the intermaxillary bone.

This bone, subservient in the bats, to the variations which distinguish the organ of smelling, possesses a movement peculiar to itself. It is raised or lowered (oscillating as if on an axis) by the upper lip, which is of a thickness and consistence adequate to the operation of directing it. Diminished as it is at its points of articulation, it cannot participate in the fixedness of the other bony parts.

It is, doubtless, in consequence of the domination of the surrounding organs, that the intermaxillary bone is so remarkably small. It does not project beyond the canines, whence it happens that the upper jaw is shorter than the lower, and appears almost truncated. Another result of this is, that the incisors of the two jaws do not correspond, and resting edgeways, their summits are never worn, which remain with two lobes above, and three indentations below.

The nasal cavities of the Nycteres, when first examined in the cranium, might be considered as of no great depth, because the bones which circumscribe their extent are very narrow. The lower or palatal lamina, is not extended beyond the second molar, and the external plate, or as it were, nasal maxillaries are quite rudimentary. But we form a very contrary opinion of these nasal cavities, when we see them provided with their soft parts. The hinder or back nostrils open considerably beyond the point where the maxillary bone is terminated, and the external conduits have their large entrances filled, and apparently incumbered with lobes and cutaneous appendages. A fold of skin originates from the middle of each conduit. A lobe, which is formed like the head of a nail, and is nothing but the cartilage of the nostril, is seen at each side, and unites with the interior fold in hermetically sealing the nasal orifice. The only effort on the part of the animal, necessary for this operation, is simply to wrinkle or knit up these parts, or, perhaps, only to abandon them to their natural elasticity.

The cavity of the nostrils is prolonged behind with the forehead. But what is not less remarkable, is the size and channelled form of this peculiar part. This gives to the nycteres the sombre and savage physiognomy by which they are characterized.

The forehead is, in fact, extended beyond its habitual dimensions, by means of the bony laminæ which originate from the sides of the os coronalis, and unite together on the vertex. The canal or longitudinal scissure which results from the projection of these crests, extends upon the nostrils.

The forehead undergoes these strange metamorphoses, probably, to supply the extreme smallness of the nasal apertures, and may be, perhaps, a sort of funnel where the odorant fluids are collected. The edges of this scissure are bristling, with long and abundant hairs, which fill it.   But this is not so when the labial muscles raise the opercula, distend the interior folds, and partly open the nasal conduits.   These edges, by the tension of the skin are turned upwards, and with them, the long hairs with which they are furnished.

Nostrils, which are habitually closed, and which to communicate with surrounding bodies, require an act of volition in the animal, and the consequent exercise of certain muscles, are, doubtless, a characteristic which possesses in itself no ordinary degree of interest to the observer of nature.   The nycteres derive no small advantage from it; they are thus enabled to establish their dwelling in places from which other animals are repulsed by powerful or pestiferous exhalations.   It is doubtful, however, whether this singular and inverse disposition of the nasal conduits be meant to preserve the animal from the inconveniences of infectious odours alone.   Like every thing else, of the kind, it supposes some corresponding modification elsewhere, and we shall find, on investigation, that to this general law of nature, the nycteres form no exception.

The faculty of flying, in the bats, very naturally led men to the notion of comparing them with birds.   The latter were observed to possess much more ease and grace in this style of locomotion, owing not only to the superior perfection of their direct organs of flight, but also to the power they possess of inflating themselves with air, and thus diminishing their specific gravity.   It did not appear likely that a similar faculty would be discovered in the bats, whose pulmonary functions are, in fact, so very different from those of birds.

It is, nevertheless, true, that aërial vesicles have been found in the nycteres, similar to those in birds, but still larger, and that the animal can fill them at what time and to what extent it pleases. But, as may well be imagined, the nycteres convey air into these vessels, by virtue of a peculiar mechanism, and by means of an organization, which with all its anomalies, is still derived from the primordial type of the class to which they belong.

The results of so new a mode of organization are, unquestionably, worthy of research. The means which produce it are perfectly simple.

The skin has no adherence to the body, except in some places where it is retained by a celular texture, very flaccid, and separated. The air is introduced through this, and remaining between the skin and flesh, gives to the animal that appearance which is observed in veal when blown up in the butcher's shops. There are none of those threads, or cellular tissue, to be found, except in the neighbourhood of the conduits, and on the sides of the thumb. Thus, the skin is completely raised on the back, the chest, and the abdomen. The nycteres are thus immersed in a bath of air, or rather placed in a sort of muff, which this elastic fluid forms around them.

However extraordinary, such a fact as we have been describing, may seem, it yet appears to detract nothing from the essential character of the mammiferous type. Nor is this essential character any more affected by the mode in which this singular and extensive cellular system is inflated.

At the bottom of each cheek-pouch is found a small aperture, and it is simply by means of this, that the aërial sac communicates with the mouth. The animal, in opening its nasal cavities, causes the circumambient air to enter and inflate its chest. On the other hand, in a moment after, by abandoning the nasal membranes to their natural elasticity, and, at the same time, keeping the mouth closely shut, it forces the gas, which has been respired, to return into the cheek-pouches, and thence into the large aërial sac.

Although there is, at the entrance of the sac, a sphincter which is very apparent, yet it is not by means of it, or at least by means of it alone, that the return of the air is resisted. There are large valvulæ on the neck and back, which are charged with it.

The air proceeding from the sphincter, takes the front of the ear in its way, passes into the sinus of the forehead, from whence it reaches the vertex, the occiput, and the upper part of the neck. It is there that it is discharged into the great sac.

Thus, the nyctere manages exactly like the tetrodons, (*Tetraodon lineatus*, L.) It carries, at pleasure, a mouthful of air in its sac, then a second, and so on. It breathes exactly as we ourselves can do, in the very same manner, in fact, that we do breathe, with this difference only, that it breathes into its mouth, the cavity of which, is then without any external passage. It then becomes a true bladder, within which, the trunk is, as it were, deposited. Thus inflated, it assumes a spherical form, and in this state, the animal bears no indistinct resemblance to a balloon, to which wings, a head and feet have been attached.

More fortunate than the tetrodon, which cannot have recourse to a similar operation, without reducing itself to an inert mass on the face of the waters, the nyctere preserves all its faculties, or to speak more truly, it augments their energy, by becoming more light, and more capable of velocity, in the act of flying.

It was natural to suppose that such singular anomalies in the olfactory conduits, would influence another system of organs, and perhaps occasion considerable changes elsewhere. We find, in fact, that this extensive sac modifies, or rather procures for the nycteres, a most valuable appendage to the respiratory organ. If this same apparatus, which is so well adapted to their system, be not, in truth, the motive of those modifications of the nasal cavities, and will not give an explanation of them completely satisfactory, at least, it cannot be denied that reciprocal and necessary relations must exist between all these parts.

The distinctive characters of the nycteres are confined to the differences which we have now detailed. The other teeth of these bats, canine and molar, resemble those of the vespertiliones. The same is true of the abdominal viscera. The teguments only, present a greater extent of surface. The ears are longer than the head, without the auricula, which borders the meatus auditorius, being augmented in a similar proportion. This extent of the tegumentary system is peculiarly conspicuous between the legs, where the interfemoral, or as we may, in this case, term it, the *caudal* membrane, surpasses, in both its dimensions, the length of the animal. The compass of the wings, however, and the peculiar size of each of those organs of volition, present nothing very anomalous or extraordinary. The osselets of the fingers are in the smallest number ever found among the bats. One, (the metacarpian,) constitutes the index, and the others are formed of three pieces, that is to say, of the metacarpian, and the two phalanges.

The last vertebra of the tail is bifurcated, a singular kind of separation, found in all the nycteres, and not existing in any other division of the bats.

There was but one species of nyctere known, and described, in the time of Daubenton, his " *campagnol volant,*" (the *vespertilio hispidus* of Linnæus,) and bearded bat of Pennant. The nyctere Thebaïs, differs from this, as also does another species, brought from Java, by M. Leschenault de la Tour.

The dimensions of these bats are not the same. The nyctere of Daubenton, is as thirty-eight in length, from the head to the origin of the tail; the nyctere Thebaïs, as fifty-four ; that of Java, as sixty-seven.

The ear is larger in the Egyptian species, and the fur is shorter, and tufted. The fur of this nyctere is clear brown above, and ash-colour underneath. In the nyctere of Daubenton, the tints are almost the same, only that on the back there is an approach to reddish, and to a dirty white under the belly, where there is also a mixture of fawn-colour. The Javanese species has the upper parts of a bright red, and the lower of a reddish ash-colour.

The nyctere, which was first described, had been brought from Senegal, and we find that all the species inhabit the warm countries of the old continent.

It seems probable, that two species exist in Senegal. Daubenton at least, has described two varieties, both of which had been sent to him by Adanson. The second, which, however, he established by a stuffed or dried specimen, differed from the first, in having the whitish-colour of the lower part of the body mingled with an ashen-tint, and in there being no reddish on the membrane of the wings. The cranium, and principal bony parts of the same individual, were inspected by M. Geoffroy St. Hilaire, and they do not agree either in dimensions, or in some details of form, with the nycteres of Daubenton, or the nyctere Thebaïs.

We shall now proceed to the VESPERTILIONES. This name was employed, at first, to designate the small number of bats which were known to the oldest systematic writers. Brisson was the first who restrained its acceptation, and applied it solely to such of these mammalia as have four incisors in the upper jaw, and six below. This, naturalist, as we have noticed before, established for the other species of this family, a new genus, under the name of Pteropus, to which he assigned this distinctive character : " Incisive teeth, to the number of four, in each jaw."

Such was the classification to which Erxleben, in 1777, seemed desirous to conform ; but as he wrote at an epoch when the discoveries of Daubenton had considerably augmented the number of the bats, and proved them to be susceptible of many more differences in that very relation in which Brisson considered them, he found himself embarrased by these rich materials. Erxleben, without hesitation, had entirely adopted the principles of the French naturalist, and had made as many generic divisions as the discoveries of science had presented him with new types. But this was an innovation which he did not dare to adopt, and, consequently, he destroyed the true character of the genus vespertilio, at least, as far as its first definition extended, by placing in it all the bats which had, more or less, than four incisors ; in fine, all those to which the characters of the genus pteropus were not found to agree.

Linnæus, who was acquainted, at first, with a very small number of bats, united them in a single group, under the name vespertilio. It was only in the last edition of his *Systema Naturæ*, that he deviated from this arrangement, by separating from the bats, the *leporinus*, or hare-lipped bat, to bring it (for no very sufficient reason, apparently,) into the order of glires, under the name of *noctilio*. This great man, too much occupied in establishing the broader bases of his zoological classifications, often neglected the subdivisions, of which they were susceptible. The bats present a remarkable example of this. It might really be almost imagined that he knew nothing about them ; for, in the first instance, he attributed six incisors to these animals ; a description which is applicable to none : and when, in his later editions, he corrected this character, it was only to extend to all of them the characters of a few species, namely, those of the pteropus of Brisson. Later systematic naturalists, struck with these inconsistencies and the embarrassing consequences to which they led, recurred to the notion of a single genus. They established, however, some subdivisions founded on the number of the incisive teeth. But this was done less with the view of grouping the bats according to the order of their common relations, than to procure the means of determining the species with more rigorous precision. It was unfortunate that this sacrifice made to the advantages of a good classification, failed even to produce its intended effect, inasmuch as the observations relative to the number of incisors which are quoted in these writers are for the most part incorrect.

The result then, in fact, proved to be neither a good classification for the families, nor an exact method for arriving at the determination of the species. The disorder which reigned on those points in later systematic works, is sufficiently proved by the difficulty, if not, indeed, impossibility, of employing the characters given in such works for the purpose of recognising the living objects themselves.

Under such circumstances it was absolutely necessary to revise in some degree the labours of preceding naturalists on the subject of the bats. This task has been undertaken and most admirably performed by M. Geoffroy St. Hilaire, to whom we owe the substance of most of our observations on the cheiroptera. He set out upon the principle, that the cheiropterous tribes admitted of subdivisions perfectly in conformity to nature, and he has particularly succeeded in re-establishing the genus vespertilio, such as it was formed by Brisson, in explaining and enforcing the grounds on which it rests, and describing the species of which it is composed.

These bats, which comprise nearly all the cheiroptera found in Europe are remarkable for the largeness of their head, their short muzzle, naked ears, and the existence of an auricula or tragus, which originates from the centre of the auricular conch; for nostrils without membrane or appendage ; and for a very long tail. They are capable of a very permanent and extensive flight, comprising within the two extremities of the wings a length four or five times as great as that of the body. The surface of their wings is moreover materially augmented by the interfemoral mem-

brane, which extends upon the tail, and takes it in altogether. The vespertiliones have also two mammæ, situated
on the breast, pretty near the arm-pits. The tongue is soft ;
only in one species it is possible to perceive some papillæ
at the basis. All the fingers included in the membrane of
the wings have neither nail nor unguical phalanx.

These characters would be sufficient for the purpose of
recognising the vespertiliones. But to those are also joined
the consideration of the teeth. These are arranged as in
the lemurs. The same number and same disposition of the
incisors prevail in both : four in the upper jaw, separated
in pairs, six in the lower very closely approximated to each
other. These teeth are not worn, they preserve their tops,
so that the upper are always cylindrical and pointed at the
extremity, and the lower divided into two lobes and cut as
it were in scissures. The intermaxillary bone is formed of
two portions not joined in front. As the upper incisors
form but a very narrow lodgment in this bone, they are
always remarkably small, and are easily disengaged from
their beds.

The canine teeth, to the number of two in each jaw, are
the same as in all the bats.

The cheek-teeth, on the contrary, have a form exclusively
proper to the vespertiliones. There are from four to six
on each side, according to the species. The anterior are
conic. The others with large coronets are bristling with
points. The lower ones are furrowed on the sides. The
upper, twice as large, exhibit moreover a coronet with an
oblique edge, insomuch, that they partly displace the lower
when the jaws are closed. These large teeth moreover are
hollow at their centre. In both rows they are respectively
engrained, and present on inspection a general appearance
from which it may easily be judged that they appertain to
insectivorous animals.

Such are the characters which agree without exception

to the several species of vespertiliones. The indicating characters of this sub-genus may be given thus: Incisive teeth, four above, six below, nose simple, ear with auricula or tragus.

We might be tempted to believe after what has now been said, that the sub-genus vespertilio including species so closely approximating to each other, might occasion some difficulty to the naturalist in making a rigorous specific determination. In truth we can but very seldom make use of the consideration of colour, a character to which we are obliged so often to have recourse in zoology, as all these bats are more or less brown or reddish. Nevertheless when we come to examine them attentively, we are astonished to find that they present so many appreciable differences. Their physiognomy varies ad infinitum. Their ears and auriculæ are in proportions extremely different in the different species.

This mode of considering them will furnish the characters of each.

The *vespertilio murinus*, or common bat, is that species which has been known in Europe from the earliest period. The figures found of this animal in Johnston and Edwards are by no means good. It had been compared with no species excepting the oreillard or great-eared bat. From that circumstance it derives its name of *vespertilio major* in Brisson, and its character, "ears smaller than the head," by which description Linnæus only meant to oppose it to the smaller species, in which the ears are nearly as large as the body. Linnæus also changed its name of *major* into that of *murinus*, in consequence of Brisson's observation that it had hair of a mouse-coloured gray.

By the following character it may be always recognised. Oblong ears of the length of the head; auricula shaped like a demi-heart: fur ashy red above, grayish white underneath. The murinus has the head moreover tolerably long, the forehead narrow and incurvated, and the cerebral case oblong. Its fur is of two colours, dark ashen at the origin of the hairs and red at the points above, and white under the belly.

Preserved specimens of this species are sometimes of a tolerably bright red, but this is probably owing to the liquor in which they are kept. A numerous assemblage of these bats were found in the church Des Grands Jesuites, in the Faubourg St. Antoine, in Paris. They were of all ages. The young had a shorter muzzle, the fur somewhat coarser, and of a tint in general more bordering on ash-colour. The males did not differ from the females except in the fact of the colours being generally a little brighter.

In the last expedition of MM. Peron and Lesueur to the Austral regions was discovered a new species of the murinus. These gentlemen sent to the French museum two individuals exactly alike, and much larger, and of a clearer colour than our European murinus. The back was of a clear yellowish-ashen, and the belly of a decided white. We do not know where they were found.

The bat of Carolina *vesp. Carolinensis,* is smaller than the preceding ; but in other respects there is a strong resemblance. The ears and auriculæ are of the same form and the same relative dimensions. The fur is likewise of two colours, dark ash-colour in the origin, and marron brown at the points. The extremities of the fur underneath are of a yellowish colour as it approaches the belly. The ears are provided with hairs for more than half their length, and there is a small portion of the tail not enveloped in the interfemoral membrane. These peculiarities, joined to the difference of colour, have seemed sufficient to the French naturalists to establish the non-identity of species between this and the common bat. The proportions of the cranium give additional strength to this opinion. The forehead is shorter and larger in the bat of Carolina.

This species was first described by M. Geoffroy. It was sent to France by M. Bose, who procured it in Carolina. It is excessively common in that country. Its characters may be summed up : oblong ears, of the length of the head, partly covered with hair, auricula in the form of a demi-heart. Fur brown marron above, yellowish underneath.

The *Noctule* (*Vesp. Noctula*) might itself admit of another very natural subdivision by the approximation to each other of those species which have a mutual resemblance in the number of incisors and false molars. In this point of view, the Noctule and the Barbastelle, (of which we shall speak presently,) deserve to be united, and distinguished from all the other Vespertiliones, inasmuch as they have, in the upper jaw, four incisives and four false molars; and in the lower, six incisors and four false molars, a combination exhibited by no other group of the same family, nor, indeed, to any of the Cheiroptera hitherto discovered. The upper incisors are separated in pairs, rounded, pointed, and a little curved, and the first of each intermaxillary incisor is larger than the second. The first false molar is a very small anomalous, rudimentary tooth, concealed at the basis of the canine. The second is a regular tooth, and very large. The inferior incisors are trenchant, and uniformly situated on the arc of a circle. The two false molars, in this row, are regular.

In pursuance of our plan of avoiding all superfluous details of organization, we shall confine ourselves here to such as are peculiar to the species in question. In the mammiferous tribes, generally, the movement of the wrist is performed by rather an exterior turning of the axis of the bone of the arm ; but in the bats generally, the carpus, instead of revolving below, in front of the fore-arm, does so on the side, and in this position, the thumb becomes the external finger, and the calcaneum is prolonged into a long apophysis to sustain the interfemoral membrane in flight.

The head of the Noctule is large, the muzzle obtuse and without hair, the cheeks prominent and rounded. The ear is large and naked, and the eye is placed immediately beneath it: the mouth is remarkably open, which gives a very peculiar expression to the physiognomy of the animal. Its organs of sense also, considered in detail, present characters exclusively peculiar to the species.

The eye very small and round, has thick eyelids, which surround it like swellings, and the upper one is surmounted in the front with a wart; the iris and the pupil being both black, the form of the latter is not distinguishable. The nostrils very much separated from each other, are open on the sides of a flat and large muzzle, of a glandulous appearance, but not very distinct from the naked parts which surround it. The orifice of these nostrils is circular, and is terminated behind by a tolerable large sinus, which has an upward direction.

The upper lip is entire, and in the middle part of the lower, we remark a semicircular portion, which is furnished with an uniform and black skin, which appears to be organized differently from the neighbouring parts. The mouth has no cheek pouches, and the tongue, which is covered with soft and fine papillæ, is rounded at its extremity, and divided transversely in the middle by a sort of swelling that looks like a second tongue, and the anterior border of which is furnished with a rank of papillæ, or soft and conical fringes; this swelling is itself covered with soft papillæ, and terminated by two large glands rounded, and partly flattish. The ear is large, rounded at its extremity, of a breadth equal to its length, and is particularly remarkable by an elongation of the external edge of the helix, which is continued underneath as far as the commissure of the lips, and still more in a smaller ear, which seems to be an appendage to the anthelix; it is placed in front of the conch to protect the auditory conduit, and apparently to direct the sounds thither by repercussion. The conch in the middle part of the helix, has four or five folds, formed by the motions it makes for the purpose of bending back upon itself.

The body, with the exception of the muzzle, the ears, and a considerable portion of the membrane of the wings, is covered with hairs, remarkably soft and fine, the true character of which it is not very easy to determine. These hairs are longer, and form a thicker covering on the shoulders, the back, the flanks, the breast, and belly, than on the head and tail. The colour is of a golden-brown, deeper on the head, and paler on the lower parts. The naked parts are of a violet-black.

When the noctule is on the ground, the fore-arm is drawn in towards the body, and the wrist, as well as the thumb, rests upon the earth; the folded fingers are extended at full length along the external edge of the fore-arm, and are concealed under the membrane of the sides. In this situation, the two last phalanges of the third finger are somewhat bent, as is also the last phalanx of the second. The two others remain entirely extended, and the tail is bent back under the belly, from the second or third vertebra. When this animal walks, or rather drags itself along, these various parts preserve the same positions, and strange to say, they are the very same when the noctule, in a state of the most absolute repose, is suspended by its hinder feet with the head downwards.

The individuals upon which the above observations were made, were found in the hollow trunk of a poplar tree, by M. Saulnier; they were ten in number, in the upper part of this cavity, which extended to a considerable height in the trunk. These animals flew about the tree, and re-entered the hollow with inconceivable rapidity, and without justling, though the entrance was very narrow. This species is also found in old buildings, beneath the roof of churches, &c. The individuals just mentioned, having been placed in a large box with some minced-meat, paid no attention to it; all their anxiety appeared to be to hook themselves, and hang head downwards, as speedily as possible; oftentimes they fastened, in this way, on one another, and would then utter a sharp and hurried cry, through vexation. Finally, at the end of a few days, they died of inanition.

M. F. Cuvier, to whom natural history is so highly indebted, has favoured us with some details relative to the Barbastelle, (*Vespertilio Barbastellus*,) with which we shall take the liberty to enrich our pages, but in as compendious a form as possible.

The indvidual, which this naturalist examined, was taken in the steeple of a church ; being shut up in a glass case or press, furnished with several shelves, it traversed its whole extent, getting through the smallest passages, and finished by retiring into its most obsure corner. In standing, it placed on the ground the entire sole of its feet, and its wrist as well as its thumb; the other fingers, with the membrane which unites them, were raised in a contrary direction to the fore-arm, and preserved, by this position, from friction against the ground. The tail was bent back underneath, and the membrane, which enveloped it, folded in such a manner as to fill the least possible space. In walking, the limbs were raised alternately as with other quadrupeds ; the fore-arm was carried obliquely forward, and followed by the hind-foot of the opposite side. The fingers generally remained united, as in standing; but sometimes they were triflingly separated, as if for the purpose of maintaining the equilibrium of the animal. Sometimes, too, the nail of the thumb was crooked, apparently

L 2

for the purpose of dragging forward the posterior part of the body with more effect; and, at other times, it remained extended and unemployed. The repose of this animal consisted in suspending by the hind-feet, with the head downwards, and a vertical superficies seemed indispensably requisite for this position. When it wished thus to suspend itself, it would stop at a convenient place, fix its thumbs where the nails of its hinder-feet should be hooked, and for this purpose, the slightest inequality was sufficient; then, fastening itself strongly, it would detach one of its thumbs, and revolving its body, would carry the foot of the same side to the spot which the thumb had just occupied. The nails being once properly fixed, it would let go its other thumb, which movement, by leaving the body to its natural weight, carried the head downwards, and brought the second foot alongside of the first, where it became hooked in its turn.

When this bat was desirous of flying, being on a horizontal surface, it would spring perpendicularly to the plane on which it rested, and suddenly extend its wings, which sustained it, and left it sufficient time to make some motions to enable it to rise. If it was suspended, it would quit its hold, fall, unfurl its wings, and fly.

This Barbastelle thus passed eight days, going from one place, where it was suspended, to another, to suspend itself afresh. It was especially during the night that it quitted its retreat and repose. It remained, during this time, without taking any nourishment, though meat cut into very small pieces had been put into the case where it was shut up. At last, however, in full day-light, it fell upon the meat, which it entirely devoured. When a piece was too large, it would fix it to the ground with its wrist, and cut it with its cheek-teeth; but if these teeth got engaged in the meat, or any morsel attached to them, it would not use its feet to get rid of the embarrassment, but would seek for some projecting spot, against which it would rub its muzzle. After the time above-mentioned, it was observed to eat no more, and it soon died. Cleanliness appeared a very peculiar characteristic of this animal ; with its hinder-feet it would rub all the parts of its body, and cleanse its nails, fingers, and the membrane forming its wings, very dexterously, with its mouth. There was very considerable vivacity in the motions of its head, appearing to indicate very lively sensations ; and its projecting cheeks, its mobile nose, and great ears, exhibited partial movements of astonishing variety, which gave to its physiognomy an expression extremely singular in so small an animal.

The system of dentition, as far as it respects the molars, is the same in all the insectivorous bats. In the Barbastelle, the upper incisors are four in number : the first is larger than the second, and bifid. There are six incisors in the lower jaw, and all indented. There are two false molars on each side of the two jaws, so that, in all, this animal has four-and-thirty teeth. The first false molar, in the upper jaw, is so small, and so completely concealed at the basis of the canine, that it is scarcely perceptible ; for this reason, it escaped the researches of Daubenton, who gives but eight upper cheek-teeth to this bat. On each side of the muzzle is a large tubercle or cushion, surrounding the nostrils in part, and behind which, is the eye. The eye itself is so small, that it is perceived with difficulty in the midst of the hairs which surround it, and it is impossible to distinguish exactly the form of the pupil. The nostril is bored at the extremity of a furrow, which has the form of a V, one of the branches of which, is much larger than the other. The lips are entire ; the tongue is smooth, and there do not appear to be any cheek-pouches. The most extended sense of this animal is evidently that of hearing. The external conch of the ear, is of a most disproportioned magnitude ; it extends, in front, as far as the middle of the forehead, is united by its basis to the opposite conch, and in the rest of its length, it approaches the other ear so closely, that at a front-view, it is impossible to perceive any of the back part of the head, and there appears to be nothing but muzzle. The same organ, on its other side, extends below, as far as the edges of the jaws, so that the eye is actually enclosed by it, and from its internal basis springs a lobe or second ear, in the form of an elongated leaf, which is entirely free, and seems destined to augment, if not, the sensibility of hearing, at least, the effect of sounds upon the auricular nerves.

As to its wings, &c., it is unnecessary to go into any details. The membrane of the flanks, also, includes the tibia, and extends from thence to the extremity of the tail, which is nearly one-third of the length of the whole body, so that when the animal extends its arms and hands, and its hind-legs horizontally, and erects its tail, it has not only wings, but a large parachute, which contributes not less than the wings, to diminish the specific gravity of its body. It is not easy to conceive why the men who have attempted to fly, and who have often paid so dearly for their folly, did not endeavour to construct their apparatus for flying after the model of the bats, than after that of the birds. There are certainly many more relations existing between man and the cheiroptera, than between him and the feathered race.

The specific rank of the Barbastelle was first established by Daubenton ; before his time, it had been confounded with the other bats. Buffon's figure of this animal is pretty good, but the fingers are not long enough.

The Serotine (*V. Serotinus*) is another vespertilio of Europe, discovered, described, and named by Daubenton. It is found abundantly in Paris and its environs. The hollow trees, in the Bois de Boulogne, enclose them in vast numbers. Its magnitude is the same with that of the murinus.

It may be recognised by its oval and triangular ears, shorter than the head. The auricula arched, and with a summit, large and rounded. The hairs short; fur bright-red above, more dull underneath.

It must not be confounded with the preceding species on account of the similar disposition of the ears. It differs in the auricula, which is shorter, arched, and terminated by a large sort of head. Its ears are also more extended in front, their anterior border falling in a line over the eye, while, on the contrary, it is removed back a trifle in the noctule. Besides which, the head of the serotine is shorter, the forehead and muzzle considerably larger, and its hair gives out a brilliant reflection from the upper part.

The Pepistelle (*V. pepistellus*) is another French bat, the knowledge of which is owing to M. Daubenton; it is the smallest of all. It resembles the noctule so strongly in colour and proportions, that one might almost be tempted to consider it as a juvenile specimen of that species. It differs, however, from the noctule, and that essentially. Its ears are of a triangular oval shape, shorter than the head; the auricula is almost straight, and terminated by a roundish kind of summit. Its hairs are long; its fur of a blackish-brown above, and brownish fawn-colour below. Thus the pipistrelle differs from the noctule, not only in size, but also in the auricula, which instead of being large at the base, and pointed at the extremity, approaches more to the configuration of that of the serotine. The colour of the hairs is a decided brown-black, which proceeds from the hair being black within, and fawn-colour only at the point. The pipistrelle is also remarkable for its long tail, which is very nearly as long as the body. Its cranium is also considerably remote in form from those of the noctule and serotine. The cerebral case is larger, more convex, and projects more beyond the forehead, and the occiput is rounded, and without any crest.

It is not uncommon to find it in the day-time on the ground, but this always happens when it is remote from any culminating point. It suffers itself to be taken without resistance, being overwhelmed by the fatigue of its fruitless efforts to resume its flight, and regain its habitation.

There is a variety of the pipistrelle found in Egypt. It is usually more of an ashen-colour ; the points of the hair are of that tint.

The vespertilio auritus, or great-eared bat, is designated by Brisson, under the name of *vespertilio minor*, because it is much smaller than the other. But as its ears are of an excessive length, Linnæus changed its name into that of *auritus*, desirous of translating the French name *oreillard*, given to this bat, by Daubenton. Its ears almost as long as the body, and re-united in the front, easily cause it to be recognised among all the known species. It is but small in size ; the muzzle is tolerably large ; its nostrils present a singular peculiarity. After the nasal apertures, such as they are found in all the vespertiliones, are two holes, or rather two small purses. The ears are joined in front, a little way up ; their interior edge is folded backwards ; there are some hairs bordering the length of this fold, exactly in the manner in which the eye-lashes are fixed on the human eye-lids, ranged upon a single line. At the lower part of this edge, is a small fold, which cuts it in an angle of about 60°, and then proceeds, inwardly, towards the origin of the auricula. The tail is also very remarkable by its considerable length. The membrane of the wings is the same which extends between the legs, and that to an equal degree as in the wings ; these last, even in their extension, are somewhat folded or wrinkled, in consequence of the numerous tendinous threads which are spread in the interval between the upper and under membrane.

The fur of the auritus is greyish-brown above. and ash

colour underneath. M. Geoffroy found a bat of this kind in Egypt, apparently smaller than the European muritus, and which bordered more upon the red.

The slope-eared bat (*vespertilio emarginatus*) is another of the bats of Europe. It escaped the researches of Daubenton, and of all the naturalists who succeeded him. It is, however, very common both in England and France. M. Geoffroy gives it the name of *emarginatus*, in consequence of a very marked slope at the external edge of its ears, and he thus characterizes it: " Oblong ears, of the length of the head, and sloped at their external edge ; auricula, awl-shaped ; fur reddish, gray above, ash-coloured below." This is the first instance we meet of this long and narrowed auricula, shaped like an awl, but it is a character of almost all the foreign vespertiliones. This species might be confounded with the pipistrelle, though it is somewhat larger, in consequence of the great resemblance of physiognomy. It also resembles the common bat in the two colours of its fur. It is not only towards the point that this hair is grayish-red, but it begins to be so from one half of its length ; hence, the general tint is uniform. Nevertheless, as it is long and tufted, when brushed back, some spots of dark ash-colour are perceived, which is the colour of the other portion of the hair. Under the belly, the extremities of the hair are of a dirty white.

The *vespertilio pictus, Kirivoula* or striped bat, is one of the most anciently and best known, though a foreign species, and belonging to India. Seba was the first who gave a figure of it. Daubenton afterwards described it, and gave a much better print. Pallas also described an individual, which he had seen in Holland ; he was wrong, however, in attributing to it eight incisors in the lower jaw. The Kirivoula, as Daubenton had already described it, has but six. Kirivoula is the name of this bat, at Ceylon ; it is also found, according to Seba, at the island of Ternate.

Its epithet, *pictus*, is given from the yellow streaks which proceed from the carpus, and extend over the fingers. Its head is large, the muzzle slender. The forehead is arched, and the front of the cranium considerably incurvated; but this is not very visible in the fresh state, as this part is covered by long hairs, in which the ears are also partly concealed.

These, though oval, have a small point above. The auricula is subulated, very narrow and long ; the fur is of a very beautiful colour, being of a most brilliant golden-red. The membrane of the wings is of a fawn-coloured brown, and is radiated with yellowish streaks the whole length of the fingers.

The rough-tailed bat (*vesp. lasiurus*) inhabits Cayenne. Schreiber and Pennant have given a tolerable description of it. It is about the size of the emarginatus. Its ears are oval, shorter than the head. The auricula narrow, and shaped like a demi-heart ; the fur varied with yellowish and red ; radii of a grayish-brown, proceed from the carpus, and extend over the fingers. The hair above is yellow for almost its whole length, and cinnamon-colour at the point. The belly is yellowish.

The Timor bat (*vespertilio timorensis*) is owing to the labours and researches of M. M. Peron and Lesueur. The ears are large, of the length of the head, and joined together by a small membrane ; the auricula like a demi-heart. It is brownish-black, and ashy in the under parts. The fur is long, soft, and thick.

The Bourbon bat (*vespertilio Borbonicus,*) comes from the island of Bourbon. It may thus be characterized : Ears triangular oval, half as short as the head; auricula long, formed like a demi-heart ; fur red above, and white below. Its head is short and large ; muzzle inflated, and nose projecting.

The Senegal bat of Pennant (*vespertilio nigrita*) was called, by Daubenton, the flying marmotte. Seba was the first who gave to two species of the Indian bats, the names of flying dog and flying-dormouse. Daubenton adopted this sort of nomenclature for the purpose of giving an idea of the relative sizes of these animals, not because there was any resemblance between them and the species whose names they received in part.

The ears of this vespertilio are oval triangular, one-third the length of the head. The auricula is long, and terminates in a point; it is of a fawn-coloured brown. The muzzle is broad and thick, the lips long, but neither inflated nor varicose. The end of the tail is disengaged beyond the interfemoral membrane. Daubenton gives it but two upper incisors ; there are two others, but very small. It is a native of Virginia, and was made known by Adanson's voyage to Senegal.

The *vespertilio maximus* is the largest of this tribe. Its ears are oval, shorter than the head ; the auricula subulated, and the muzzle long and pointed. The fur is brown ; marron above, clear yellow on the sides, and dirty white under the belly. It inhabits Guiana.

The *vespertilio Noveboracensis*, New-York bat, has been described by Pennant ; it seems to belong to this genus, though this clever naturalist says that it has no upper incisors ; Pennant, probably, saw an imperfect individual. The ears are short, large, and rounded. It is remarkable for a white spot at the origin of the wings. It is about the size of the noctule.

The (*vesp. lasiapterus*,) lasiopter bat of Shaw, in size, form of head, and colour, strongly resembles the serotine ; but it differs in having the membrane of the wings covered with hair internally for about one half its length.

The (*vesp. villosissimus*,) hairy bat, is a species of Paraguay, found in d'Azzara, and called villosissimus by M. Geoffroy, because the hair is longer than in all the other bats of that country, and extends even over the interfemoral membrane.

It is of a whitish-brown, and there are some radii of the same colour on its wings, after the fashion of the pictus.

The *vespertilio ruber* and the *vespertilio albescens* are two other species of the same country, and described by the same writer. The first is of a cinnamon-colour above, and rose-colour below. The second is blackish above, and of an obscure brown in the under parts. It appears as if powdered with white under the belly, because the points of each hair are of this colour; the whitish-tint increases more and more behind. The ears of both these species are remarkably sharp, and the auricula are subulated, *i. e.* awl-formed.

The word GALEOPITHECUS is a compound of two Greek words, signifying cat monkey. It was appropriated by Pallas to the *lemur volans* of Linnæus, which had been designated by travellers under the names of flying cat, flying monkey, flying dog, &c.

The galeopitheci are at present but very imperfectly known. M. F. Cuvier has given us the best account of them. All we learn from those who have seen them living is, that they hang suspended by their hind-legs from the branches of trees ; that they feed on insects, and probably on small birds ; that they move with difficulty on the earth's surface, but climb trees with surprising facility, and spring from one to the other supported, as by a parachute, in their passage by the membrane spread around their body; and that they are crepusculous animals, active only during twilight.

More is known of their organization than of their mode of life. The various vulgar names applied to them have reference each to their general physiognomy, in which may be traced traits of similarity to the cat, the monkey, especially the lemur and the dog. The largest species known, if indeed more than one be really described, is not bigger than a young cat; but it is longer and thinner than any of the felinæ, and in this respect approximates the lemurs.

Their teeth are also nearer to those of the lemur than any other animal, though they present some strongly-marked differences. Of the six incisives in the lower jaw, the two external are indented like a comb, especially the intermediate; the four upper incisors are very small, but the two external are the largest; the canine teeth are very small, and sharp pointed. The cheek-teeth, $\frac{6.6}{5.5}$ are sharp pointed, and the anterior are very like the canine teeth.

Their extremities are entirely enveloped in the membrane. The four feet have each five toes, parallelly disposed, furnished with long strong sharp semicircular nails. All the fingers are attached by the membrane, beyond which nothing is to be seen but the nails. The tail also, which is long, is included in the membrane.

This singular appendage springs from under the throat, and extends to the fingers of the hands, from which it passes on to those of the hinder extremities, and so to the extremity of the tail, so that when the animal spreads its limbs to make a leap, it covers a space much larger than its body, which lowers it gently, in the manner of a parachute, and assists also in lengthening the extent of its leap, though it does not appear to be of much service in aiding the ascent, being calculated simply to break or ease their fall.

We know but little of the senses of these animals, and and still less of their modes of generation. Their eyes are large and prominent, the nose is simple; the tongue is soft; the ears are not very large, and the fur on them is soft and silky; they have no whiskers, and their skin, particularly of the hands and feet, is very soft; their mammæ are pectoral.

The situation of the galeopitheci in natural arrangement

seems hardly yet determined. Linnæus and Pallas joined them to the lemurs. Geoffroy St. Hilaire attached them to the cheiroptera, or rather made them intermediate between the lemurs and bats. Our author, as we have seen, has placed them at the end of the cheiroptera, considering them apparently as more approximated by their organization to the omnivorous mammalia than to the quadrumana. Illiger makes them the first family of his *volitantia*, the cheiroptera of the Baron. It appears, says M. F. Cuvier, that the true place of these animals in their natural order is between the lemur and bat, whether we consider them with Linnæus as the last of the former, or with M. Geoffroy at the head of the latter.

The oleck, red galeopithecus, (*lemur volans*, L.) whose description is sufficiently stated in the text and table, is said to emit a strong and disagreeable odour, though its flesh is perfectly palatable. The Pelew Islanders call it oleck, which should be adopted in preference to its factitious but more scientific appellative. The figure of the animal is from a specimen in a Mr. Bullock's late museum, with a young one attached to the teat. The figure in Audebert's History of Monkeys and Lemurs, like most of the rest in that splendid work, is very good and accurate in its detail.

The varied galeopithecus of Audebert is smaller than the preceding, of a darkish brown, varied with white spots on the legs. It seems likely to be a young individual of the preceding species.

Seba's figure has been treated as of a distinct species, under the name of the galeopithecus of Ternate. The fur is reddish-gray, soft, like that of the mole, deeper in shade above than below, with some white spots on the tail.

Whatever may be the real number of species of this singular family, the distinctiveness of those above enumerated seeme at present uncertain.

C.Hamilton Smith Esq.ʳ del.

Griffith. sc.

THE COLOCOLO.

F. COLOCOLO OF MOLINA ?

4.

The family of the cheiroptera, or bats, is, without question, the most abundant in species of any subdivision of the Mammalia. By a reference to the organs of mastication, sense, and motion, it can fortunately be divided into numerous groups, without which it would be next to impossible to study these singular animals with any prospect of success.

The ancient authors busied themselves but little in distinguishing the different species of bats. To Daubenton (so recently as 1759,) we are indebted for our knowledge of the noctule, though one of the commonest species. It may, also, be safely asserted, that many more accurate observations will be necessary, before we are in complete possession of its history; and the same remark is equally applicable to all the other bats whose habits, structure, and disposition, have been as yet but very superficially studied. In truth, little is known concerning them beyond their mere zoological characters. Their natural history, properly speaking, has been made the subject of very few observations. There are, however, but few of the mammalia whose peculiarities would better repay the attention of a curious observer, inasmuch as their extraordinary organization obliges them to a mode of existence, if possible, still more extraordinary. The extreme difficulty of sustaining them in a state of captivity will, for a long time, hinder us from becoming acquainted with their true nature, and the advantages which they derive from their organization for procuring subsistence, escaping from their enemies, and propagating their kind. Before all the phenomena relative to these fundamental points of the natural history of any animal can be collected, a large number of individuals of each species must be subjected to attentive observation.

We have been a little more diffuse on this curious family, on account of the paucity of information in the English language on the subject, than our allotted space will allow for the numerous other branches of the animal kingdom, which we hope will be considered not merely as venial, but as acceptable. It is principally from the works of M. Geoffroy St. Hilaire, whose name so often occurs in these pages, that the substance of our essay has been extracted.

Teeth of Insectivora.

With short canines.

1

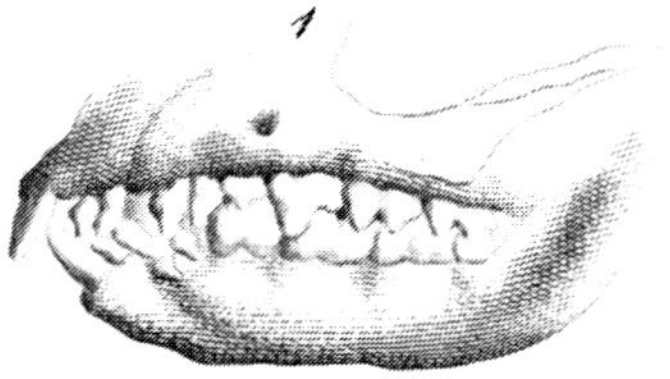

With long canines.

2

Cheek teeth

3

4

5

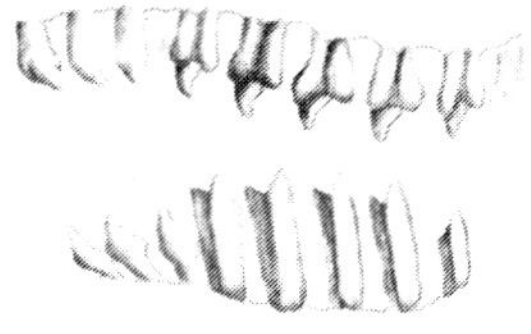

6

7

1 Hedghog.
2 Mole.
3 Shrew.

4 Desman.
5 Chrysochlore.
6 Tenrec.

7 Tupaia tana.

London: Published by C.B.Whittaker, March 1.1827.

Supplement on the Insectivora.

This name, in its literal signification, might be applied to a very extensive assemblage of the animal world. Some of the Lemurs, several of the smaller Simiæ, and many of the Rodentia or Glires, are truly insectivorous. We have also seen that most of the Bats are principally sustained on insects. Our author, however, in his classification, restricts the application of the term to a particular tribe of mammiferous Carnassiers, which forms the second family of that order. In this family he comprehends the genera, or rather subgenera, Hedgehog, Shrew, Desman, Scalope, Chrysoclore, Tenrec, and Mole ; which, like the Cheiroptera or Bats, have cheek-teeth, bristling with conic points. Their mode of existence is not only nocturnal, but for the most part subterranean, whence Illiger has named his corresponding order Subterranea. It may be remarked, that a similar formation of the cheek-teeth, namely that with sharp tubercles, obtains in all insectivorous animals, even though they do not belong to the family we are about to describe. This conformity of the teeth with the nutriment of the animal is a very general law of nature.

The family of Insectivora, as we have seen in the text, is conveniently divided into two tribes, distinct from each other by the position and relative proportion of their incisive and canine teeth. The first having long incisors like those of the Rodentia, and short canines sometimes called lateral incisors, ambigui, or false canine teeth ; and the second with small incisive and longer canine teeth as in the Quadruma Carnivora, &c. The former tribe includes all the subgenera of this family except the Tenrecs, Moles, and Condylure of Illiger, which compose the latter.

We have accordingly given a figure of the general system of dentition in these two tribes, to which are added examples of the variations in the cheek teeth, which, though

different from each other, are of the insectivorous character in all the species.

The HEDGEHOG, which forms the first sub-genus of the
insectivorous family, has been also called in popular language the *Urchin*, a word obviously derived, as well as the
French word *Herisson* from *Erinaceus*, the name by which
the Latins designated our Hedgehog, as the Greeks termed
it *echinos*. From the species to which this name has been
applied, it has been extended to others bordering thereon,
and thus it has at length become a generic appellation.

The Hedgehogs appertain to that family of the insectivora, which subsist for the most part on little animals, on
insects, and on fruits. The cheek-teeth of this genus, with
few exceptions, are distinguished by conical points, which
are mutually intergrained together. They are plantigrade,
dig into the earth to conceal themselves, pass their lives in
a state of repose approaching to lethargy, and escape from
their enemies by means of the obscurity in which they envelope themselves. Their gait is heavy, and their intelligence very limited. They may be so far domesticated as
to be brought up in gardens, where, without doing any
mischief, they may prove of considerable service in destroying many hurtful insects. It is said that their flesh is good
for eating.

The Hedgehogs have on each foot five toes, armed with
nails proper for the purpose of digging. The soles are
naked, and provided with projecting tubercles, which cover
a soft skin underneath. The ear is rounded, and of a very
simple structure. The eye is small with a round pupil.
The nostrils, which extend considerably beyond the lower
jaw, have their openings on each side of the muzzle, the
external edge of which is fringed. The lips are entire, the
tongue soft, and there are no cheek-pouches. The upper
parts of the body are covered with thorns or bristles. On

the under parts they are flexible, and partake more of the character of hair, though rather stiff. A few are found with something of a woolly character, and the upper lips are furnished with mustachios.

These animals have at the extremity of each jaw two incisors similar in form to canine teeth, and of which they may make the same use as the Rodentia do of their incisors, or the other Carnassiers do of their canines. Those of the upper jaw are separated from each other; those of the lower are approximated and nearly touching. Behind these first incisors in the upper jaw, are found, on each side, two small teeth with a single root, which have the form of the small anterior cheek-teeth called by our author false molars, though implanted in the intermaxillary bone. After these come the false molars themselves, separated from the last-mentioned by a small vacant interval. They are three in number. The first, which is the largest, has two roots; the second has but one, and the third again has two, and a small internal protuberance beside. The true cheek-teeth then follow, to the number of four. The first has three tubercles, one on the external side, large, sharp, and trenchant, the two others on the internal side, but smaller. The second and third of these teeth resemble, with the exception that the latter is not quite so large. They have all four tubercles of equal size, terminating the four angles of a square. The last is slender, situated obliquely in relation to the others, and not unlike a false molar. In the lower jaw there are three small teeth, with a single point and a single root, following immediately the large incisives. After this comes a first molar with two principal points, and terminated by a small protuberance. The second and third resemble. Three points form their anterior part, and two their posterior. The first are disposed in the form of a triangle, the second are beside each other transversely. The last molar, which is the fourth, and is very small, pre-

M 2

sents in front a small protuberance, and a forked point behind. The molar teeth of the two jaws are opposed crown to crown, so that the anterior part of those below correspond to the vacuum above, and the posterior part of these last to the vacuum below. We are not satisfactorily acquainted with more than two species, of the first of which, the Erinaceus Europæus, we shall now speak.

There are some animals, and among others the Hedgehog, which manifest in an eminent degree the care and attention bestowed by nature on the preservation of living beings. In the majority, indeed, of mammiferous animals, the continuance of their existence appears mainly to depend upon that kind of equilibrium in the animal world, which properly constitutes the economy of nature, and not so much on any direct means afforded them for that purpose, or any express provision for their security and preservation. Those which attack, those which defend themselves, those which pursue, those which fly, and those which have recourse to concealment for refuge, are reduced in each country from their reciprocal relations to a number pretty nearly fixed, and which cannot alter with respect to one division without some consequent and corresponding alteration with respect to another. Thus, for example, if any fortuitous circumstances should conduce among us to the multiplication of the mammiferous Carnassiers, the natural consequence would be that the species on which they subsisted must become the victims of this unusual increase, or they must themselves perish from misery and hunger. It would be impossible for either, by their own peculiar activity, or any internal resources, to resist the imperious control of circumstances in such a case. The state to which they would be reduced would be the inevitable operation of a law founded on unalterable conditions, and involving within itself the truth of evidence and certainty of result which belong to mathematics.

But the Hedgehog is one of those animals which form some exception to this general law. It is by no means so exclusively submitted to the influence of those causes which surround it. Its means of defence are to a certain extent independent of the number of its enemies. Notwithstanding its weakness, it bids defiance to their power, it braves their attacks in all the consciousness of security, and finds the means of escape, of shelter, of resistance, in the resources which it has received from the bounty of nature. The thorns in which it possesses the faculty of self-envelopment, and which radiate round the circle of which its body forms the centre, constitute a rampart, before which the most powerful and voracious animals, which might otherwise be tempted to prey upon the Hedgehog, must infallibly be obliged to retire.

It is not easy to conjecture for what reason so especial a favour has been conferred upon an animal, which by its littleness, its silence, the obscurity of its life, and the fewness of its wants, would have been sufficiently concealed from all eyes, and sheltered from every enemy. One would imagine, from its provisions of defence, that it was destined to perform some important and necessary part in the grand economy of nature. But its habits, its appetites, and its instincts, do not differ materially from those of the other Mammalia, much more ill-provided against the approaches of danger, but which, nevertheless, contrive to preserve their existence for the prescribed period of its duration. The Moles, the Shrews, and all the omnivorous Rodentia, subsist, like the Hedgehog, on worms, insects, roots, and fruits.

The majority of them, too, like this animal, shun the light of day, and conceal themselves in retreats of obscurity and silence. Nor is the Hedgehog distinguished for more intelligence than the rest, for in all of them the power of perception seems limited to the faculty of distinguishing among the small number of causes which are influential on their being, the hurtful from the advantageous. Why the Hedgehog, therefore, should be thus peculiarly favoured by nature, is a question which apparently we have no means of resolving. It is not included, most probably, in the circle of proximate causes, the only ones which can form the legitimate objects of human knowledge, and which are placed within the reach of human ability.

The nature and constitution of the Hedgehog are obvious to an observer at his very first glance at the animal. Its heavy form, short limbs, and plantigrade motion, indicate at once the small proportion of its agility, the weakness of its intelligence, and the obscurity of its mode of existence. The animal, in fact, is almost always in a state of concealment. It is usually found at the foot of trees, in those hollows left between them by the roots, and for which the moss forms a slight and second covering, or else in the vacuums which are found in heaps of rocks or stones, or in deserted rabbit-holes. In retreats like these it passes its gloomy days, and never sallies from its obscure abode until the congenial shades of night approach. It then proceeds, with a slow and measured pace, in search of its food, which principally consists of snails, worms, and other animals of the same description. It also feeds on sweet and succulent fruits; but it is not true, as has been most absurdly asserted, that it carries them off upon its prickles. This supposition is not merely devoid of truth, but even of probability; for the animal could have no means of detaching the fruits which were thus fastened on its thorns. It is also during the obscurity of night that the Hedgehogs seek to satisfy their sexual wants. They have the faculty of lowering their prickles, and of smoothing them to a level with their body, so as to be able to perform many operations without the least difficulty, which would be out of their power if the prickles were always rectangular with the body.

It is in the commencement of Spring that the want of

re-production begins to be felt by these animals, and the young are born in the course of the month of May. The period of gestation is not precisely known. The young come into the world covered with small prickles entirely white, and their reddish skin may easily be perceived between them. Their eyes are closed, and what is very remarkable, their ears also. This is a very singular peculiarity, and not observable, as far as we are aware, in any other species of the mammiferous animals. Some of the Bats, as we observed in our last number, possess the faculty of voluntarily closing the ears; but this character of having them closed at and a little after birth, is exclusively confined, we apprehend, to the Hedgehog kind. The length of these young Hedgehogs is rarely above two inches, and the tail is very short. In a little time, their prickles increase both in size and number, and become coloured, and towards the Autumn, they differ little from the adults, except in magnitude.

When the Hedgehog has grown to its full dimensions, it is about nine inches in its greatest length, and all the upper parts of its body are covered with thorns, gray at their origin, then of a brownish black, and finally terminating in a white point. Its head, and the circumference of its ears, are clothed with harsh and brownish hairs, and the upper part of its fore-paws and all the lower parts of its body are covered with whitish hairs. The paws themselves, the extremity of the muzzle, and the tail, are very nearly naked, but some rudiments of weak mustachios may be seen on the sides of the upper lip. The prickles are retained in the cuticle only by a very small pedicle. The eyes are simple, extremely projecting, and the pupil is round. The ear, small and rounded, is also of a very simple construction. The tragus and antitragus cartilarge, and very closely approaching to each other, so that the interval between them is but a small cleft above the auditory nerve.

They are also surmounted by a sort of valve, which may be considered as a very extraordinary development of the internal point of the helix, and which closes the ear completely above when the animal approaches the anterior part of the conque to the posterior ; an operation which is always performed whenever any foreign body advances to touch this organ. The nose extends considerably beyond the jaws. It is terminated by a snout or muzzle, the posterior part of which is divided by five or six scollops, which compose a kind of fringe work round it. The orifice of the olfactory conduit is open at the lower part of the nostrils, and consists in a furrow, which following a curved line, rises towards the upper part of the muzzle. The lips (as they are indeed in the whole genus) are entire, and no other accessory organ is found within the mouth.

The organs of locomotion consist in plantigrade feet, which have five toes, armed with very long nails. The relative length of the toes in the two feet proceeds thus in a decreasing ratio :—the middle toe, annular, index, small toe, and thumb. The sole is furnished with three tubercles at the end of the four longest toes, and one extremely large towards the middle. The palm has also three tubercles at the base of the four first toes, then two more at its lower part, a large one and a smaller. All these parts are covered with a skin extremely soft and susceptible of the most delicate sensation of touch. The tail is very short, and usually folded back upon the genital parts. The teats, to the number of five on each side, extend from the arm-pit to the groin.

The description of the teeth we have already given in our observations on the genus, and it is unnecessary to repeat it here. We have also given every thing that is known concerning the manners and habits of the animal, and usages to which its organs are appropriated. We shall only add that the faculty which it possesses of enveloping itself completely in its thorns, when it rolls itself up into a ball, consists in the muscles of the skin, which, once extended beyond the head and paws, contract and enclose the body in a sort of purse.

Some authors have spoken of two species of the Hedgehog among us, one characterized by a more obtuse muzzle than the other. But this distinction has not been confirmed, nor does it apear to be founded on any very exact observation *.

Our Hedgehog is met with throughout all Europe, excepting in the very northern parts. It was well known by the ancients, and the majority of figures which have been published, from the time of Gessner to our own days, are tolerably faithful, and communicate a pretty just notion of the animal. In all methodical catalogues it is known by the name of *Erinaceus Europæus.*

The second species of the Erinaceus, is the *long-eared Hedgehog (Erinaceus Auritus,)* called also by some writers the Siberian Urchin. It inhabits the Eastern regions of Asiatic Russia, near the lower parts of the Volga and the Ural, and beyond the lake Baikal. It has also been found in Egypt by M. Geoffroy. Those near the Volga and the Ural are considerably smaller than the European species, but those beyond the lake are of a larger size. Its muzzle is short. The ears are distinguished for their size, from which character the animal derives both its popular and

---

* The fact is, that these can only be considered as varieties of the *Erinaceus Europæus.* The first has received in French the name of *Herisson Pourceau* (Swine Hedgehog), from the form of its nose, which is elongated like the snout of a Hog. This is the most common variety. The other is much more rare. Daubenton could not discover it in the space of ten years, and M. Desmarest declares that he never saw it either alive or dead. It is called the *Dog-hedgehog,* from the form of its nose, which is proportionally shorter than in the first-mentioned variety. Its prickly armour is also less extended, the tail is longer and more slender. The hairs are thicker, stiffer, and of a deep red. The only figure of it has been given by Perrault, *Collect. de l'Acad. des Sciences,* tom. III., 2d part, p 41.

scientific appellations. The upper part of the body is covered with slender brown spines, with a whitish ring at their base, and a yellowish one at the point. They are not joined in tufts or stalks at their extremity to the root, but separated singly, and are smoothed back when the animal is in a state of repose. The nostrils are denticulated like the crest of a cock. The limbs are longer and more slender than those of our European Hedgehog. The tail is rather shorter, conical, and almost naked. The hair is in general much finer than that of the Erinaceus Europæus ; in the limbs and belly it forms a fine whitish fur. The muzzle is furnished with four rows of whiskers. The colour of the tail is of a yellowish white. The iris of the eye is bluish. The female has usually two litters in the year, and brings forth from six to seven at a birth. It hybernates in holes a few inches below the surface of the ground. It feeds on insects, and can eat cantharides and such vesicatory insects, which would destroy other mammalia, without sustaining any inconvenience ; thereby evincing an instance of that harmonious agreement and consistency, which however prevalent, universally and in more familiar instances, too generally escapes observation. The powers of stomach of the Hedgehog are equally adapted with its teeth for insectivorous regimen, and for such insects even as are destructive to others differently adapted ; were it otherwise, its mental impulse and its physical capabilities would frequently induce its destruction. It grows remarkably fat, especially towards the Autumn, rolls itself up when afraid of any thing, and closely resembles the European species in all its habits and manners. We present a figure of this species.

The Hedgehog with pendent ears (Erinaceus Malaccensis), is usually considered as belonging to this genus. Gmelin, however, was doubtful whether to class it here, or among the Porcupines. He placed it, notwithstanding, among the Hedgehogs, on the authority of Brisson. M. Desmarest

THE GREAT EARED HEDGE HOG.

ERINACEUS AURITUS. Gm.

London, Published by G.B. Whittaker, Dec.ʳ 1824.

gives it but a conditional place among them, as he seems to consider that its general appearance, and the form and length of its spines approximate it more to the Porcupines. We are not yet sufficiently acquainted with its system of dentition to assign it a determinate place in the Baron's classifications of the animal kingdom. Indeed all we know of the animal is from the figure, and very short description given of it by Seba. Its eyes are large and brilliant. The ears almost naked and pendulous. Its spines are from five to six inches in length, and are variegated with white, black, or reddish colours. There are soft hairs between those spines. The hairs of the under part of the body are red. Of its peculiar habits little or nothing is known. From this species was said to be procured the stone called *piedra del porco*, which was formerly held in high estimation for its medicinal virtues, and is probably a kind of bezoar. The habitat of this *Erinaceus*, or perhaps *Hystrix*, are the islands of Java and Sumatra, but principally the peninsula of Malacca.

In addition to the species which we have now enumerated, some naturalists admit two others into the genus Erinaceus. The first is the Hedgehog of Siberia, called *Erinaceus Sibiricus*, by Erxleben, Brisson, and Klein. It presents no very material points of dissimilarity to our Hedgehogs, except that its ears are flat and short, its nostrils are not fringed, its spines are red, and their points are of a golden yellow. The lower parts of the body are covered with hairs of a clear ash-colour, and slightly shaded with yellow. Sonnini considers it only to be a variety of the European Hedgehog. It has been rarely seen, and we are not aware of any existing figure of it.

The second is the *Earless Hedgehog (Erinaceus Inauris)*, also called the Hedgehog of America. It is known only by a figure and short description given by Seba. The ears have no external conque, and the spines are ash-coloured, approaching to yellow. The fore-part of the head, the belly, and the limbs, are covered with silken and whitish hairs. Those on the upper part of the eyes are of a deep brown, those of the temples long and blackish.

According to Seba, who seems to have been deceived respecting this point, it inhabits Dutch Guiana, and is supported by fruits, roots, herbs, and the larvæ or eggs of ants. Its flesh is white and delicate, and the inhabitants are said to use it as food. D'Azzara suspects, and apparently with good reason, that this animal is the same which he has mentioned under the name of Couy, and which belongs to the genus HYSTRIX

The next family of the Insectivora, according to our author's arrangement, are the SHREWS (SOREX), in French, MUSARAIGNES. These little animals have been known at all times in Europe, and compared, from their littleness and the meagreness of their limbs, to spiders. This resemblance seemed to form so decided a trait in the conformation of the animal, that from it its French name is obviously derived. Pliny speaks of it under the appellation of *Mus Araneus*, and from this, in French, was formed first *Musaragne*, then *Muserain*, and finally *Musaraigne*.

The first part of this appellation *(Mus)* indicates that the Shrew was originally supposed to belong to that numerous group of the smaller Rodentia. This notion was so implicitly received, that we find no variation except in its specific *araneus*, which, however, the greater number of naturalists adopted. Some Italians changed it for *Cæcus*, and Gessner gave the Shrew the name of *Moschius*, on account of the peculiar musky odour which it exhales.

It was soon, however, observed, that the Shrews formed a group of animals entirely isolated, and bearing small relation to the Glires. Upon this it was thought necessary to establish the new genus, *Sorex*, and the necessity of so doing was more especially recognised, after Daubenton had explained and taught naturalists thoroughly to appreciate the distinctive characters of this genus, in a memoir which he inserted in the Transactions of the Academy of Sciences, in 1756. He also made known, at the same time, two distinct species.

The Shrews, in fact, have no relation with the Rodentia, and consequently none with the Mice. Their jaws are completely furnished with teeth. They are destitute of a cæcum, and even of large intestines. Their ossa pubis are not united, and their head, which is excessively elongated, gives them quite another physiognomy. It is not easy to reckon their teeth, and at the same time to distinguish with precision their different kinds ; for all that we can learn on the subject, after the most attentive examination, is, that the anterior teeth, or if it must be so, the incisors, are the longest ; that the lateral teeth are the shortest, and that those in the hinder part of the mouth are furnished with conical points.

A single Sorex, and one of the most common, the fur of which is of a grayish red, was the only one known until the year 1756. Daubenton then brought to light one of those species which live near the water, and to which Erxleben and Blumenbach have given the name of its discoverer, calling it *Sorex Daubentonii*.

Some other species were also discovered in France, but at a period considerably later, in 1778. The celebrated Professor Hermann, of Strasburg, was informed of their existence by one of his pupils, whom an enthusiastic ardour for the study of natural history was at that time continually stimulating to researches after the smaller tribes of animals. This pupil, whom considerable services to the cause of science have since advanced to an honourable celebrity, and to whom, whatever may be thought of his peculiar doctrines, none can refuse the praise of persevering assiduity in the cultivation of knowledge, was no other than Dr. Gall. This disciple made known to his professor, without permitting himself to go any further, three new

species of the Sorex. The charge of publishing this discovery to the world remained with Hermann, but unfortunately was not executed for a long time afterwards, and
even then, in a very imperfect manner, the professor being
obliged to return to the task several times. Thus, for instance, he began by confiding the figures to Schreber, who
published them without any explanation. Boddaert, from
the same source, introduced three new Shrews into his *Elenchus Animalium;* and it is only in a posthumous work of
Hermann, dated 1804, that we finally discover any details
concerning these singular animals.

M. Geoffroy St. Hilaire, being thus made acquainted
with the existence of several species of the Shrew in France,
was desirous of enriching with them the collection of the
Museum of Natural History. His correspondents supplied
him not only with those already noticed, but also with several others, which had hitherto escaped the researches of
naturalists. We shall now have the pleasure of presenting
our readers with the substance of his valuable observations
on this subject, in as condensed a form as possible.

The first remark which may be made respecting this family is, that there are no genuine Shrews to be found, excepting in the ancient continent. Those which Gmelin
places in America, either belong to genera altogether different, or must, as utterly inauthentic, be blotted out entirely from the catalogue of living beings. Thus we find
that the *Sorex Aquaticus* belongs to the Baron's new genus,
the *Scalopes.* The *Sorex Cristatus* is the Condylure of
Illiger ; the *Sorex Brasiliensis* is probably a Didelphis, and
the *Sorex Surinamensis* may be suppressed altogether.

Among the species of the ancient continent there are
even some which must be withdrawn from the family of the
Sorex. The *Sorex Auratus* is the *Chrysochlore* of Count
Lacépède, and the *Sorex Moschatus* belongs to the subdivision *Mygale.* Exclusively of all these anomalous species,

the Shrews will form a very natural and as we shall presently see, a tolerably numerous family.

The Shrews are easily recognised by the elongated
and conical formation of the head, and more particularly
by the exceeding length of their nostrils. Their ears have
more width than height. The apparent shortness of their
legs arises from their placing the entire sole of the foot so
completely on the ground. The neck also appears short,
because the clavicles contribute, by their length and peculiar disposition, to approximate the head to the anterior
extremities.

To form a just idea of the true length of their heads, it
will be necessary to consider them when divested of their
softer parts. We shall find that the skull presents a very
sufficient capacity for the lodgment of the cerebral masses.
It is remarkably wide between the fossæ temporales, a fact,
the existence of which would not be conjectured from a
contemplation of the head with its integuments, but which
is owing to the absence of the zygomatic arch.

The jaws of the Shrews are completely furnished with
teeth, to the number of twenty-eight or thirty. The two
central teeth in each jaw, or the incisors, properly speaking, like those of the Glires, are the largest and strongest;
those above have a sort of spur or protuberance near the heel
of each tooth, while those below grow out horizontally from
their alveoli for some distance, and then take a curve upwards, where, in many species, they then become of a
bright-brown colour. The three succeeding teeth in the
upper jaw, (sometimes four,) and the two in the lower jaw,
on each side, may, perhaps, be either termed lateral incisors or false canines; they are shaped like ordinary canine
teeth, but cannot perform the same office, being, in fact
shorter than the cheek-teeth, which are four above, an
three below, on each side, in both jaws, with large crowns,

furnished with several sharp conical points, the ordinary character of insectivorous cheek-teeth.

The feet of the Shrews are completely divided, while those of the Desmans are palmate. They are pentadactylous, and the nails are short, curved, pointed, and elevated.

It remains to distinguish the species of so natural a genus, by characters, not of the highest order it is true, but still by characters very certain and important, as the reader will be very easily convinced, from our subsequent observations.

Since we are assured that this little sub-genus of animals is composed of several species, it becomes necessary to inquire what is the common Shrew, and to consider as not yet determined, the *Sorex Araneus* of Naturalists. It would be natural to have recourse to Hermann for the characters of this species, who has given them, comparatively, in reference to the head species, which he described. But in his article on this subject, in his " *Observationes Zoologicæ*," he found himself involved in so much doubt and obscurity, that he has actually declared that he did not precisely know the animal of which he was speaking. The " Tresor " of Fabre, who attributes a red fur to this species, increases our perplexity concerning it tenfold. The description, indeed, of Fabre, which applies to another Sorex, must be put altogether out of the question.

The fur of the common Shrew, is a of a mouse-coloured gray, paler underneath, and bordering on fawn-colour, in some individuals of a smaller size, though perfectly adult. In others, a little stronger and larger, it partakes more of brown. The first kind is supposed, by M. Geoffroy, to be characteristic of the males, and the second of the females We own, that we should have hazarded a different opinion. The hairs, in their entire length, are ash-coloured, excepting at the point, where they are reddish. The little points of the hair, under the belly, and tail, are white. The tail is flesh-coloured, in the individuals first above-mentioned, and brown in the second. These differences may appertain to distinct races, or even to distinct species. This is a point not very easily decided, and such points must often recur where the distinctive characters are not of a high order.

A third has been observed by the naturalist we have referred to, with tail one-fourth longer than that of the two fore-mentioned animals.

Finally, individuals are found naked on the flanks, and exhibiting, on the sides, instead of hair, a white spot, of an elliptical form. These circumstances, our Naturalist seems to think, distinguish them as mothers suckling their off-spring. But he puts this question in an interrogatory form.

The tail is covered with short hairs, it is tolerably full-shaped, demi-rounded, or to speak more correctly, sub-quadrated, or triflingly squared. Its four sides are rather projecting, but as the lines in angle, which separate them, are clearly perceptible, we cannot admit that the tail is entirely round. It is more so, certainly, in living individuals, and in specimens recently dead, some difference, in this respect, can be perceived. These observations, however minute they may appear to persons superficially acquainted with Zoology, are, nevertheless, to be insisted on, as necessary to the precise determination of a species, which demands exactitude, in a very peculiar degree, satisfactorily to distinguish it. The lips and feet are flesh-coloured, and thinly sown with a few short and whitish hairs.

Beside the three varieties we have noticed, Albinos are occasionally found.

The Water-Shrew of Pennant, Shaw, and the English Naturalists, which we rather prefer, after Erxleben, calling Daubenton's Shrew, *Sorex Daubentonii*, is larger than the preceding. Its tail is also longer in proportion. The tail and the limbs are better furnished with hairs, and the muzzle is a little more blunt.

It is generally found in the neighbourhood of small rivulets, from which circumstance, Daubenton, its first describer, took occasion to call it the *Water-Shrew.* There are others, however, of this genus, which live in marshy and aquatic places, and this renders preferable the appellation given to the animal, by Blumenbach and Erxleben, namely, that of *Sorex Daubentonii.* Hermann changed it into that of *Sorex Carinatus*, having remarked that the upper part of the tail bears some resemblance to the keel of a vessel. Before him, Pallas had substituted the name of *Fodiens*, which is also the name under which the Shrew of Daubenton is inscribed in Gmelin's catalogue.

The colour of its fur, blackish above, and a pure white underneath, is peculiar to itself. The white of the lower parts extends over the sides, being raised nearly above the level of the thighs. It may also be easily distinguished by a white spot, and by the ferruginous colour of the extremities of the teeth.

M. Marchand, presented a specimen of this Sorex to M. Geoffroy. He found it in the stagnant waters, and beheld it maintain a combat of more than half an hour's duration with a frog, which it had seized with its paw.

The ears of this species are furnished with three small valves, by which it is able, almost hermetically, to close the passage, a provision, as is also that of the stiff hairs bordering the feet, which accords with its aquatic habits, and is serviceable to it in the water.

The next species to be noticed is the *Sorex Tetragonurus.* This is one of the new species described by Hermann. Daubenton, in 1791, having had some knowledge of the labours of this Professor, from the *Elenchus Animalium* of Boddaert, inserted all the Shrews, known at that period, in a systematic table, which he formed for the " *Encyclopédie*

Méthodique," and which is printed in that collection, at the head of the Anatomical System of Animals, by Vicq-d'Azir.

The Tetragonurus is a little smaller than the Sorex Araneus, and the tail much longer. The head is larger, and the muzzle more slender. The teeth are brown at their extremities ; there are two false canines more in the upper jaw, and all of them are smaller, and of an equal volume. Its ears, without being entirely concealed within the hairs, are shorter, and less apparent than in the common Shrew. Its fur is blackish above, and a brown ash-colour below. Its tail is most decidedly squared, each face being altogether plane, and terminating suddenly in a very fine point. The lower part of the face exhibits a slight furrow. This species lives nearly in the same places as the common ; it is found in barns, particularly in the country, and sometimes within walled gardens.

The next species of the Shrew is that which is called, by Hermann, *Sorex Constrictus,* and by Daubenton, *Plaron.* The only specimens ever possessed by Hermann, consisted of a litter of seven young ones, whose teeth had not yet appeared. No longer assisted by the labours of Dr. Gall, he found himself unable to procure an adult individual, and, consequently, he remained in some doubt respecting the reality of the species. The eminent Naturalist, to whose labours we are so deeply indebted for accurate information, respecting many tribes of animals, (M. Geoffroy,) had the good fortune to have access to many adult specimens of this species, which enabled him to complete its verification and description.

The Sorex Constrictus is equally found in Germany as well as France. It has been described, figured, and extremely well-coloured, by M. Bechstein, in his work on Zoology, printed at Leipsic, in 1801. He gave it the name of *Cunicularis,* an appellation, objectionable from

N 2

its want of novelty, and, also, because the habit which it indicates, namely, that of digging burrows in the earth, is common to all the Shrews, who invariably do so, when they find no cavity of the kind ready made for their reception.

This Shrew is about the size of the common one, but its tail is a little longer than the latter's. Its muzzle seems stronger, and strait stiff hairs, extended over the cartilages of the nose, give it a shorter and blunter appearance. The ears are entirely concealed in the hair with which all their external parts are completely covered.

On the inspection of the crania of this and the preceding species, it was found that the cerebral case was sensibly larger, and less bulbous in the *Constrictus,* and the forehead was more arched in the *Tetragonurus.* The teeth of both are similar. The *Constrictus* has, as well as the other two, additional canines in the upper jaw. All these may be safely considered as so many characteristic traits, having been ascertained in individuals taken at the same age.

The form of the tail of this animal has given rise to the specific appellation of *Constrictus.* It is flat at its origin, narrow, and exhibits an appearance that seems to be the result of compression; while in the remaining part, especially towards the middle, it is thick, as if swelled out, and round, except at its extremity, where it is again flattened, and where the hairs unite in a point, like those of a painting-brush.

The fur is tolerably long, and soft to the touch. The hairs, for a considerable portion of their length, are blackish, and red at the points. The belly is of a grayish-brown, and the throat is ash-coloured. The hair, of recent growth, is somewhat of a clearer colour than the old. In the first kind, underneath the belly, the points are gray, but in the hair, which is ready to fall, they are red.

This Shrew is also remarkable for the copiousness and equality of its hair, dispositions, which impart to its fur, a velvety appearance, and an uniformity and harmony of tone. The feet of the *Constrictus* are covered with hair like those of the preceding species. The peculiar formation of the tail would seem to indicate that it lived in elevated situations ; Dr. Gall, however, found the litter of seven, which we mentioned above, in a meadow which had just been mown, at a small distance from a rivulet.

The *Sorex Leucodon (White-toothed Shrew)* is another of the species described by Hermann. It seems probable that the specimen seen by the Professor was but young. Its dimensions are much the same as those of the Shrew of Daubenton, excepting that the tail of the *Leucodon* is shorter. Its toes likewise are a little thicker, the nails are shorter, and the eyes larger. Its tail is not precisely rounded, but rather partakes of that approximation to a square which characterizes this organ in the common Shrew. Its incisive teeth, notwithstanding its name, cannot be considered as entirely white, except in its earlier age. In adults their extreme point is tinged with brown. Inaccurate or imperfect observation could have only given rise to its specific appellation.

The fur of this species is a much more distinctive character than any other that we have yet cited. Its back is brown ; its belly and (what is not observed in the other Shrews) its flanks are white. Elsewhere, as in all the other species, it is the points of the hairs alone which are of this latter colour; the rest is ash-coloured. The upper part of the tail is the colour of the back, and the lower part the colour of the belly.

The *Striped Shrew (Sorex Lineatus,)* is found in the environs of Paris. Its form is more lank, and its muzzle longer and finer than those of the preceding species. Its tail is round, and strongly partaking of the keel shape underneath. Its fur is generally of a blackish brown, the belly is more pale, and the throat is ash-coloured. By two other characters it may be easily recognised amid the little group of the Shrews; the first is a narrow and white line which extends over the forehead, and departing from the front, disappears gradually over the nostrils; the second is a white spot upon the ears. The hairs which compose this spot grow from the inside of the auricular conch, and border the two small lobes which we noticed in describing the first species. Similar hairs may be seen, but shorter, and more scanty, in the other Shrews.

The incisors are brown towards their extremity. From the form of the tail, it seems likely that this species is aquatic, and seldom removes far from moist situations.

The *Sorex Remifer* or *(Oared Shrew)* is considerably larger than any of the preceding. M. Geoffroy regards it as a distinct species. He saw three different specimens.

This Shrew differs from the preceding by its proportions, especially by those of the muzzle, which is very blunt and short. It is generally more clumsy in its figure, but bears a resemblance to the last in the colours of its fur, which is, however, of a brown black, considerably deeper in its shade. The belly is of a brownish ash-colour, and the throat of a clear ash-colour. The same spot is observable upon the ear, but not the stripe across the forehead.

The extremities of the teeth are of a ferruginous brown. The peculiarity, which among other characters, more especially distinguishes this new species, not only from the striped Shrew, but from every other, is the very singular form of the tail. In its first half it is a perfect square, having each face entirely plane, except the under one which is furrowed. From the end of this furrow originates in the other half a keel, which is prolonged in proportion as the tail diminishes in breadth. The tail ends in a compressed and flattish form, which bears no indistinct resemblance to the shape of certain kinds of oars.

It seems likely that all the Shrews which visit the water participate more or less in this sort of organization ; and it is more probable that this conformation determines the habits of these little animals, and the preference which they give to marshy grounds, rather than any disposition of the hairs of the toes, as has hitherto been believed.

The *Shrew* of India (*Sorex Indicus*) has been described by Buffon, in the seventh volume of his Supplement, and figured in his 71st plate.

It is considerably larger than the Shrews of Europe, but notwithstanding this, it perfectly resembles them in all essential characters, such as the teeth, toes, and length of muzzle. Its hair is altogether extremely short, and of a grayish brown, tinted with reddish at the top, because the point of each individual hair is of this colour. All the teeth are white. The round tail indicates that it is a terrestrial animal ; and Buffon, in fact, informs us that it inhabits the fields, from whence it sometimes comes even into the houses. Its presence is soon betrayed by the strong odour of musk which it exhales.

According to M. Geoffroy, the *Sorex Murinus* ought to be referred to this species.

The *Cape Shrew (Sorex Capensis)* must not, according to Geoffroy, be confounded with the preceding animal. It is no doubt approximated to it in size, in the colour of the teeth, in the rounded and thick form of the tail, in the magnitude and nakedness of the ears, and in the musky odour which it exhales. But, nevertheless, there are differences which appear to be essential. No Shrew has a longer and slenderer muzzle, and its tail which is but one half shorter than the body, is in proportion much longer than that of the Indian Shrew. The tail is likewise of a different colour, being red, and quite contrasted with the colour of the fur ; its surface is covered with hairs that appear shorn, with a few silky hairs intermixed.

The entire fur is ash-coloured, excepting on the back,

where there is a slight sprinkling of fawn. The sides of the mouth are reddish.

This species is not entirely new. It is the same which has been designated by Petiver under the title of *Sorex Araneus maximus Capensis*. He has given a bad figure of it, which was copied by Valentin. Burmann has also mentioned it in his work on the animals of the Cape.

According to MM. Peron and Lesueur, this species lives in caves. They take wonderful pains at the Cape for the destruction of this animal, as it is extremely troublesome, both on account of the mischief which it occasions, and the powerful odour which it exhales.

The *Rat-tailed Shrew* (*Sorex Myosurus*) bears a considerable resemblance to the preceding: For some time M. Geoffroy was inclined to consider it as an albine variety of the Sorex Capensis, for it approaches it in size, in the magnitude, and nakedness of the ears. There seem, however, to be essential differences between the two animals, which may justify their reference to distinct species.

The tail of the Myosurus is longer and, more especially, much thicker than that of the Cape Shrew. The muzzle, on the contrary, is much shorter and singularly inflated on the sides. The limbs are strong, the feet thick, the ears very large, the hairs of the tail more dispersed, and the silky ones more numerous and longer.

This animal is entirely white.

The *Sorex Myosurus* was described by Pallas in the Acts of Petersburgh, in 1781, and figured. In the same plate he gives the figure of another Shrew, which he considers the male of his *Myosurus*. It may be observed, however, that the appearance of this last is considerably different from the *Myosurus*; the head thicker, the tail shorter, and the fur altogether of a brownish black.

M. Geoffroy gives us a figure of the skeleton of this Shrew, on which it may be observed, that it has two dorsal vertebræ, and two ribs more than the common ecies, having fourteen altogether. The Shrew of Daubenton has but one less, thirteen.

Some others are mentioned in the table as unauthenticated and uncertain.

We shall now speak of the sub-genus DESMAN (MYGALE). The Desman is in the same predicament in relation to the Shrews as the Ondatra to the Field-mice. It is larger, and more necessitated to the adoption of an aquatic life.

This genus has been known since the year 1605, and yet there has been no classification of the Mammalia in which it has not received a different allocation. It was first given by Clusius under the denomination of *Mus aquaticus exoticus*. Aldrovandus soon reduced this title to the simple name of genus and of species, calling the animal *Mus aquatilis*. Klein afterwards ranged it among the Dormice, giving it the name of *Glis moschiferus*. Hill and Brisson confined themselves to the name of Musk-rat, under which it had been so long confounded with the Ondatra. Finally, what had been a mere conjecture of Brisson, who had given too exclusive an attention to the form of the tail, was decidedly adopted by Linnæus. The Desman passed among the Castors, with which it remained confounded in the tenth and twelfth editions of the *Systema Naturæ*, under the name of *Castor Moschatus*.

Some original labours appeared soon after. Such were the descriptions of this species by Buffon, Gmelin, and Guldenstadt. But little resulted from their researches to enrich the natural history of the Desman, except with better figures than any before published.

The true affinities of this species were not ascertained and fixed until the year 1781. Pallas, after having established and discussed all its characters, finally replaced it among the Shrews, and his *Sorex Moschatus* was adopted by all subsequent writers on natural history. But this notion of the affinity of the Desman, which Pallas put forth as something new, was not altogether so. The fact had evidently been suspected by Charleton in 1677, when he designated this identical animal under the name of *Sorex Moscovitus*. The Desman, in fact, in the natural order, must follow close upon the Shrews. But at the same time it must be kept separate from them, and not confounded with them, as Gmelin and Shaw have confounded it in their catalogues. The general relations which connect these beings together do not hinder them from differing in some very essential parts. Their teeth do not present a similar appearance, nor are they alike in number. The toes are free in one set of animals and palmate in the other, and their nostrils are so much unlike, that the name of horn might be not inappropriately applied to that of the Desman.

From these differences, the necessity of classing the Desman apart from the Shrews was felt by the Baron, as otherwise a hiatus must have been left in the catalogue of animals not easily filled up. Accordingly, in the first volume of his Comparative Anatomy, in a list of genera and species which terminates it, he proposes the name of Mygale for this new genus.

There are several considerations, independent of the distinctions just noticed, which must oblige us to accede to the justness of this arrangement. On the one hand the number of genuine Shrews has become so very considerable, that it is necessary to reject from the group every animal whose attributes do not come strictly within the limits to which we have confined its definition. On the other hand also, the Desman, in consequence of the acquisition of a new species found in France, was clearly indicated as the centre of a little tribe, for the further augmentation of which nothing more in all probability was wanting than a belief in its existence and in the plurality of its species, and a little more attention to the characteristic traits of each. In fact, the acquisition of a species, to which we have alluded, gave to the new genus all the sanction which it required. But independently of that, the principal characteristics, which we shall now notice, would have furnished a sufficient motive for its establishment.

The cranium of the Desman exhibits as much analogy with that of the Mole, as with that of the Shrews. It does not terminate in so fine a point as the cranium of the Sorex. The bones of the nose are more elongated, and there is no want of zygomatic arch. The rising branches of the lower maxillaries are more elevated. It has two additional incisors in the lower jaw. The upper incisors have totally a different form from those of the Sorex; they are large, and of a sugar-loaf form. The small canines are conic, and double the number of those of the Shrews, being six on each side. The Desman, in fine, has forty-four teeth, as well as the Mole, which are distributed in the following manner :

		Lateral Incisors or false Canines.		Cheek Teeth.			
Upper jaw,	2	—	12	—	8	=	22 }44
Lower jaw,	4	—	12	—	6	=	22 }

The orbit is not more apparent than in the Mole, because the eye is equally small.

The nasal conchs are prolonged so much as to present the appearance of a horn. Their length is equivalent to that of one-half of the cranium. They decrease insensibly after leaving the muzzle, and then gradually enlarge towards the nasal apertures. This horn, or proboscis, is as mobile as that of the Elephant. Pallas speaks of its suppleness and agility, and, at the same time, describes the muscles which regulate and impel its movements.

Another difference, not less important, and which, like the preceding modification, results from the necessity under which the Desman is placed, of constantly inhabiting the water, is the total absence of the external ear. The Shrews, as we have observed above, are always provided with it.

Finally, the completion of this system of organization, by virtue of which the Desman, renouncing, as it were, the attributes of a quadruped, comes to partake, with the fish, their natural element, is found in the transformation of its organs of locomotion into genuine oars. Its limbs are extremely short, and partly involved under the teguments of the trunk, and the toes, which terminate them, are connected by membranes. The tail, in fine, as well as every other part, is accommodated to the same system, being flattened on the side, and contributing, by this conformation, to facilitate the operation of swimming.

This co-relative disposition of all the parts of the Desman ; this concordant adaptation of all its forms, to constitute it an aquatic animal, exercises, as may be well supposed, the most imperious control over its habits and propensities. The Desmans, in truth, pass the most considerable portion of their lives in and under the water. Never, of their own accord, do they seek a dry place; and if they proceed from one pond to another, it is only when they meet with subterraneous channels, or ditches filled with water, to conduct them.

They choose, by preference, ponds, lakes, all kinds of dormant waters, but more especially inundated places, surrounded by elevated banks, as their general habitations. They make there a sort of burrow, the entrance to which is under the water. From this entrance, they commence their operations; they dig on, gaining, by degrees, in height, elevating their work in multiplied and lengthened windings, so as to embrace, at times, an extent of more than one-and-twenty feet. Thus, there is but one part of their burrow under the water; there they live, either singly, or in a monogamous state, according to the season.

They do not fall into a torpid state during the winter, a circumstance which exposes them to an inevitable and serious evil. The ice, which is formed along the surface of the ponds, imprisons them under water, and within their burrow, and it would appear that they then endure a state of the most cruel torment; for, if there are any fissures or holes through which it is possible for them to respire, they run thither to gain a little place, on a level with the water, to thrust out the extremity of their proboscis; should they fail in this, they can only exist on the small quantity of air contained within their burrow. A trifling number thus survive, but the others perish by suffocation. Their attempts to release themselves, which we have just noticed, are more numerous in proportion to the duration of the cold season.

The Desmans seldom come to a level with the water, except in the rutting season; then they grow bolder, proceed along the channel of the river, or collect in the water-plants, or attempt to climb along the shrubs that border the water-side.

It has been asserted, that they feed on the roots of the *nymphæa*, and on acorns, of which they lay up stores; but Pallas never found any thing in their stomachs except the remains of larvæ, and of worms.

The epithet *Moschatus* has been given to these animals, in consequence of the strong musky odour which they exhale. This odour is so powerful and penetrating, that the flesh of pike, and other fish, which have chanced to feed upon the Desmans, becomes tainted by it.

In the *Desman of Russia, (Mygale Moscovita,)* the tail is shorter than the body. The form of this appendage is very remarkable. At its base, it is compressed; soon after, it becomes cylindrical, somewhat swelled, and increases rapidly, in a bulbous form; but, at a little distance, it begins to decrease again, which diminution continues insensibly to its extremity: the more it diminishes, the more it becomes vertically compressed. It is, also, like the tail of the Beaver, thickly set with scales between small intervals, which are filled with short and isolated hairs. Some scales are also found on the upper part of the toes.

The fur of this Desman is much esteemed. It is com_posed, like that of the Beavers, of long silky hairs, and a kind of soft and marrowy felt concealed beneath. The fur is of a brown colour, paler above, and deeper on the flanks. The belly is of a silvery-white.

The *Desman of the Pyrenees*, (*Mygale Pyrenäiea,*) The discovery of this new species is owing to M. Desrouais, Professor of Natural History, at the central school of Tarbes, in France. It is one-half smaller than the Desman described by Pallas. The tail is neither compressed at its commencement, nor inflated beyond that point. It diminishes gradually and insensibly towards its extremity, is cylindrical in three-fourths of its length, and vertically compressed in the remainder. It is covered with short flat hairs, that are almost entirely adherent. Its nostrils are twice as long as in the preceding species. The front toes are but half enveloped in the skin, and the external toe of the hind foot is also nearly free. The fur is the same as that of the last species, as to the nature of the hair, which is silky, and as to its felty basis. But its colours are somewhat different. All the upper part is of a Maroon-brown, the flanks grayish-brown, and the belly silvery-gray. There is no white part upon the face, as Pallas relates, to be the characteristic of the Desman of Russia.

This species has, as yet, been found only at the foot of the Pyrenees, in the neighbourhood of Tarbes. The very great distance between the places where these two Desmans are found, is additional reason for believing in the diversity of their species.

The Scalope is the next sub-division of the insectivorous

family. The two upper incisors are large, plane, and towards their extremities are shaped leaf-wise. The lower ones are conical, erect, tolerably long, and separated one from the other, and in the interval a little forwards, are two very small ones. In the upper jaw there is an inter-dentary space after the incisives. But in the lower jaw the teeth which replace the canines follow immediately. They are conical, and proceed gradually increasing in size towards the bottom of the jaw, where they change into molars, with a coronal provided with sharp tubercles. The muzzle is considerably elongated in the form of a horn, and cartilaginous at its extremity. The eyes are extremely small, and the external ears are altogether wanting. The fore-paws are very short and broad. The toes, to the num-ber of five, are united as far as the last phalanx. The nails are very long, flattened, linear, and proper for dig-ging. From the thumb (or that toe which may represent it), they proceed increasing as far as the third toe inclu-sive, while the two others diminish, and the external one is the least of all. The hinder feet are very small, very slender, and are armed with small and sharp nails. The tail is short. The whole body, the general form of which is more elongated than that of the Moles, is covered with very soft fur.

The Scalope lives in the same style as the Moles, and digs like them subterraneous galleries. It is peculiar to North America, and is most usually found along the banks of rivers. This animal was classed for some time with the *Radiated Mole* of America, of which Illiger has formed his genus *Condylura*. But its characters must separate it from that animal.

The only species known is the *Scalope of Canada (Sca-lops Canadensis)*. This is the *Sorex aquaticus* of Linnæus, and the *Talpa fusca* of Shaw. It is of the same size as the European Mole, nearly about seven inches in its total length.

The general colour of the head, when considered at first view, is fawn-coloured gray, but examined separately, each hair is of a mouse-coloured gray at its base, and fawn-coloured gray at the point.

We may notice as somewhat singular that M. Fred. Cuvier has described, in the twelfth volume of the Annals of the Museum, the teeth of the Desman, under the name of this animal. It is apparently a typographical error in the title of the article, and in one passage in the article itself.

The genus or subgenus of the CHRYSCHLORE was founded by the Count Lacépède on a single animal which resembles the Moles in its mode of life, but differs from them in many other respects, and principally in the character of the teeth. In the upper jaw are two strong and sharp incisors ; in the lower are four, two similar to those above, and which exactly correspond with them, and two more, extremely small, placed between the others, and of no great apparent utility. The molars are nine in number in the upper jaw. The three first have each but a single point, and are all similar to each other ; the other six are tuberculous. Their general form is that of a triangle, each angle of which possesses a tubercle. The most acute angle is that on the outside of the jaw, and at its base springs an isolated tubercle tolerably strong. The last of these teeth, much smaller than the others, is little else than a slender plate, in which, however, the general form of the other molars may be recognised. The lower jaw possesses but eight molars. The three first are single pointed like those in the upper jaw, and the five others have likewise a triangular form with tubercles. They are more slender, and the acute angle is outside. All these teeth are separated by *an interval equal to their thickness,* and it is in the vacuum left thus in one jaw that the teeth of the other are inserted when

THE CAPE CHRYSOCHLORE or SHINING MOLE.

TALPA ASIATICA. *Seba.*

Published by G.B.Whittaker. Sept. 1826.

the mouth is closed. The Chrysochlore, is, we believe, the only example in the animal world, of teeth being opposed by their anterior and posterior faces.

The toes of the front feet are three in number, and the external one, enveloped altogether in a nail formed for digging, is of a monstrous size. The hind feet have five toes, and the external one is the shortest. These animals have no tail, nor any external conch to the ear. According to Seba, neither the eyes nor mammæ are perceptible and he also says that the nostrils are situated on the anterior part of the muzzle, as in the Hog. The other parts of the organization have not been yet described.

The *Cape Chrysochlore (Talpa Asiatica)* is smaller than the common Mole, but it has the same forms and very nearly the same physiognomy. It also lives beneath the earth, in burrows, the arrangement of which is not known. It digs with its fore-feet, which are armed with very thick nails, and the force of which is moreover sustained by a peculiar bone found in the arm below the cubitus. But what distinguishes this animal from all the other Mammalia is the brilliancy of its fur, which exhibits metallic reflections, changing from green to a bronze red or golden yellow, and which reminds us of the dazzling plumage of some tropical birds, or the splendid hues of many of the scaly tribes. The female, as Seba hath it, does not differ from the male, except that the hairs of the muzzle and head are shorter and more yellowish, and those of the belly exhibit greater brilliancy of reflected radiance. It is from this golden colour that this sub-genus has received its name.

The Chrysochlore is a native of the Cape of Good Hope. Seba had announced it as original in Siberia, and this error was adopted by Buffon and Linnæus. Brown, in his Zoological Illustrations, was the first who pointed out the true habitat of this animal.

The *Red Mole (Talpa rubra, Lin.)* This species is

known only by the description and figure given of it by
Seba. It is only, therefore, in consequence of the resemblance of its paws with those of the Chrysochlore that it has
been admitted into this genus. Seba it is true, says that
this Mole has but four toes on the hind feet; but as he
has given but four to his Siberian Mole, which he compares to his red Mole, it is probable that in both species
the external toe, which is remarkably small, had escaped
his notice. If we may judge by the figure given by this
author, this species has a tail. We shall conclude this
notice of the Chrysochlores with his description of the
Talpa rubra.

This mole is of a red, bordering on a clear ash-colour.
It resembles considerably the common Mole, only that the
fore-feet are formed differently, being only divided into
three toes, the first of which is armed with a nail very large
and long, pointed, and a little curved. The middle toe is
the smallest, and its nail also. The third is also very small.
The hinder feet are divided into four toes, armed with
nails almost equal.

We shall now proceed to the MOLES, the most interesting subdivision of the insectivorous family.

The genus *Talpa* of Gmelin consists of four species,
which are thus named,—*T. Europæa, T. Asiatica, T. longicaudata,* and *T. rubra.* Two of these, the *Asiatica* and the
rubra have, as we have just seen, been separated from the
Moles by later naturalists to form the new subdivision of the
Chrysochlores. The *Talpa longicaudata* is too little known
to render it possible that we can refer it with precision to
any established genus; and, in fine, the *Talpa Europæa* is
the only species which has been observed with sufficient
accuracy to admit of an exact detail of its characters.

In the genus Sorex of Linnæus and Gmelin, two species
are found, in which the fore-paws are conformed, or very

nearly so, like those of the European Mole, and in consequence of this have been placed by some writers among
the Moles. One of these we have already described,
namely, the *Sorex aquaticus*, of which the Baron has made
his genus *Scalope*. The other is the *Sorex cristatus*, of
which Illiger formed his genus *Condylure*, which the Baron
has, in the " Règne Animal," at least, suppressed. It will
be proper, however, in this article on the Moles to present
our readers with the observations opinion of M. Desmarest
on the subject,

M. Cuvier, in the " Règne Animal," established the characters of the Mole on the inspection of one proper to the
European species ; and he suppressed this genus of Illiger,
because, from the inspection of its teeth, he was assured that
it was a true Mole, and not a Sorex, as we have seen in
the text. This opinion was received by M. Desmarest, until
M. Lesueur sent a *Sorex cristatus* from Philadelphia to
Paris ; and on inspection of this subject, M. D. became convinced that it ought to be considered as a separate genus.
It presents characters, according to him, altogether peculiar to itself, and which separate it equally from the Moles
and from the Shrews. We shall first, however, treat of the
true Moles, and then the Condylures after M. Desmarest.

The Talpa is defined as a genus of the insectivorous
Mammalia of the second tribe, or that which contains species with four large canine teeth separated from each other,
and having between them small incisive teeth ranged upon
a single line.

This genus is thus characterized: six upper incisors,
small, vertical, and nearly equal in height, the intermediate
ones being broader than the lateral ; eight small lower
incisors, arranged archwise, and rather on a declivity ; two
canines in each jaw, which surpass the incisors, and are
triangular and compressed, the upper ones being larger
than the under. There are seven molars on each side of

the upper jaw, and six below, with coronals provided with sharp points. The head is elongated, and terminated by a kind of snout; eyes extremely small. No external ears. Short limbs with five toes on all. The fore limbs are stronger than the others, and are terminated by hands extremely large, which have the palm always turned outwards or backwards, the lower edge being trenchant, and the toes (or fingers) joined as far the basis of the nails, which are long, strong, and sharp. The hind-feet are more slender, the toes more feeble, more separated, and provided with nails of moderate size. The tail is short and but slightly covered with hairs. There are six abdominal mammæ.

The Moles exhibit anatomical characters not a little remarkable. Their head is extremely elongated and somewhat flattened above. The cervical ligament possesses exceeding strength. The bones of the anterior extremities are angular, and so very thick, that theirt ransverse diameter is hardly exceeded by their length. The two bones of the fore-arm are attached. The clavicles are extremely strong. One very elongated bone of the carpus communicates solidity to the under edge of the hand. The motive muscles of these extremities are enormous; above all, the pectoral, which are attached upon a very large sternum composed of five pieces, and which, like that of bats and birds, has a central ridge considerably developed. The pelvis is fact, extremely narrow. The pubes are not joined by a symphysis, which fact, according to the observations of M. Breton, allows the displacement of the vulva at the period of parturition, and its movement in front of the pelvis. Without this arrangement it would be impossible that the young could come forth, as they never could do so by the ordinary way, in consequence of the narrow diameter of the pelvis. The stomach is membranaceous, and of an elongated form. There is no cæcum; the liver has three lobes, and the gall-bladder is round.

The Moles are animals eminently adapted for digging. All their muscular force is situated in the levator muscles of the head, which is to them a kind of lever, and, in their hands, which act as spades. Placed upon a solid ground, these animals move slowly; but when they are permitted to dig, they speedily disappear.

The Moles are insectivorous; but besides earth-worms, and the larvæ which they meet with in their subterraneous passages, they also feed on the roots of certain plants, and more especially on those of a tender kind.

Hearing and touch are senses which they possess in a state of high development, but the others are less perfect.

After the removal of the *Chrysochlore, Scalope,* and *Condylure,* (if this last indeed be to be removed,) there will remain but a single species in the genus TALPA, which will be the *Talpa Europæa,* or our common European Mole. For the substance of some curious observations, respecting this animal, we are obliged to M. Sonnini.

The Mole, properly so called, is one of the commonest animals we possess, and one of the most hurtful in cultivated lands. It appears that this animal was not known to the ancients, who have been very wrongfully accused of having fallen into the gross error of supposing that the Mole had no eyes. Aristotle, it is true, in two places of his History of Animals, repeats this assertion. But the researches of modern times have ascertained that this illustrious naturalist was perfectly right in refusing the organs of vision to the Mole of his native country, to the σπαλαξ or ασπαλαξ of Ancient Greece. There does, in fact, exist, in this country, a little subterraneous animal totally deprived of sight: Naturalists have become acquainted with it, but comparatively, lately, and have designated it under the appellation of the *Rat-Mole.* They have been obliged to confess, after many ages of injustice towards the ancients, that these last had truth altogether on their side, with regard to the Mole known in Greece, and had correctly observed that this animal was not only completely blind, but did not possess even the smallest rudiment of an external eye. This, indeed, is not the only error into which a too great precipitancy of judgment has led many persons, in modern times, who are more prone to censure than observation *

It had been, in truth, not a little singular, if men so accustomed to the observation of the works of nature, had not perceived the eyes of the common Mole, which, though very small, and a little concealed, are yet sufficiently apparent. The skin which surrounds them, as well as the hairs by which they are partly concealed, may be removed at the will of the animal, to permit it to perceive objects when it is above ground, while, at the same time, they preserve it from being dazzled by any glaring light. This skin, and these hairs, on the other hand, form, occasionally, a complete covering for the eyes, and prevent them from receiving any shock or injury when the Mole is at work in its subterraneous galleries. At such times, the or-

---

* Almost all the fertile lands of Europe are inhabited by the common Mole. There are none, however, in Ireland, and few are to be found in Greece, where their place is supplied by the *Spalax* or *Rat-Mole*. Aristotle and Pliny have reported that the Moles, which had been brought to Lebadia, in Beotia, refused to dig the ground, while in the neighbouring territory of Orchomenos, they turned up all the cultivated lands. It must be remarked, here, that Lebadia is a very mountainous country, while Orchomenos is more flat.

We should be deceived, if we considered, as genuine Moles, the little animals, of which Spallanzani met a numerous colony, in certain beech forests, near the Lake Seaffajolo, situated on the most elevated summit of the Apennines. (Travels in the Two Sicilies.) It is easy to recognise, from his account, that those animals, to which he gives the name of Moles, have no relation whatever with this genus, and are, in all probability, field-mice.

Pallas saw, repeatedly, the species of the common Mole, in the Canton of Kouschwa, not far from the Tyrol mountains. Those Moles are larger in breadth than the European, and almost all the individuals are white.

gan of vision is of no use, and that of smell alone can
direct the animal in its darkened paths. If a living Mole
be plunged into the water, the teguments which cover the
globe of the eye, immediately dilate, and leave the organ
perfectly uncovered; because, under those circumstances,
the animal has occasion for the use of all its faculties to
extricate itself from impending danger. In other respects,
the eyes of the Mole, which are not imbedded in their orbits, like those of the majority of quadrupeds, are somewhat of the shape and bigness of a grain of mustard-seed,
and appear like two black and glittering points.

The muzzle is elongated, mobile, and pointed very nearly
like that of a swine. It is an instrument extremely well
adapted to facilitate the labours of the Mole; for while
the animal, with its fore-paws, removes the soil, the snout,
furnished with powerful muscles, and a small bone, raises
the earth, and prepares the passage, through which the
body is to go. The muzzle is terminated by two large
nostrils, which advance a little beyond the opening of the
mouth. There are twenty-two teeth, of which we have
already given the detail, in each jaw. The tongue is long,
and not very unlike that of the Carp. The mouth, being
opened by the motion of the snout, a small membrane,
placed below the upper lip, and which descends over the
lower, hinders the earth from entering. Relatively to the
magnitude of the Mole, and its mode of life, the organ of
hearing is perfect, though less so than that of smell,
which is extremely delicate. There is no external conch to
the ears; the meatus auditorius is concealed by the hairs
which surround it. It is cartilaginous, and descends obliquely, as far as the cavity of the os petrosum, to which
it adheres, by many small membraneous fibres. The orifice
does not exceed, in diameter, the quill of the feather of a
pigeon's wing; and a small membranous valve, which is
raised and lowered like the eye-lid, and the mechanism of

which may be perceived by shaving the head of the Mole, closes the aperture, at the will of the animal, to prevent its being obstructed by earth or sand.

All the feet are divided into five toes. The fore-feet, which have the form of hands, are broad, and placed obliquely, so that the palm is always secured, and the toes, (or fingers,) armed with flat and strong nails, are directed externally, and downwards. The hinder-feet are much smaller than the fore. The tail is short and scaly, like that of rats. The anus is exceedingly prolonged, from the origin of the tail. The body, thick and muscular, is covered with skin, which adheres strongly to the flesh, and is well furnished with close, soft, and silky hairs, insomuch, that the body of the Mole is not unlike a velvet pincushion, the two extremities of which are formed by the pointed muzzle, and the short and round tail.

The left, or upper, orifice of the stomach, is surrounded by a fibrous wing, destined to bind this viscus together. Severinus found a transverse line, which attaches in some way, and separates the pylorus. Other anatomists have not made the same observation. The liver is divided into four lobes, though sometimes there are but three; and some anatomists have even discovered five. Their colour is a reddish-brown.

The gall-bladder is observed with difficulty, and does not contain much liquor. There are five lobes to the lungs. The heart is of an elongated form, and situated entirely on the left side. The spleen, which immediately adheres to the stomach, resembles that of a dog, precisely in its conformation.

But it is the parts of generation which are chiefly remarkable in the Mole. " Nature," says Buffon, " has been munificent, indeed, to this animal, in bestowing on it, as it were, the use of a sixth sense. It possesses a remark able apparatus of reservoirs and vessels, a prodigious quantity of seminal liquor, enormous testicles, the genital member of exceeding length, and all secretly concealed in the interior of the animal, and, consequently, more active and vivid. The Mole is, in this respect, of all animals, the most advantageously gifted, the best organized, and must, of consequence, possess the most vivid sensations."

The ordinary colour of the fur of the Moles is a fine black ; but various colours are found, which constitute the following varieties :

The *White Mole.*—A variety first noticed by Wagner, in his Natural History of Helvetia. It is sometimes seen in Lorraine, in Switzerland, and in Holland, but more commonly in climates farther north. Pallas tells us, that the Moles of Siberia are almost all of them white. Razoumousky describes one of these Moles, which was caught near Lausanne. It was generally of a dirty-white, bordering on fawn-colour, but with reflections of a more lustrous white, according to the light in which it was viewed. There was a little red on the throat and belly, and some spots of the same colour on the head. The end of the muzzle and the nails were of a blood-red. This animal was smaller than the common Mole.

The *Ash-coloured Mole,* of which some German writers speak, seems to be nothing but an aged individual. The black colour of the Mole lightens, and grows gray with age.

The *Citron Mole,* which is of a fine citron colour, is not met with, says M. De La Faille, except in a portion of the territory of Alais, and its colour is pretended to be the result of the quality of the soil which it inhabits.

The *Fawn-coloured Mole,* according to the same writer, is only found in the country of Aunis. Its fur is of a clear red, bordering on the fawn-colour towards the belly, without any spot or mixture. Some have been seen, the upper part of which was fawn, and the lower of a lustrous white.

The *Pied Mole.* Seba asserts, that it has been found in

East Friesland. It is longer than the common Mole. Its skin is marbled on the back, and under the belly, with white and black spots, in which may yet be distinguished an intermixture of gray hairs, as fine as silk. The muzzle is long, and furnished with long and stiff hairs.

Authors have mentioned some other varieties, with which we do not think it necessary to trouble our readers.

Very handsome furs, and light and warm coverlets, have been manufactured from the skins of Moles, as well as very fine and handsome hats. Scandal will have it that, at one period, French ladies, of a certain age, were wont to use the same material for false eye-brows; if so, they could only have succeeded in deceiving those who put forth the imposition. In ancient medicine, many virtues have been attributed to the different parts of the Mole, both in the cure of disease, and the recovery of beauty, which had suffered from the relentless ravages of time. These sorts of medicaments have been long abandoned, and to enumerate them would be but loss of time. In Thrace, according to the report of M. Sestini, it is still believed that the skin of the Mole possesses the capacity of curing a defluxion of the eyes.

Of all animals, the Mole is, probably, the most advantageously gifted by nature. With the exception of sight, which is the weakest of all its senses, because it is the least exercised, its other organs possess very great sensibility. Its hearing is remarkably fine, its touch delicate, and its sense of smelling most exquisite. Its skin is fine, and it always maintains its " embonpoint." Its fore limbs are terminated by hands rather than feet. Its strength is very considerable in proportion to the volume of its body; and it possesses an address, in addition to its vigour, that accurately directs the employment of all its faculties.

The Mole exhibits an admirable degree of industry in constructing the habitation to which it retreats. It passes its life under ground. If ever it abandons its asylum, it is but for a few moments, and only for the purpose of seeking some commodious soil, and the moment it has found it, it sets to work immediately. It closes the entrance to its retreat, for it fears the open air as much as the open day. It equally avoids mud, and a hard or rocky soil. It chooses, by preference, prepared and cultivated lands; but if the water should surprise it, it hastily quits its abode for some more elevated situation. The overflowing of rivers and streams is the greatest scourge of the Moles, and the most certain and natural means of diminishing their number. These animals change their habitations according to the variations of the atmosphere. During the winter, and the rainy season, they remain in elevated situations. In summer, they descend into the valleys, and if the drought continues long, they will take refuge in cool and shady places, along the banks of ditches or streams.

There is no animal more accustomed to labour than the Mole. Its means of subsistence are dispensed through the very bosom of the earth, and it is continually occupied in searching them out. Long alleys, usually parallel to the surface of the soil, and in depth from four to six inches, constitute the evidence of its laborious life. A skilful miner, it forms its galleries with equal art and activity. Sometimes it only raises the superficies of the soil, and sometimes it digs deeper, according to circumstances and temperature. All the roads which it opens have channels of inter-communication. According as it digs, it throws out the earth which it detaches, which produces those little domes of ejected earth, called Mole hills. If, while engaged in its excavations, it should happen to be disturbed, it does not attempt to fly, by issuing from its galleries, but buries itself in the earth, by means of a perpendicular tunnel, to the depth of nearly two feet. If its channels of communication be disarranged, or the heaps of earth which it has formed, it comes instantly to repair them. The Mole is said to pant and blow, when with its muzzle and paws it pushes the earth to a mole-hill, or when it forms a sort of oblong vault of moveable earth in the place where its track has been intercepted.

The male of this species is lustier and more vigorous than the female. Its labours are easily recognised from the volume and number of the hillocks which it raises. Those of the female are smaller and less numerous. Those of the young are small, imperfect, of a zig-zag form, and the channels or trenches which terminate each are nearly on a level with the surface of the soil. It has been observed that the hours of labour with the Mole are sunrise and sunset, noon, nine in the morning, and nine at night. These animals are less eager at their work in winter than in summer. Their activity is diminished during the frosty season, but they do not fall into a torpid state as me writers have erroneously imagined.

At such periods they seek the warmest places, such as the beds of gardens, and as soon as the cold becomes less rigorous, they resume their work and remove the earth as before. When the Moles begin to work, that the thaw is not far distant is a common observation among the inhabitants of the country.

The nutriment of the Moles consists in tender and succulent roots, the bulbs of the *Colchicus*, also worms and insects. Their season of love is the early spring. The powerful means of propagation which nature has bestowed on them animate their union with the liveliest ardour. The male and female have accordingly much attachment for each other, and the latter has a peculiar tenderness for her young. She prepares for them before-hand a particular retreat, which Buffon has described with equal elegance and truth. " This abode," says this admirable painter of nature, " is constructed with singular intelligence. The females commence with removing and raising the earth and forming a vault of tolerable height. They have partitions, a sort of pillars, at certain intervals. They then beat and press the earth, mix it with roots and herbs, and render it so hard and solid beneath, that the water cannot penetrate the vault, in consequence of its convexity and firmness. They then raise a hillock below, to the summit of which they bring grass and leaves, to make a bed for their little ones. In this situation, they find themselves above the level of the soil, and also, consequently, sheltered from any ordinary inundations, and at the same time secured from the rain by the vault which covers the hillock on which they repose. This hillock is pierced throughout with many sloping holes, which descend low and extend on all sides, like so many subterraneous roads, through which the mother can sally forth and seek subsistence for her young. These subterraneous paths are firm and beaten, extend to about a dozen or fifteen paces, and all proceed from the retreat like the radii of a circle. There may be found, as well as under the vault, the bulbs of the colchicus, which are apparently the first food given by the mother to her offspring."

In the interior of the Moles' nests are to be found leaves, grass, and the skins of those bulbous roots just mentioned. The little ones may be found there in the beginning of the month of March. At first they are quite naked and red. It is an old remark that more males than females are produced. They bring forth twice a year, generally four or five at a time, whence it is that the young may be found almost at all times, from spring to autumn. The exact period of gestation is unknown. Out of the season of sexual intercourse, and of the care of the young, every Mole lives an isolated, retired life, seeking its sustenance by continual labour.

The Mole, however destructive to agriculture, repays us to a certain degree by her services. Hence some would proscribe and utterly annihilate the race, while others would rather promote their increase. If she destroys the roots and seeds which are the care of man, she destroys worms, and insects, and noxious weeds, which are alike inimical to his industry; even the heaps of earth she raises on the surface, when spread, become serviceable as manure, particularly to meadow-land. However detrimental, therefore, these animals may be, especially to gardens, it is difficult, where they are not at least locally excessive in number, to ascertain the balance of mischief they may do when they have been credited for their beneficial offices.

It is justly observed, that to annihilate a species is an usurpation, and an abuse of the power of man, and which would, we cannot doubt, be detrimental to his interests in the end, were he to succeed in his exterminating projects. Nothing surely is made in vain, and to destroy the balance of nature in the immensity of her work is to violate her laws, and consequently to lead to confusion and mischief. To a certain extent it may be the duty of man, as the agent of Omnipotence, to limit the undue multiplication of some species, but not to annihilate any. Such undue multiplication indeed, in a great measure, is provided for without his interference; and species, noxious in certain particulars, are generally the means of checking the increase of others equally or still more extensively mischievous.

The *Tenrecs*, or *Madagascar Hedgehogs*, have four upper incisors bent, and six trenchant and lobed laterally in the lower jaw, a canine tooth on each side, and six cheek-teeth in each jaw. The first of these is a compressed isolated false molar, but the five following are genuine insectivorous teeth. Their body is covered with spines like those of the Hedgehog; the head is elongated, the muzzle excessively pointed, the eyes moderately large, the ears short and round, and they have no tail.

The anatomical characters of the Tenrecs approximate

tooth se of the Hedgehog. Like that animal, they have clavicles, and are destitute of the great intestines and of a
cæcum ; but they differ in the fleshy panicle, which is not
organized so as to envelop them as it were in a purse. The
Fenrecs have the tibia and perona distinct, but they are so
intimately connected as not to be able to play on each
side.

We know but little of this sub-genus. They dig in the
earth and remain there during the day, and are said to
sleep for three months in the year, not indeed in winter, but, on the contrary, during the greatest heat of sum
mer, as Buffon informs us, on the authority of Bruguière.

The Tenrec, properly speaking, the Tenrec or Tanrec of
Buffon, is about eight inches long and formed a good deal
like a Hedgehog, except that it is rather more elongated.
The spines, which cover all its upper parts, are yellowish
at the root and black in the remainder, the longest not exceeding an inch ; they form a sort of tuft above the head.
The back, crupper, and sides of the body are covered with
silky hairs of the same colour as the spines ; rough yellowish hairs cover the throat, breast, belly, and legs ; the
ears are very short ; the muzzle very elongated, of a
brownish colour : the nails are crooked, strong, and calculated for digging.

Their flesh, like that of most of the insectivorous animals, is unpleasant to the taste ; notwithstanding which,
as Buffon informs us, it is eaten by the Indians with much
avidity.

The *Tendrac* (*Erinaceus Setosus, Gm.,*) is smaller than
the Tenrec, not exceeding 5 inches in length. Its body is
covered with silky flexible hairs, of a pale yellow colour
the top of the head, and upper part of the neck and shoulders
are furnished with numerous small strong spines, those behind the head, particularly, elevated almost into a tuft ; they
are white at their base, and deep red toward the point ;

the belly and paws are covered with long hairs, annulated, and very rough.

The *Varied Tenrec*, considered erroneously by Buffon, says M. Desmarest, as a young Tenrec, is about four inches long. Its whole body is covered with silky hair and 'spines mixed together, so coloured that the back, of a blackish brown, is marked by three longitudinal lines, of a yellowish white ; the middle one streching from the lip of the muzzle to the anus, and the two lateral beginning at the ear, and terminating on the flank. The paws and under part of the body are yellowish white.

The CONDYLURE is stated by our author, as we have ob served in a note on the text, to be, by its dentition at least, no other than a Mole. In this, we should have implicitly followed him, did he not seem to treat the same species as distinct in his " Ossemens fossiles ;" and as we have not the means of referring to the teeth themselves, we shall append M. Desmarest's description of the animal, which he considers to be distinct from the Moles ; and, indeed, as intermediate between the two tribes of the insectivorous family.

The genus Condylure, says he, belongs not to the second tribe of the family insectivora, but by its characters approaches more to the first, in which are included the species provided with two long incisive teeth, and canines shorter than the molars. This genus, as we shall see, may be arranged between the two already admitted.

It is thus characterized: In the upper jaw there are six anomalous incisors implanted in the maxillary bone, the two intermediate are very large and contiguous, occupying the whole extremity of the jaw, furrowed, trenchant, a little obliquely, having the angle by which they touch each other more salient than the external angle ; the fol lowing incisors on each side touch the intermediate, and resemble long canine teeth, being conical, a little triangular at their base, where there are two very small tubercles, one before and the other behind ; the external or lateral incisors, which are the smallest of all the teeth, are simply conical, slightly compressed, and a little bent backward at their point, and placed at some distance from the incisors, like canine teeth. There are seven cheek-teeth on each side, the three first being the smallest and apart from each other ; the four following rather larger. But we shall not pursue M. Desmarest further, in relation to the upper cheek-teeth, as they seem to present the ordinary character of insectivorous teeth, making in all twenty in the upper jaw.

The lower jaw is very slender ; it has four incisors, flat, close, furrowed longitudinally, with the upper end of the furrow wider than the rest ; the lateral incisors are partly bedded horizontally on the intermediate, and rise a little on the external edge ; five teeth with many lobes follow on each side, which may be considered as false molars, as distinct from each other as those of the upper jaw, the first being larger than the others, and resembling in that particular only a canine tooth. The true molars in this jaw are only three. In all in this jaw also twenty.

Hence, continues M. Desmarest, the Mole and the Condylure differ in their dentition as follows :

Condylure, incisors $\frac{6}{4}$, canine, $\frac{0 0}{0 0}$, cheek-teeth, $\frac{7 7}{8 8}=40$ *

Mole, $\frac{6}{8}$, $\frac{1 1}{1 1}$ $\frac{7 7}{6 6}=44$

The Condylure has the muzzle very much elongated, very wrinkled, provided with a bone at the snout, and furnished at its point, in the species, at least, the most known with a naked disk, which encloses in its centre the opening of the two nostrils, and the edge of which is furnished with twenty cartilaginous moveable points granulated on their surface.

* If we consider the second incisive a canine tooth on account of its strength, it will then be, incisors $\frac{4}{4}$, canines $\frac{1 1}{0 0}$, cheek-teeth $\frac{6 0}{8 8}$.

Like the Moles, their neck is not distinct. The anterior paws are very short; the hands large, naked, scaly, with the lower edge trenchant, though in a less degree than in the Moles. They have five short fingers united as far as the second phalanx with long straight nails. The hinder feet are longer than those of the Mole or Scalope, and longer by a third than the anterior; they are weak, and the toes are deeply divided; the nails not so long as in the hands, more bent and sharp at their points. The tail is very slender, with the vertebræ a little salient. It is about one third the length of the body; the skin which covers it is divided into transverse folds, moderately close and scaly, and between the scales spring a few rough hairs.

Their fur is short, fine, soft, and silky. Their eyes very small, and so concealed in the fur, that it is with difficulty they can be found. The ears are totally destitute of an external conque. They have long rough mustachios, elevated parallel with each other, and directed forward towards the muzzle. The eyebrows are indicated by three or four similar hairs, which direct to the situation of the eyes. The palms are perfectly naked.

The first, and indeed the only species known with certainty, is the *Sorex Cristatus* of Linnæus, the Radiated Mole of Pennant. It was first described by M. de la Faille, in the Memoires of the French Academy for 1769. To the generic characters already detailed, it may be necessary only to add, that the fur is of the same blackish-gray, velvety colour as in the Mole, and that the rays round the nose are of rose colour. By enveloping and enclosing the nasal conduit, or by a contrary action, the Condylure is enabled to draw together or to open and spread these rays like the calices of a flower. The Condylure is in a small degree less than the Mole.

The Condylure is an inhabitant of Canada and the United States, having very much the manners of the Mole.

1

2

Griffeth

THE DESMAN OF THE PYRENEES.   THE CONDYLURE OR RADIATED MOLE.

MYGALE PYRENAICA. Geoff.   *TALPA CHRISTATA. Cuv.*

London Published by G. & W. B. Whittaker Sept.r 1824.

De la Faille observed that the vertebræ of the tail in this animal, were particularly prominent and distinct, which induced Professor Illiger to name it Condylure, from κονδυλος *nodus*, and οὐρη *cauda*, a name by no means satisfactory.

The long-tailed Mole of Pennant, is the species which Illiger cites as the example of his genus Condylure. The tail is half the length of the body, and as Pennant makes no mention of the cartilaginous rays round the nose, probability strongly indicates that it is not allied to the Sorex Cristatus of Linnæus. It is known only by Pennant's description, and its insertion, as a second species, is conditional, until more particulars can be ascertained relating to it.

Before we conclude the insectivora, it is necessary to notice an additional sub-genus, formed by the discovery of certain animals in the island of Sumatra and Java, which bear a considerable resemblance to the Squirrels, in external form, but approximate closer to the Shrews, on the one hand, in the system of dentition, at least of the cheek-teeth, and to some of the quadrumana on the other, in the characters of its incisors, prominence of the eyes, elongation of the tarsus of the hind foot, and habit of life. They may, consequently, be considered as bearing some affinity with these three genera of animals, but as the teeth are the most leading character, they are placed in the insectivorous division of our author's system.

M. Diard, in 1820, first proposed the formation of this new genus, and having given its characteristic analogies, bestowed on it the name of SOREX-GLIS. Sir Stamford Raffles published a summary description of two of these animals, under the generic title of *Tupaia:* these were the *Tupaia Tana* and the *Tupaia Press;* and a description of a third, namely, the *Sorex-Glis Javanica,* or *Tupaia Javanica,* as well as the Tupaia Tana, may be found in the valuable work of Dr. Horsfield, on the animals of Java. The name *Tupaia* is generic with the inhabitants of Sumatra, and is common to the Squirrels, and to these insectivora, inasmuch, as it is more peculiarly descriptive of the external form and general physiognomy of the animals. The admitted rules of natural history will scarcely permit us to retain it, as its local acceptation seems too indefinite.

One of those species, the Tupaia Javanica, of Horsfield, is called the *Bangsring*, and *Sinsring* by the Javanese. This animal lives in the forests, on the most elevated trees, where it subsists on insects, small birds, &c. Little has been collected concerning its habits, but we may presume that they bear much analogy to those of the last sub-divisions of the quadrumana, more perhaps than to those of the insectivora, properly so called, which generally live and seek their sustenance in a state of subterranean obscurity, or at least sheltered in retreats inaccessible to the light ; while the animals in question seem to pursue insects, small birds, &c., about the tops of trees. It has all the teeth, says M. F. Cuvier, of the insectivora : its incisors exhibit the same anomaly observable in all the genera of this family ; nor is there less irregularity in its canines, than in those of the Hedgehogs, Shrews, Chrysochlores, and Desmans. As for the molars, we find there the most exact resemblance with the above-mentioned genera. We have added Dr. Horsfield's figure of these teeth to those of the other sub-divisions of insectivora.

The head of the Tupaia Javanica is oblong, rather depressed, and very gradually attenuated to a conical muzzle, which is somewhat compressed laterally. The nose is obtuse and naked. The nostrils slightly curved, and pierced laterally. The upper jaw a little longer than the under. Slight mustachios. The eyes are very large and prominent, equidistant from the mouth and crown of the head ; the pupils are circular: The ears, says Dr. Horsfield. offer some peculiarities both in their disposition and form. They are externally provided with a large helix, which being margined in the upper part, passes in an angle to the sides, where a well-defined anthelix runs parallel to it ; and between both, patches of short hairs are scattered, without regularity: the tragus is of moderate size, and naked, reflected, in part, over the meatus auditorius externus, and is calculated to cover it over entirely, whenever the economy of the animal requires this organ to be protected. The antitragus is naked, and occupies a considerable portion of the auricular cavity. The ears are situated far behind.

The feet are plantigrade, and terminated by five toes, armed with slender and sharp nails, which are raised, and appear not to wear in walking, though not retractile. These toes increase in size, in the following order: the thumb or great toe, the fifth, the second, the fourth, and the third. In the hinder feet, the fourth toe is the longest. The nose is terminated by a muzzle, divided in the centre by a furrow   The ears are large, and provided with several tubercles ; these are rounded, and situated on the side of the head, and do not exceed it.   The structure of the eyes, of organs of taste, and of generation, is unknown. M. Diard says, that there is a cœcum, large eyes, four ventral mammæ, long tongue, and simple stomach.

All the body is covered with thick and soft hair, uniformly brown, except that it is a little tinted with yellow in the upper parts, each hair being terminated with one or two black and yellow rings. The lower parts, under the jaw, the throat, breast, belly, and internal face of the limbs, are of a yellowish-white, and a white straight line proceeds from the lower part of the neck, and terminates cn the middle of the shoulder.

These hairs appear to be of a twofold nature, but the woolly are the most numerous. The silky, which exceed the others in length, seem all to terminate in a black point, and are rare in all parts. Those of the limbs and muzzle are much shorter than the others, and those of the tail, the longest of all, are divided underneath, like the barbs of a quill, as in the Squirrels. The skin of the naked parts, *i. e.*, of the soles of the feet, and of the ears, is flesh coloured.

Another insectivorous member of this new sub-genus, (*Sorex-glis*) is the *Press*. The Press has all the generic characters of the Bangsring, and the general physiognomy of that animal. It is unnecessary, therefore, to say anything concerning its organs of mastication, motion, and sense. We shall stick to its specific characters.

The upper parts of its body are a beautiful brown maroon, deeper on the back than on the flanks. The hairs are, some black, with a fawn-coloured ring in the midle ; others reddish fawn, with a black ring. The brown maroon is the predominant colour on the back and flanks, because the reddish hairs are found there in the greatest numbers. But further back, the black hairs becoming more numerous, give to this part of the body a deeper and more obscure tint. The tail is of a grayish-brown, in consequence of the white rays upon its hairs, which, for the remaining part, are black.

The breast is a whitish-yellow, which colour is mingled with gray on the abdomen, and internal face of the limbs.

The colour of the head is nearly the same as that of the tail, but the rings are yellow, and more numerous. The ear is entirely covered with black hairs.

It measures between six and seven inches from the extremity of the nose to the root of the tail.

We have observed that this new genus approximates three distinct and different genera. In the first place, continues Dr. Horsfield, this genus agrees with the animals of the second family of the insectivora in the elongated form of its rostrum, and in certain peculiarities of its dentition. In the latter, it is most nearly allied to the genus Mygale. Mygale is also the only genus among the insectivora which agrees with our animal, in having, in both jaws, single well-defined canine teeth ; but the incisors of Tupaia differ essentially from those of all other animals of this family. As far as regards the rostrum of Tupaia, it should be observed, that although it is long and tapering, the upper jaw projects beyond the lower, and is not extended into a naked proboscis, which constitutes a peculiar character as well in Sorex and Mygale, as in other sub-genera of this family. Of other characters, it should be noticed that the eyes, in most of the insectivora, are minute, or scarcely perceptible, while in Tupaia they are large and prominent. The structure of the external ear is also entirely different in our animal; in Sorex, for instance, in which this organ is most developed, the antitragus is enlarged to such a degree as to close the meatus auditorius, while the helix has a similar disposition, and can be folded over it as a double membrane. In Tupaia, on the contrary, it is in the tragus, which is so constructed as to afford a covering to the external passage, while the antitragus is a simple eminence in the cavity of the ear; the helix constitutes a narrow border, forming an extensive circuit about the auditory passage, and can only be elevated to regulate the admission of sound, without affording an additional covering as it does in Sorex. Tupaia is further destitute of these glands, which in Sorex and Mygale are placed in the sides of the body, or at the root of the tail, which by their odoriferous secretion, constitute a very peculiar character.

The form and habit of the body, the length and structure of the extremities, and the broadness of the tail, give to Tupaia a physiognomy entirely different from that of the insectivores hitherto named, and which have led to its association with the Squirrel.

Notwithstanding, however, these aberrations from a common type, the teeth, and influential characters of these newly-discovered animals, are all, generally speaking, insectivorous, an dictate their place in artificial arrangement. Similar inclinations to deviate may be observed in certain species throughout all the genera.

The tail is of a grayish-brown, in consequence of the white rings upon its hairs, which, for the remaining part, are black. The breast is of a whitish-yellow, which colour is mingled with gray, on the abdomen, and internal face of the limbs. The colour of the head is nearly the same as that of the tail, but the rings are yellow, and more numerous. The ear is entirely covered with black hairs. The size of this animal is about eight inches, (French measure,) from the end of the muzzle to the origin of the tail. The tail is five. The head is two inches long, and the middle height is three inches, six lines.

Sir S. Raffles has mentioned this animal in his catalogue of the animals of Sumatra, published in the thirteenth volume of the *Linnæan Transactions*. He found it tamed, and living in a domestic state. It was free, would run through all the house, and would come of itself at every meal, for fruit or milk. In its wild state, this author avers that it lives on the fruits of the *Kayo-gadis*. But we may safely presume that it preys on insects, and other small animals. Its organization is a certain proof of this fact. Resembling the rest of the insectivora, in the organs of digestion and mastication, it cannot differ in its appetites, though it may in its habits, which may depend more exclusively on the structure of the brain than on that of the teeth or intestines. Its habits are the same as the Bangsring. It is diurnal, and lives in trees, in the thick forests of Sumatra, which it traverses, like the Squirrels.

Sir S. Raffles refers this animal to his genus *Tupaia*, and gives it the Latin name *ferruginea*, in consequence of its colour, which is like that of the red oxide of iron.

Supplement on the Carnivora.

QUITTING the insectivora, we come to those races that are more or less decidedly carnivorous; that will prey upon animals of a larger size, or gorge on carrion. Many of them unite, to the sanguinary appetite for flesh, the most cruel and unmitigated ferocity; while others exhibit little or nothing of the murderous instinct, except under the goading influence of indomitable hunger. Some also subsist much more exclusively on flesh than others, while many are found capable of being supported almost entirely on a vegetable diet. Among these last, are some of the first species which will come immediately under our review, in the brief additions which we shall make to the text, on the family of the

Plantigrades.

Among those, the first genus is that of the URSI or BEARS, and the first of which we shall speak, pursuing the order of the text, is the *Brown Bear of Europe,* or as it is sometimes called, the *Brown Bear of the Alps and of Norway,* the *Ursus Arctos* of Linnæus.

The Bears, with a brown fur, approaching more or less to black, on the one side, and, on the other hand, to fawn, or even a fairer hue, are so very numerous, and have been so much confounded together, that it is impossible to decide whether they belong to many, or are only varieties of a single species. All the critical discussions which have been entered into, with the view of throwing light upon this question, have proved nothing but the impossibility of attaining the proposed object. Authors are not agreed upon the point, many even contradict themselves, and recent observations, have not been sufficiently accurate or multiplied, to reconcile these conflicting authorities. Brown Bears appear to have been found in all parts of Europe and

Asia, in the Molucca Islands, Mount Atlas, and the western regions of North America.   The only mode of fixing the the analogies existing between animals which inhabit such a diverity of climates, and are exposed to such varied influences, is evidently to describe and represent, with exactness, such specimens as are discovered, so that we may be able to approximate and compare them together, in all the details of their organization.   Accordingly, we shall here present to our readers, M. Frederic Cuvier's description of two Bears, which came under his immediate inspection ; one was a Brown Bear from the Alps, which was adult ; the other, a very young Bear from Norway.

The proportions of the first (in French measure,) were :
From the extremity of the muzzle to the buttocks, 1 ft.
7 in. 6 lines ; from the end of the muzzle to the occiput, 3 ft. ; its mean height was 2 ft. 1 in.

It was covered all over the body with very thick, long, and rather soft hair, generally of a maroon brown, deep on the shoulders, the back, thighs, and legs ; but on the sides of the head, ears, and flanks, tinted with yellow.   On the paws, this hair became short, and nearly black ; as also on the muzzle, where, however, it retained rather more of the brown colour of the head.

The circle surrounding the pupil of the eye was of the same colour as the fur.   The sole of the hind feet was entirely naked, and marked with four folds, which correspond to the divisions of the toes.   These last were separated from the sole by some hairs, and each of them was furnished with an elliptical tubercle.   The palm of the fore-paws was naked only on its interior half ; but behind, there was a naked and rounded tubercle, surrounded by hairs.   There were three folds on the naked part, two which corresponded to the two internal toes, while the two external ones were embraced within that part circumscribed by the third fold.   But this part was also divided by a fold, cutting it obliquely from front to rear, and from

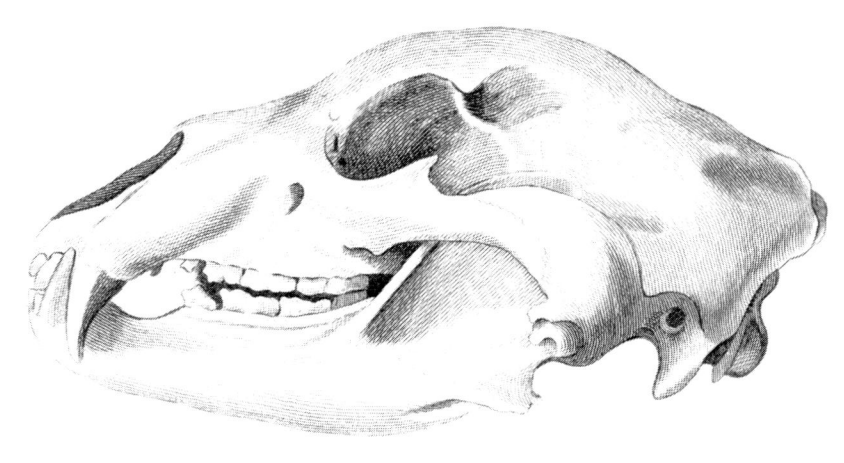

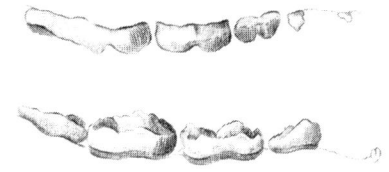

TEETH OF THE GENUS URSUS.

SPECIES — THE BROWN BEAR OF EUROPE.

London. Published by C. B. Whittaker Dec.r 1824.

the outside to the inside. The toes were also furnished
with elliptical tubercles. Each foot had five toes, armed
with strong and trenchant nails. On both feet, the middle
toe was the longest; the others went on diminishing gradually. The eye was circular, small, and without any accessory organ. The nostrils opened in front of a glandulous muzzle, and passed over its sides, bending convexly,
and forming a section. The external conch of the ear was
very simple, and rounded. The tongue was soft, narrow,
and long. The lips were very extensible, and the muzzle
participated in their mobility. The incisors were six in
number, in both jaws, and the canines were as in the other
carnassiers. There were five molars in each jaw; two
very small and pointed ones in the upper jaw, and three
very large and tuberculous ones. In the lower jaw, one
very small, and four large and tuberculous, like those
which were opposite. These are the genuine triturating
teeth. In the upper jaw, these teeth proceed in one increasing proportion, from the first to the last; in the
lower, the last but one is the largest. The one preceding
that is less than the last, and the first is the smallest.
This animal drank by suction, was sustained on vegetable
substances only, which agreed with it very well, and it ate
but a very small quantity in comparison with its size, six
pounds of bread being found sufficient for it, while ten
pounds of meat are usually given to a lion at a meal. Its
walk, posteriorly, was altogether plantigrade, and all its
movements were heavy and embarrassed. It was extremely
malicious, and slept during a considerable portion of the
day. Without being in a lethargic state, during the winter,
it was observable that it ate considerably less at that season
than any other. This animal had lived a long time in the
pits at Berne, from whence it was brought when the French
conquered Switzerland, to the Museum at Paris, and in
1819, it had lived six years in the pits of the menagerie.

The Norway Bear was presented to the royal menagerie

of Paris in 1818. It was five weeks old when it was received there, and its only food was milk. Its proportions, three months after this period, were as follows, in French measure:

The head from occiput to muzzle, 0 ft. 7 in.
Body from occiput to buttocks, 1 ft. 4 in.
Its height in front,   -   -   - 1 ft. 1 in.
In the hinder part,   -   -   - 1 ft. 0 in. 6 lines.

Its whole body was covered with a crisp and very thick hair, excepting the muzzle and paws, of a very uniform umber-brown. No trace of white hairs was visible. The organs of sense and motion, in this young animal, were already conformed, similarly to those of adult Bears. It differed in nothing from the Alpine Bear just described, and even resembled it in disposition. Though young, on its first arrival at the French menagerie, and forced by its weakness to obedience, it evinced no small degree of malice ; and always attempted to bite when it met with any opposition. Since that period, the wickedness of its disposition increased.

This young Bear was particularly fond of sucking its paws, during which operation it always sent forth a uniform and constant murmur, something like the sound of a spinning-wheel. This appeared to be an imperious want with it, and it was surprising to observe the ardour with which it commenced the operation, and the enjoyment which it seemed to derive from it. The belief, which once so gene-rally obtained, that these animals, during the season which they pass without eating, and surrounded by snows, support themselves by sucking their paws, seems not utterly without foundation. In truth, every natural action must have a tendency to some useful end, though it has not been observed that the Bear extracts anything from its paws by the act of suction. After all, it is more probable that Bears lick their paws, as cats do, from a love of cleanliness, or merely in consequence of some pleasing sensation which nature has attached to the act, for inexplicable reasons, rather than for sustenance.

Both sexes retire in the winter, and the period of parturition with the female is in spring, after a gestation of seven months. She produces from one to five at a birth.

We have observed that Bears feed chiefly on vegetable substances, and may consequently be considered as semi-carnivorous. They will, notwithstanding, at times destroy the poultry, &c., in a farm-yard, and sometimes they subsist on fish. They are also very fond of honey, and notwithstanding the clumsiness of their conformation, exhibit a considerable degree of agility in mounting trees in search of it. Their mode of fighting is chiefly by seizing their adversary, and squeezing him between the arms and breast until they have deprived him of life. They never attack man except under the influence of severe hunger ; and it is reported that, when so pressed, they will associate together in search of animal food.

Major Hamilton Smith (to whose researches natural history is so deeply indebted) made a drawing of an European Bear at Dresden, which seems to be, if not a distinct species, at least a strongly-marked variety. It was about four feet high at the shoulders. The physiognomy differed from that of the common Bear. The ears were small and round, and the facial line was greatly depressed at the junction of the nasal and frontal bones. The colour was a fiery yellow on the head and back, passing into chestnut and red on the sides and hams. The belly and paws were brown, and there was a dark streak upon the nose spreading into branches towards the orbits. The general form of the animal was extremely clumsy.

The same gentleman also took a drawing of a specimen at Buda, in Hungary, of a Bear which appeared about forty years ago on the shores of the Danube, in Upper Hungary. This animal was uncommonly large, and had proved excessively destructive to the cattle, consequently every effort was made to seize or destroy it. But shot appeared to take no effect upon him, and when hard pressed he would swim to the other side of the Danube, and resume his depredations there until chased back again. In this manner he was fairly hunted into Lower Hungary, having travelled most of his way by water. From Semlin he was chased beyond Belgrade, but the Turkish peasantry drove him back, and it was many months before he was killed. Besides the peculiarity of his excessive bulk, his colour was purpurescent, and several balls were found lodged in his skin. Although the Bear is not uncommon in Hungary, his extraordinary colour and bulk excited so much curiosity, that he was stuffed and preserved at Buda.

Before we quit the European Bears, it will be necessary to say something of the *Black Bear of Europe.* Though the Baron speaks with some uncertainty on this point in the present work ; yet in the " Ossemens Fossiles," he seems pretty well decided as to the distinction of species. It is our duty, therefore, to present to the reader the substance of his opinion on the subject.

All the terrestrial Bears of Europe he thinks reducible to two species, different in their general forms, and more especially in the conformation of the crania ; and one of those species he considers divisible into many varieties, founded on the character and colour of the fur.

In the first of these (which we have already noticed) the upper part of the eranium is arched in every part. The forehead forms a part of the same curve which prevails from the muzzle to the occiput. It is arched also from right to left in the same style as in its length, and there is no clear distinction between the forehead, the middle portion of the parietal bones, and the temporal fosses. The sagittal crest only begins to be sensibly marked very near the occipital.

In the other species, the frontal portion is flattened and even concave, especially crosswise. The two ridges which separate it from the fossæ temporales, are strongly marked, and form behind an acute angle prolonged into a very elevated sagittal crest, which is not furnished until it meets the occipital.

Of the first of these species sufficient has been said: of the second, the Baron says, that he never saw more than a single living individual, which he afterwards dissected. It was of considerable magnitude ; the skin was of a brown black, rather rough, partly woolly, and long, especially on the belly and thighs. The upper part of the nose was a clear fawn-colour, and the remainder of the muzzle of a brownish red fawn. This the Baron believes to be the Bear to which naturalists have given the name of the Black Bear of Europe, and it must not be confounded with the Black Bear of America, whose fur is black, pliant, and shining. The peculiar and flattened form of the cranium can be perceived through the hairs which cover it, quite sufficiently to distinguish the animal from the common Brown Bear.

The Baron has seen skeletons of the same species, which, with some unimportant deviations, preserved in the main the characters above described. He is unable to point out the strict habitat of the animal, or any of its variations as to shades of colour or other accidents. He is certified, however, that the characters we have noticed are not the result of age or sex ; for he possesses crania of the other species of both sexes, and equally adult with those which he has seen of the second.

Judging from the form of the cranium, the size of the temporal fossæ, the strong points of attachment furnished by the crests to the crotaphite muscles, the Baron is persuaded that the Black Bear of Europe is more decidedly carnivorous than the other species. If the contrary opinion has generally prevailed, it is owing to the confusion of this species with the Black Bear of America, which in its native country appears to live for the most part on fruits and fish. But the fact is that all the Bears are omnivorous; and in the menageries, all of them, even the Polar Bear, to which so marked a character for cruelty has been given, are fed with bread alone, and that without sustaining the least detriment. For more than twenty years in the French menagerie, the Bears have been supported on no other regimen. In truth, the cheek-teeth of the Bears, flat and tuberculous like those of Man and the Simiæ, and never trenchant like those of Lions and Wolves, would lead us à priori to the conclusion, that they were destined to make use of every kind of aliment.

A third species, according to the Baron, is the *Black Bear of America*, which, though bearing some affinity with the last, is nevertheless very clearly distinguished from it by characters of sufficient certainty.

The osseous head is shorter in proportion to its bulk, and the zygomatic arches, being less convex and less separated from the cranium, leave consequently less volume to the crotaphite muscle. This explains, to a certain extent, the fact attested by all travellers, of the milder disposition generally evinced by this species. On the other hand, its forehead is arched like that of the Brown Bear, not flat and concave as in the black, and yet the temporal crests are well marked, and meet in sufficient time to form a sagittal, which occupies as great a portion of the cranium as in the Black Bear of Europe.

It may be remarked that in both species, as indeed in all the Carnassiers, the sagittal crest increases in length with age, because the crotaphite muscles grow more bulky, and produce more marked impressions. This, however, does not affect the question of the distinction laid down above, between the Black and Brown Bear, as the latter at no period of life possesses a long sagittal crest. All the fur is black excepting on the muzzle. The skin is first covered by a very copious wool of a reddish black, and the hairs silken, pliant, reddish at their origin, and finally of a brilliant black, entirely conceal the first, from which results a very thick and luxuriant fur. The muzzle is of a fawn colour, more or less grayish on the sides of the mouth. In some individuals white fur is found on the breast. In youth it is more of a chocolate brown, and at a certain age it is covered with a gray down before it assumes its fine black colour. The young are of a clear, uniform, cinereous gray.

Buffon at first regarded the Black Bear of America as a simple variety of the Brown Bear of Europe; but afterwards, having seen an individual of this species at the menagerie of Chantilly, he formed a juster notion of it, though he did not make it a distinct species. It was Pallas who first recognised its characters, and since his time the species has been adopted.

The generality of travellers who have visited North America speak of this Bear. By amalgamating the substance of their various reports, a natural history of this species has been obtained, very nearly as minutely detailed as that of the Bear which inhabits our European mountains.

The Black Bear of America, in its mode of life and subsistence, exhibits much affinity with the Brown Bear of Europe. It inhabits the recesses of the forests, and the wildest tracks of the country, and seldom approaches cultivated and peopled regions, except when the rigour of the season deprives it of all sustenance in its ordinary quarters. It feeds on fruits, roots, insects, flesh, and even fish, of which last it is reported to be extremely fond, and it descends to the borders of lakes, and to the sea-shore in pursuit of its favourite diet. It never attacks the larger animals or man, except when severely pressed by hunger. Its movements, like those of the Brown Bear, are heavy and awkward, but it climbs trees with facility, and swims well. In

its excursions it always follows the same paths, which thus
in time become so well beaten, that the Indians follow them
to hunt the animal in his retreats.

In our climates, it is generally the season which determines the epoch and the duration of the retirement of the
Bears. The case is not precisely similar with the Black
Bears of America. When the winter commences in the
most northern parts, the Bears which inhabit these regions
abandon them, and betake themselves to a less rigorous climate, where they remain as long as the season obliges them.
They choose a shelter, either in the trunk of some hollow
tree, or under the projection of some jutting rock. They
furnish it with dry leaves, and soon fall into a lethargic
sleep, from which nothing awakes them but the return of
spring. They descend no further to the south than the latitude of the Floridas, and to the west they proceed as far
as the Pacific Ocean.

About the month of June, the Black Bears are in heat.
During this period they grow excessively thin, and the Indians will not touch their flesh ; they are also much more
dangerous to meet at this time than any other. Gestation
lasts about six months, and in January or February the
cubs are born. They are about six or eight inches long,
covered with hairs, have the eyes closed, and are devoid of
teeth. Their nails, however, are very much developed.
The gray tint, we have mentioned, of their fur continues for
the first year, and they are suckled for six months. The
moulting takes place in spring and autumn, and all the
hairs fall almost about the same time.

The hunting of this species was formerly much more productive than at present. The fur of these animals was
formerly preferred by the Indians, but since the Europeans
have established themselves in the northern parts of America, the hunting of the Bear has been neglected for that
of the Castor. The flesh, however, is still much sought

after, especially that of the feet, which is in high request, and the fat is a perfect *bonne bouche* for the savage hordes of these countries. The chase of the Bear is accompanied among the Indians with many superstitious observances, which are minutely detailed by Father Charlevoix; but an account of them would be better adapted to illustrate the history of uncivilized man, than that of the animal at present in question.

The Black Bear of America does not appear 'to possess the same degree of docility and intelligence as the Brown Bear of Europe. He does not minister like his congener to the purposes of curiosity or amusement, and is never exhibited dancing to the sound of the flageolet and tambourine. M. F. Cuvier remarks, that among the different species of Bears presented in the pits of the Parisian menagerie to the curiosity and whims of the public, the Black Bears of America seemed to profit the least by the education which they might have thus received, and were found less to engage the interest and attention of the spectators. They attained, however to the comprehension of certain signs; they lay down, rose up, and turned to the right or left at the word of command. But the Brown Bears did much more; and *Monsieur Martin* (of whom all Europe has heard) has acquired no less celebrity by his intelligence and ad dress than by his cruelty.

In the organs of sense, motion, and generation, these Bears resemble the brown, as also in their gait and habit of body. Their voice resembles groaning, more or less violent according to the strength of their sensations.

Among many adult individuals which the Baron observed, two resembled each other entirely. They were male and female; the muzzle was a deep brown above, and a grayish fawn colour at the sides. A small fawn-coloured spot was in front of the eye; all the rest was of a fine shining black. A third, which died of illness, had the hair a little more brown and less smooth, and the spot on the eye was less marked. A fourth was of the finest black, without any such spot. The muzzle was brown above, and the edges of the lips whitish. Two whitish lines occupied the region of the sternum between the fore-legs, forming the resemblance of an H. This Bear the Baron regards as an individual variety.

Our author remarks a fifth, which he considers a still more marked variety. The black is remarkably fine; the muzzle a clear fawn colour. A white spot is on the top of the head, and a white line commencing on the root of the nose proceeds on each side to the angle of the mouth, and continues over the cheek to a large white space mixed with a little fawn, which occupies the entire throat, and of which a narrow line descends upon the breast. It is the *Ours Gulaire* of M. Geoffroi.

The Baron also thinks that the Yellow Bear of Carolina is a variety of the same species. This is scientifically termed the *Ursus Luteolus*. We shall not venture to assert, in contradiction to the authority of the Baron, that this Bear forms a distinct species, but assuredly it is a very strongly-marked variety. Major Smith took a sketch of one at New York; the specimen was semi-adult. He does not consider that there is sufficient proof of its being a distinct species. In the specimen drawn by the Major there was a greater convexity of forehead, and a sharper nose than in the Black Bear. This comparison was easily made, as the two animals were chained very near each other. The ears of the Yellow Bear stood more back, were not quite so large, and the physiognomy was very different*. Both were remarkably tame. Although the Yellow Bear cannot be affirmed to be specifically different, yet it is certain that there is a distinct race of these animals. They

---

* It must be remembered that this specimen of the Ursus Luteolus was but *semi-adult*. P.

NORTH AMERICAN BEAR.

URSUS FEROX, Lewis and Clark   *U. CANDESCENS*, Hamilton Smith.

T. Landseer del. et fecit, Tower

London, Published by G.B. Whittaker, March 1, 1827.

were formerly common in Virginia, and they are still abun-
lant in North Western Louisiana, where they are called
White Bears, and are said to feed chiefly on honey, on
acorns of a large size, wild berries, &c. Both this and the
Black Bear are far from disliking animal food, especially
the flesh of the Hog, which they probably prefer on account
of the fat, and the facility with which they can overtake
the animal. The young Fawn is likewise hunted by these
Bears; but notwithstanding their acute smell, they cannot
follow its track, as the musky secretion in the hoof, which
occasions the scent, does not take place until the Fawn becomes adult.

The *Cinnamon Bear* in the Tower appears to be of the
same race as this Yellow Bear. The Major, however, remarks, that it is a received opinion that the Black Bears
occasionally produce white or fawn-coloured cubs; and
after all the Yellow Bear may be nothing more than an
albino variety, such as are constantly springing up in the
human and many other species.

There is another of the American Bears, which, from all
accounts, we have every reason to consider as a distinct
species. It exceeds, in size, ferocity, and strength, both
the last-mentioned Bears, and seems, in truth, to be an uncommonly fierce and cruel animal. This is the *Grisly Bear*,
(*Ursus candescens*, Smith.) Lewis and Clarke, in their travels, have given a number of interesting adventures, hair-
breadth escapes, and surprising anecdotes, relative to the
North American Bear; but as they have not stated any
satisfactory anatomical particulars, and as there is much
uncertainty, and some confusion, in their work, it is
extremely difficult to determine as to the identity or distinctness of species in the several individuals which they
mention. Many species of this genus, according to them,
are to be found in the Arctic regions of America, endowed
with great strength, and corresponding ferocity. But it

seems most probable, that these are individual varieties, distinguished by different shades of colour. Yet, still we think that this Grisly Bear, which differs so considerably in size and ferocity from the rest, may, at least, until we have more complete information on the subject,be set down, provisionally, as a distinct species.

These American travellers, from actual observation, and also from the information which they derived from the Indians, seem to be of opinion that there are two species of the Bear in the New World, distinct from the common Black Bear of America. The first is the White or Grisly Bear, under which, they include the pure white, the deep, and the pale grisly red, the grisly dark-brown, and, in short, all those in which, be the ground-colour what it may, the extremities of the hairs have a white or frosty appearance. The second species consists of those individuals, in which the black or reddish-brown is intermixed with a few entire white hairs, or which have a white breast, or are of an uniform bay, brown, or light reddish-brown. Where these two species abound, the common American Bear does not inhabit. The last-mentioned of the two species of Lewis and Clarke, may be referred to those varieties of the *Ursus Americanus,* mentioned by the Baron, in his " Ossemens Fossiles," and which we have already presented to the reader.

Major Smith has made two drawings of these animals; one was from a stuffed specimen at Philadelphia, which was sent from the Missouri country, by Messrs. Lewis and Clarke. The animal was clumsy and compact, and the hairs on the neck and back were tipped with white. Another is from a specimen brought from Hudson's Bay, which was in the Tower. This last was more slender, and better proportioned than the other, very active, and exceedingly fierce. It was three feet, three inches in height, from the shoulder, and appeared to be adult. The teeth of these

specimens were similar, and both wanted the usual number of the small cheek-teeth, which immediately follow the
canine teeth, in the common species; this last, however,
cannot be insisted on as a distinctive character, inasmuch
as these teeth are so liable to fall in all the ursine genus.

The second of these seems to be the black variety mentioned by Lewis and Clarke. Difference of colour alone,
more especially in the animals of very northen climates, is
no sufficient criterion of the distinction of species. Until
more decided and anatomical characters are pointed out,
we think that the safest conclusion is, that there is but one
species of this ferocious Bear, which may branch into two
principal varieties, and each again subject to slight variations of colour. Major Smith has, indeed, very happily
applied the epithet *candescens* to all of them ; for, let the
main colour be what it may, there is always observed a tendency to whiteness in greater or less degrees. From the
ferocious character of this animal, it has also received the
epithet of *Ursus Ferox.*

" On one occasion, Captain Lewis, who was on shore with
a hunter, met two White Bears. He says, " of the strength
and ferocity of this animal, the Indians had given us dreadful accounts : they never attack him but in parties of six
or eight persons, and, even then, are often defeated, with
the loss of one or more of their number.

" Having no weapons but bows and arrows, and the bad
guns with which the traders supply them, they are obliged
to approach very near to the Bear ; and as no wound,
except through the head or heart, is mortal, they frequently
fall a sacrifice if they miss their aim. He rather attacks
than avoids a man ; and such is the terror which he has
inspired, that the Indians, who go in quest of him, paint
themselves, and perform all the superstitious rites custo-
mary when they make war upon a neighbouring nation.
Hitherto, those we had seen, did not appear desirous of

encountering us; but although, to a skilful rifleman, the danger is much diminished, yet the White Bear is still a terrible animal."

On approaching these two, Captain Lewis and the hunter fired, and each wounded a bear; one of them made his escape; the other turned upon Captain Lewis, and pursued him seventy or eighty yards, but being badly wounded, he could not run so fast as to prevent him from re-loading his piece, which he again aimed at him, and a third shot, from the hunter, brought him to the ground.

Another instance is recorded, by these travellers, of the tenacity of life in this species. An individual received five balls through his lungs, and five other wounds; notwithstanding which, he swam more than half across a river to a sand-bar, and survived more than twenty minutes. He weighed between five and six hundred pounds, and measured eight feet, seven inches and a half, from the nose to the extremity of the hind feet; five feet, ten inches, and a half, round the breast; three feet, eleven inches, round the neck; one foot, eleven inches, round the middle of the fore-leg; and his claws, five on each foot, were four inches and three-eighths in length. A specimen of this species is now in the Tower, which is engraved, to illustrate this description.

As there are very few genera among the mammalia, more natural than that of the Bears, so, by a necessary consequence, are there none of which it is more difficult to mark the specific distinctions. Accordingly, we find that the *Polar Bear*, (*Ursus Maritimus*,) had always received his proper generic appellation; yet his specific characters remained long unestablished, and naturalists were uncertain, if he should be considered as a distinct species from the others, or merely an albino variety of the common Brown Bear. Buffon, at first, as we may see, in his fifteenth volume, entertained this very doubt, having no other means

THE POLAR BEAR.

URSUS MARITIMUS . L . ? .

London, Published by G.B.Whittaker Feb.ʸ 1825.

of judging on the subject than the lights afforded him by the accounts of travellers. When Collinson, however, had sent him a figure of this bear, he then decided, from the conformation of the head, that it must be a distinct species, essentially differing from the Brown Bear of Europe. Pallas, afterwards, confirmed this opinion, by an attentive examination of a young individual of this species; and since that time, the Polar Bear has been generally admitted into systematic catalogues, under the specific denomination of the *Ursus Maritimus*. In fact, although this Bear has a general resemblance of exterior to the Brown Bear, and is also pretty similiar in the details of its organization, in the instruments of sense and motion, yet it is so different in the forms of the head and proportions of the neck, that nothing could authorize the supposition of identity of species. In the Brown Bear, the muzzle is separated from the forehead by a profound depression, while, in the Polar Bear, these two parts of the head are nearly on the same line. The front of the Common Bear is rounded, that of the Polar Bear is flat. This last has the head narrow, and the muzzle broad; the other has the head broad, and the muzzle narrow. The Polar Bear is also still further characterized, by the length of his body in comparison of his height, by the length of his neck, by the small extent of the auditory conch, and by the length of the sole of the foot, which forms, according to the Baron, one sixth of the length of the entire body; whereas, in the Brown Bear, it forms but one-tenth. The Polar Bear is, finally, distinguished by the length and fineness of his fur.

M. F. Cuvier describes an individual which was in the Parisian menagerie, in 1795. His length, from the extremity of the muzzle to the posterior part of the body, was about five feet eight inches. But he had not attained the stature of his species, in consequence of constant confinement; for their usual length is from seven to eight feet.

One which was killed by the crew of Captain Ross, and whose skin has been deposited in the British Museum, measured seven feet eight inches from snout to tail; and its weight, after a loss of blood, estimated at thirty pounds, was 1131½ lbs.

The one, of which we first spoke, was entirely covered with a white fur, composed of very long and thick silky hairs, except on the head and limbs, where the covering was short, and composed of woolly hairs, forming a thick stuff, well adapted to resist the impressions of cold. The muzzle, the tongue, the skin of the eyelids, and the claws, were black. The skin of the lips and mouth was of a violet-black. These colours did not change, and were the same at all seasons.

This Bear was fed with nothing but bread; of this he consumed only six pounds daily, and yet was always very fat. This seems to prove that, as to diet, the appetites of this whole genus are similar, and that the Bears can with no propriety be divided into frugivorous and carnivorous. If the Polar Bear, to the travellers who first met him, appeared eager for flesh, it was evidently because the high latitudes which he inhabits afforded him no other sustenance. Bears, like other animals, will choose, by preference, the species of nutriment to which they have been accustomed. This accounts for the fact, that some Bears are extremely dangerous enemies both to man and animals, while others pass them by with apparent contempt. Thus, the Polar Bears, accustomed to fish, will pass by flocks and herds without attacking them, unless they are stimulated by very violent hunger.

This Bear, of which we have been speaking, like all others of the tribe, possessed the sense of smelling in great perfection, but his powers of vision were but feeble. His hearing seemed little less developed than his sight; and nourished always on one kind of aliment, it was not probable that he had much acuteness or discrimination of taste. It would seem that these observations are tolerably applicable to the entire species, and that, with the exception of smell, all the other senses of the Polar Bears are somewhat obtuse.

This animal seemed to suffer extremely from heat, and for the purpose of cooling him, his keepers, in summer, would throw large quantities of water upon his body, which he appeared to receive with an extraordinary degree of pleasure. He was never tamed, and always would endeavour to hurt those who approached him. He did not attack openly, or with menaces, but endeavoured to wound with his paw and claws. His voice was never heard, except when he was provoked to anger, by teasing, and then it was strong and hoarse, and always of the same tone. The only influence which the winter season possessed over him, was to diminish the need of aliment. It did not plunge him into the torpid lethargy into which he would have fallen under the rigours of a polar winter.

The gait of the Polar Bears is heavy, like that of all plantigrades, but they swim well, and dive for a considerable time. In the water, they can be fatigued by a long chase. Their ordinary aliment is fish, the flesh of the seal, and all the animal substances which the sea is continually casting upon its shores. We are told, that when shoals of fish are passing, these Bears will follow them in a troop ; but, without doubt, their usual habits, like those of all their congeners, are solitary. When the winter arrives, and their time of sleep approaches, they choose the hollow of a rock, a place sunk in the snow, or an opening in a flake of ice, and there they sleep until the sun of the returning spring awakens them with his reviving beam. During this time of sleep, considerable masses of snow accumulate upon them, and preserve them from the excess of the cold, which would otherwise, without doubt, destroy them. When they sally from their retreat, in about five or six months, they seem extremely in want of nourishment, and it is not a little dangerous to meet them at that period. Captain Ross, in the account of his Voyage of Discovery to the Arctic Regions, states, that he received a message from one of the whalers, he fell in with, requesting surgical assistance for the master, whose thigh had been very severely lacerated by a wounded Bear, which had attacked and dragged him out of the boat. The animal was pierced by three lances before he would relinquish his hold ; when, disengaging himself from the weapons, he swam to the ice, and made off.

This species of Bear is found only on the shores of the Frozen Ocean, and it never descends, but by accident, from those inhospitable regions. Sometimes, in spring, when the ice is detached from the coasts, they have been known to arrive in Norway, on the floating flakes. But, in general, they are not found to establish themselves on this side of the Arctic circle. The males first quit their retreats, and it is at this period that the females are delivered of their young, (generally two at a birth,) which are nursed by them with the greatest care, to the following winter. We are assured, by travellers, that they even carry them on their backs in swimming, or when the young are tired, after the manner of Swans, and many other Water-Fowl.

Another species, which we owe to the researches of Sir T. S. Raffles, is the *Ursus Malayanus*. He observes, that " this deserves to be ranked as a distinct species from the Common Bear, and from that of the continent of India. The most striking difference is in the comparative shortness of the hair, and the fineness and glossiness of the fur, in which particular, it appears to resemble the American Bear. It is further remarkable in having a large heartshaped spot of white on the breast. The muzzle is of a

J. Landseer. del. et sc.
Tower.

THE MALAY BEAR.

FROM BORNEO.

URSUS MALAYANUS.

London. Published by G.B.Whittaker Dec.r 1824.

ferrugineous colour.   It stands lower, but is a stouter and better proportioned animal than the Common Bear *."

When taken young, they become very tame.   One lived for two years in Sir T. S. Raffles' possession, and that gentleman adds, " He was brought up in the nursery with the children, and when admitted to my table, as was frequently the case, gave a proof of his taste, by refusing to eat any fruit but mangosteens, or to drink any wine but champagne.   The only time I knew him to be out of humour, was on an occasion when no champagne was forth-coming. He was naturally of a playful and affectionate disposition, and it was never found necessary to chain or chastise him. It was usual for this Bear, the Cat, the Dog, and a small blue Mountain-Bird, or Lory of New Holland, to meet together, and eat out of the same dish.   His favourite playfellow was the Dog, whose teasing and worrying was always borne, and returned with the utmost good-humour and playfulness.   As he grew up, he became a very powerful animal ; and in his rambles in the garden, he would lay hold of the largest plantains, the stems of which he could scarcely embrace, and tear them up by the roots."

A stuffed specimen of this animal was presented, by Lady Banks, to the British Museum.

A fine specimen of this species is now in possession of Mr. Cops, at the Tower, which we have engraved, from a drawing by Mr. T. Landseer.   It came from Borneo.

The head of this individual is remarkably thick, much more so than in the other species of Bears ; the neck, also, is very short and thick, so that the head, neck, and body altogether, are nearly cylindrical.   The colour of the muzzle is dirty-yellow.   The eyes extremely small, have the iris approaching a pale-lilac colour, with a very small circular black pupil.   The snout, which is truncated, is terminated by a moveable fleshy elongation, extending beyond the

* Linnæan Trans. vol. xii. p. 1.

extremity of the lower jaw, but this elongation does by no means appear to form a foliaceous appendage. Between the upper lip and the nostrils, there seems to be a deep incision, in the form of a horse-shoe, within which the nostrils are pierced ; the upper or thickest part of the head measures about twenty-two inches vertically round ; the length, from nose to tail, is nearly three feet. This is a female.

The patch on the throat is yellow, with a tinge of red, with several irregular blackish spots.

The hasty view that could be obtained of the teeth, when the animal was occasionally induced to open its mouth, was, however, perhaps sufficient to enable us to say, that the system of dentition is ursine, and fully to evince that the animal is not edentatous, though she is exhibited under the name of the five-fingered Sloth.

Whether the differences between this specimen and those previously described, are merely individual, or whether they constitute this a variety of Borneo, we cannot determine.

The Indian Bear, at the Tower, is very full of action, though its movements may be called slow and measured. With all its muscular clumsiness, it appears to have a sort of suppleness of joint, as it assumes various and very antic postures. Its favourite position, however, appears to be that represented in the plate, of sitting on its haunches, and thrusting out its long, narrow tongue, to a very extraordinary length.

It will eat flesh and fruit, but is kept entirely on bread and milk, eating about two pounds a day.

The last of the living species of Bears which we shall describe is the *Ursus Labiatus,* placed erroneously by Pennant and others among the Sloths, under the name of the Ursine Sloth. Though multiplicity of sinonimes is as much as possible to be avoided, in zoology, from the confusion it creates, it would be highly improper to retain the English name adopted by these gentlemen, and we shall therefore take the liberty of calling this animal the *Thick-lipped Bear*.

In speaking of the Brown Bear, we remarked the difficulty of distinguishing the various species of this genus Ursus, and the state of ignorance in which we yet are concerning them. It appears evident that they are now spread through almost all the countries of the world, a fact in utter contradiction to the opinion formerly received on this subject. The Bears, indeed, appeared so essentially constituted to inhabit the coldest climates, that they were conceived to be exclusively peculiar to them, and the possibility of their existence in countries within the tropics was denied. At the present day, however, it is clearly decided that these animals belong to one of those cosmopolite genera, the privileged species of which may be found in every latitude, and can support every degree of temperature.

That Bears were to be found in Southern Asia was known for some time previous to their being admitted into our scientific catalogues. Marsden, in his History of Sumatra, tells us of a Bear of this country called *Brouorong*. Williamson, in his Oriental Field Sports, gave a figure of a Bear in the peninsula of India. Peron made a communication to the Baron of the existence of these animals in the Gattes mountains ; and M. F. Cuvier received an account from M. Leschenault of a Bear which he had seen at Java, of the middle size, and with a yellowish spot on the neck, like a gorget. This last was evidently the Ursus Malayanus, just described. It was not imagined by naturalists that one of these very Bears had lived in Europe, had been drawn, and a description of it published. This circumstance is worthy of attention, for while it demonstrates an error not likely to be renewed, it also proves the progress which naturalists have made in the knowledge of the Mammalia. This animal was not recognised as a Bear. Deprived of his incisors by the effects of age, he seemed to belong to the order of the Edentata, and the genus of the Sloths, and his figure was published as we have seen under the name of *Bradypus*. This error also shews the bigotted spirit in which the Linnæan system was applied, and how far the admission of arbitrary authority and rules into sciences which depend on observation is calculated to mislead and even paralyze the judgment. The story is worth recording :—

In the year 1790, an old individual which had lost the incisor teeth was shewn in England. Arbitrary systems so much prevailed here at that time, that Pennant and Shaw, founding their opinion on this accidental loss of the incisors, pronounced the animal to be a Sloth, and called it Bradypus Ursinus, at the same time that they could not have avoided observing that in its motions it had nothing analogous to the genus in which they placed it. Illiger, following their notions, formed of it his genus *Prochylus*.

Mr. F. Buchanan, in his Journey from Madras through the Countries of Mysore, *&c.*, published in 1807, was the first to announce that this pretended Sloth was no other than a Bear of the Indian mountains. There is in the Museum of the College of Surgeons, the cranium of the individual above-mentioned, and which has all the generic characters of the Ursus deprived of its incisive teeth, the alceoli of which, however, are perfectly visible. MM. de Blamville and Tiedman confirmed the observation of Mr. Buchanan. The first gave it the name of *Ursus Labiatus*, the second of *Ursus Longirostris*. Previous to a more particular description of this animal, we may as well notice that it is the opinion of M. Alfred Duvaucel, that there are two other species of Indian Bears. One is the Malayanus, which we have described ; the other has been seen only in Nepaul and the mountains of Silhet by MM. Wallich and Duvancel, and the Baron has given it the name of *Ursus Tibe-tanus.*

We cannot fix the characters of the Ursus Labiatus better than by presenting our readers with the substance of M. A. Duvaucel's observations concerning these three species ; and though we have already treated of one of them, yet the comparative view which this gentleman gives of the three renders their characteristics more obvious, and enables us to mark with greater precision the differences by which they are distinguished.

" The analogy which prevails between these three Bears, and the uncertainty which still exists relative to those of the old continent, cause me to hope," says M. D. to the Baron, " that you will receive with interest some comparative observations which may tend to give them a more specific character. Their difference, which chiefly consists in the conformation of crania, though less sensible elsewhere, nevertheless extends through the whole of their organization. In the feet, in the fur, in the proportions of the limbs, many characters may be recognised alike invariable and unequivocal.

" The largest of the three *(Ursus Labiatus)* has a thick and still most singularly elongated muzzle. The head is small, and the ears are large ; but the hair on the muzzle, at first smooth and even, grows suddenly rough around the head as far as the height of the ears, and completely buries them under a thick fur, and augments considerably the volume of the head. The cartilage of the nose consists of a large plate, almost plane, and possessing great mobility. The end of the lower lip, in all the specimens which I have seen, goes beyond the upper, and moves equally by contraction, by elongation, or in a lateral direction. This gives to the animal a physiognomy (as M. D. happily expresses it) of stupid animation. Its limbs are elevated, its body long, and its motions easy. These characters are more of less disguised by the length of the hair, which in old individuals almost touches the ground. Its breast is ornamented with a large white spot not unlike a horse-shoe reversed, two branches of which extend over the arms. This Bear, which appears more docile, more intelligent, and more common in Bengal than the other species, is educated and exhibited by the jugglers for the amusement of the people. It is often met in the mountains of Silhet, in the environs of inhabited places, and the general opinion of the people is that it is exclusively frugivorous.

" The smallest species is about six times less than the preceding. Its head is round, the forehead large, and the muzzle very short. The cartilage of the nostrils is rounded, and possesses very little mobility. The ears are small, but more apparent, and situated lower down than the others. The tail is scarcely visible, the fur is smooth, shining, and tight over the body as well as the head. A spot of pale fawn-colour is observable above its eyes, which disappears with increasing years. The muzzle is red, more or less deep; and the pectoral spot, equally red, presents on all the individuals an imperfect figure of a large heart. This species, though the specimens of it are rare everywhere, has yet a very extensive habitat. It is also the most delicate, and subject to the greatest number of varieties. The smallest come from Pegu ; the largest are found in the island of Sumatra, where they are very common, and it is the only species belonging to the genus which has migrated from the continent. It causes great ravages in the island by climbing to the summit of the cocoa trees to drink the milk, after having devoured the tops of the plant*. This is the Ursus Malayanus.

* Our readers will see that this description contains nothing contradictory to that of Sir T. S. Raffles, except in the colour of the pectoral spot. And that this species is subject to numerous slight variations we have the testimony of M. Duvaucel, who probably may have seen more individuals than Sir Thomas.

" The intermediate species has the muzzle of a moderate bulk ; but the forehead, not much elevated in the two preceding species, is scarcely perceptible in this, being almost on the same line with the nose. The arrangement of the hair is the same as in the Ursus Labiatus, and the volume of the head is similarly augmented by the quantity and disposition of its covering, only at the hair being a little shorter, renders this character less prominent. The ears are also very large, and the nose by no means unlike that of a dog.

" This Bear has a compact body, thick neck, and heavy limbs. But this conformation, which might lead us to assign considerable strength to the animal, does not agree with the weakness of the claws, which are one half shorter than those of the preceding species. Perhaps we may de, duce from this circumstance that this animal does not climb trees. The muzzle is black on the upper part at all ages with a slight red tint on the edges of the lips. The lower jaw is white underneath, and the pectoral spot resembles a pitch-fork, the two prongs of which, considerably apart, occupy the entire chest, and what may be termed the tail extends to the middle of the belly. This Bear was found by M. Wallich in the mountains of Nepaul, and I have seen it myself in those of Silhet. Its habitat is less extensive, and it appears to be more ferocious than the others. This Bear is the *Ursus Tibetanus.*

" It would be easy to multiply instances of difference between these species, by a minute comparison of all the details of their organization. But I presume that the inspection of my figures alone will be sufficient to remove every suspicion of their identity, and to convince you that we have in India three species of the Bear invariably black, for I have seen a sufficient number of individuals of each to certify that their covering preserves the same colour at all ages and in every season. As to the teeth, I

R 2

know little respecting them, excepting those of my second
Bear, which has at least three false molars.   It is probable
that the third species, so different from the two others, has
likewise, some anomaly in this respect, a fact of which I
cannot be assured until my specimens are dead."

Thus far M. Duvaucel.   We shall only add that the
Ursus Labiatus belongs to the mountainous parts of India
solely.   It is said to retire into caverns and holes, which it
excavates by means of its long claws, and to feed principally
on white ants, fruits, and honey ; but as little of its habits
are known with any certainty, it may rationally be presumed, on viewing the teeth, that it is as carnivorous as
Bears in general.

A disgusting story in the Oriental Field Sports, Major
Smith is inclined to refer to this species, of a poor Indian
who had his hands and arms literally ground into a pulp
by the teeth of a Bear.   If so, this would be the *Baloo*, a
name generally given to the Ursus Malayanus.   The presence of the mark on the breast, which we have mentioned,
may strengthen this supposition.

The existence of the Bear in Africa, is not so incontesta-
ble as that of the Asiatic Bear.   Pliny having found in the
Roman Annals, that under the consulship of Piso and Mes-
sala, sixty-one years before Jesus Christ, Domitius Ænobarbus exhibited in the circus a hundred Numidian Bears.
led by as many negro hunters, quotes the fact with much
surprise.   " I am astonished," says he, " at this epithet
Numidian, for it is certain that Africa produces no Bears."
Ursinus, Lipsius, and Vossius have imagined that by this
word the annalist meant Lions, as the Elephant was at first
called the Lucanian Ox ; and they have cited medals of this
same Ænobarbus, the reverse of which represents a man
combating with a Lion.   But it seems perfectly incredible
that the Romans, who, according to the testimony of Pliny
himself, had seen such numerous troops of Lions, could

have given this animal such a perverse misnomer. How,
above all, could Pliny be ignorant of this synonyme, which
must still have been in use in his time? For we find Ly-
bian Bears mentioned by contemporaneous writers—by
Juvenal,

> " Nec profuit misero quod cominùs Ursos
> Figebat Numidas."————————

And by Martial,

> " Quod frenis Lybici domantur Ursi."

And long before by Virgil,

> " —————————Acestes
> Horridus in jaculis et pelle Libystidis Ursæ."

Solinus, and among the moderns, Crinitus, Saumaise, Aldrovandus, and Zimmerman, have taken the part of the old
annalist, and maintain the existence of the Bear in Africa,
though they allow that it has been seldom found. Solinus
even asserts that it is the handsomest of the Bears, clothed
with the longest hair, and is by far the most ferocious.
But the testimony of such an author, and even that of
Strabo, who asserts that there are Bears in Arabia, need
confirmation from more modern and authentic sources.
Shaw, indeed, mentions the existence of Bears in Barbary,
but does so in a simple enumeration of animals, without
saying any thing particular about them, and indeed without
appearing to have seen them. M. Desfontaines, who made
a long stay at Algiers, and surveyed Mount Atlas with no
small degree of care and accuracy, never saw any Bears in
that country, and only mentions in a vague manner that
there might be some in the forests in the environs of the
Calle.

Prosper Alpin attributes Bears to Egypt, but such as
have no character of the animal. He talks of them as of
the size of a Sheep, and a whitish colour. None of the

naturalists on the French expedition ever saw any Bears in Egypt.

Poncet says that one of his Mules was wounded in Nubia by a Bear. But Bruce thinks that he has confounded the Arabic word *Dubbah*, which signifies a hyæna, with *Dubb*, which means a Bear. The last-mentioned traveller assures us positively that there are no Bears in any part of Africa. Dapper says that there are Bears in Congo ; but he is a compiler whose authority is not strengthened by the testimony of any traveller. It seems tolerably certain that no one has ever seen a Bear in the south of Africa.

There should be Bears in South America, if we are to attach any credit to the first describers of that country. Acosta and Garcilasso place them in Peru. But as more recent naturalists have seen none there, it is probable that the animal taken by those writers for the Bear was no other than the large Ant-eater.

The next division of the Plantigrades that come under notice, belong to the genus RACOON (PROCYON.) They have three pointed, distinct, anterior cheek-teeth, and three posterior, which are flattened ; the whole forming a continued series quite different from that of the Ursus, in which the first three are insignificant, and may be almost called deciduous. They differ also from the Bears in having a long tail, and all the teats are ventral, while in the genus Ursus there are two pectoral and four ventral teats. In running, the Racoons do not bring the sole into complete contact with the ground, but they do so when standing.

The first species, of which we shall now speak, is the *Common Racoon* (*Ursus Lotor, L.*)

One of those phenomena which are most worthy of the attention of the naturalist, and most calculated to lead us to appreciate the infinite power of the Creator, consists in the insensible and gradual changes through which the same organ will pass, by which its nature will, in some measure, be transformed, and results produced entirely different from those which constituted the object of its original destination. The organs of sense and motion offer frequent examples of this phenomenon, and the teeth of certain animals present a remarkable instance of the same. The true carnivora, the Cats for instance, have, in each jaw, teeth evidently destined, by their form and relative position, to cut, like the two blades of a scissors, the fibres of the muscles of their prey. But in proportion as the destination of an animal is less decidedly carnivorous, these teeth lose their trenchant character, and grow thicker ; and thus we, at last, arrive at a limit where they can no longer be distinguished from the tuberculous teeth, whose office simply consists in triturating the food. These teeth, when sharp and slender, are opposed face to face, but when thick, they are opposed crown to crown, so that they become truly transformed into molar teeth, and nature, in operating so considerable a transformation, has no need of making any essential change in those organs. It is sufficient for the purpose, that a very small tubercle, which is already found on the internal face of the slenderest teeth, should simply receive a more augmented development.

The Racoons are the last of the carnassiers, in which these changes of the teeth can be traced, without uncertainty. They are frugivorous as well as carnivorous. They seem, in this respect, to form the link between the quadrumana and the mammalia, which subsist on small animals, and even insects, such as the Bats, the Moles, the marsupial carnassiers, &c. They bear a near relation to the Coatis, of whom we shall presently speak. Their molar teeth are altogether similar, and the only difference between the two is found in the organs of sense, which has led some naturalists to form of these different animals but two divisions of a single genus.

In the upper jaw, on one side as well as the other, they have two tuberculous molars, one carnivorous tooth, three false molars, one canine, and six incisors; and in the lower jaw, one tuberculous, one carnivorous, four false molars, one canine, and six incisives. The eyes have a round pupil, and they offer nothing particular in the eyelids, nor in the other accessory parts. The nose extends considerably beyond the jaws, without being, at the same time, so much advanced as that of the Coatis, and it is terminated by a glandulous apparatus, at the end of which the nostrils are opened, and prolonged over the sides, reascending in a curve line. The tongue is soft, and the lips extensible. The ears are elliptical, and of a very simple structure. The sole of the feet may be an organ of touch. Its skin is extremely delicate, and it seems not improbable that these animals make use of it for feeling. Of the teats we have already spoken. The fore-feet have five toes, furnished underneath with thick tubercles, the shortest of all is that which answers to the thumb. The little toe comes next in length; then tha which is next the thumb, and the two remaining ones, which are the largest, are of equal length. They have all digging nails, long and strong; and on the palm, five very elastic tubercles are distinctly visible; one strong one towards the wrist, another at the base of the little finger; a third, at the origin of the thumb; a fourth, near the second toe; and a fifth, at the base of the two larger toes. The hind-feet are conformed precisely like the fore, as to the toes, the claws, and the tubercles. But the tarsus is longer than the carpus, and the first tubercle is farther from the heel than its corresponding one from the wrist.

These animals are plantigrade, but, as we observed in the

introduction, do not place the entire sole on the ground in walking. The gait is heavy and awkward. They can easily stand up on their hind-feet, and have the power of grasping with the fore; but this last operation is performed, not by contracting a single paw, but by putting both together. There is not sufficient pliability in the internal part of the fingers, to enable them to grasp like the quadrumana. In this mode, they often carry their provisions to their mouth, after having plunged them in water, and rolled them between their paws; an operation, the object or utility of which, to the animal, it is not very easy to conjecture. They do not see objects very distinctly in a strong light. During the day, they remain bent into a ball, seated on their posteriors, with the head placed between the thighs. It is in the night, though the eye-pupils are round, that they evince most activity, and seek their food, which consists, for the most part, of worms, insects, fruits, and roots. They proceed ferretting in all directions, and the most retired corners, and the smallest holes seem, in particular, to excite the activity of their re- earches. They climb trees with the great facility, where, in all probability, they go for the purpose of surprising the birds, and plundering their nests. They drink by suction, and water appears to be a very absolute necessity of life with them. It is said, that they frequent the banks of rivers, and the sea-shore, to catch mollusca, and fish, to which they are extremely partial. Their sense of smelling is peculiarly delicate, but not so their organs of hearing. They are commonly very fat, which, united to the propor- tions of the various parts of their body, and to the thick fur with which they are clothed, gives them a rotundity of form, very unlike what characterizes the carnassiers of higher rank, but very similar to that of the Bears. Their tail, ex- ceedingly tufted, does not appear to be of any particular use to them. They are animals easily tamed; that is to say, they soon become familiar; they even seem to look for caresses, but do not appear capable of obedience or attachment. They must always be kept chained up, to hinder them from regaining their liberty, and returning to their wild state. Captivity causes them to contract new habits, but they never lose the sentiment of independence.

The general colour of the body is a blackish-gray, paler under the belly, and on the limbs, and resulting from hairs ringed with black and dirty-white. The tail has five or six black rings, on a ground of yellowish-white. The muzzle whitish in front, has a black patch, which includes the eye, and descends obliquely on the lower jaw. Between this and the ear, on the cheeks and brows, the hairs are almost white, very long, and directed downwards. The forehead is black. On the remaining part of the muzzle, the hairs are very short; but the upper lip is furnished with long and thick mustachios. All the feet are covered, but with very short hairs.

These animals have two sorts of hairs. The woolly kind are deep gray, and very thick; the silky are ringed, as we have said, with black and dirty white. The covering which results from this combination, is very furry and soft, which must make the skin very valuable. They are found in advanced latitudes in North America. Mackenzie found them on the borders of the Red River, in 45° or 50° north longitude. They also descend pretty far towards the south. D'Azzara describes them among the animals of Paraguay.

The Racoon has been frequently brought into Europe. It is a very well known animal as to organization and character. But of its natural habits we know next to nothing. There is a letter, addressed to Buffon, on the subject, and inserted in the third volume of his supplements, quarto edition, which contains some curious details; but we know nothing of the circumstances of the re-production and development of these animals : we know not, with precision, the means which they employ in procuring their food, or in defending themselves against their enemies. In fine, the physical history of the animal is known, but of its natural history we are almost in a state of perfect ignorance. The females are smaller than the males, but in every thing else there is a strict resemblance between the two sexes.

The substance of this description is taken from that most indefatigable, and most meritorious naturalist, M. Frederick Cuvier, whose zeal in the cause of science is only to be equalled by his assiduity and judgment in its prosecution. These remarks are the result of his observations on a great number of Racoons, possessed by the French menagerie. A state of captivity is not very favourable to the development of the natural instincts or habits of any animal. Under strict confinement, and leading a mononotous life, it must always present itself under the same aspect. Animals must be seen under other conditions, to form an accurate judgment of their natural character. They must be in a state of freedom, and their various relations must be sufficiently extended, before their faculties can be completely unfolded.

D'Azzara has spoken of the Racoon under the name of Agouarapopè. He says that the female has three teats on each side. The Mapach of Mienskenberg is evidently a Racoon, but the drawing is incorrect, and the engraving very bad. Buffon has, hitherto, given the most correct figure. In possession of the editor, is a drawing of Major Smith's, of a specimen of the Racoon, of a bright-rose colour, and which, though extremely rare, we presume to be the same as the yellow Racoon, mentioned by M. Geoffroy, in the catalogue of the collection of the French Museum.

A small variety of the Racoon, with brown throat, is also mentioned in the same catalogue.

The White Badger of Brisson is likewise considered as a variety of the Racoon.

The scientific name (*Ursus Lotor*) given by Linnæus to the Racoon, is from its habit, which we have mentioned, of plunging every article of its food, if possible, into water.

The *Crab-Racoon*, (*Procyon Cancrivorus*,) is an inhabitant of South America, and is treated, by Cuvier, as distinct from the last, though they appear to have been much confounded. It is a little larger than the latter, but the tail, in proportion, is much shorter. It is of an uniform clear ash-brown colour, or, as Buffon states it, yellow marked with black and gray, the black prevailing on the head, neck, and back, and the yellow being almost unmixed on the sides of the neck and body ; the end of the nose is black ; a blackish-brown band surrounds the eyes, stretching almost to the ears, passing over the muzzle, and uniting at the summit of the head. The inside of the ears is furnished with whitish hairs, and a whitish band passes above the eyes, and there is a white spot on the middle of the forehead ; the cheeks, jaws, under part of the neck, breast, and belly, are yellowish white ; the tail is annulated with six black rings, the intervals between which are yellow mixed with gray and black.

The habits of this species are similar to those of the last. It feeds very much on crustacea and mollusca, whence it has received the epithet by which it is distinguished, though it may be observed that the other species is also fond of this kind of food, and probably to the same degree.

It is an inhabitant of South America

Shaw observed that, according to Linnæus, the Racoon has a wonderful antipathy to Hog's bristles, and is much

disturbed at the sight of a brush, which particular, relative
to a specimen kept and described by Linnæus himself, is,
by some mistake, applied byBuffon to the Coati Mondi,
and is quoted in a note belonging to the history of that animal, in his work on quadrupeds.

The Racoons produce from two to three young at a birth,
generally in the month of May. Their fur is used by hat-
ters, and is considered as next in merit for this purpose to
that of the Beaver.

The next subdivision of our author embraces the COATIS,
the first species of which is the *Red Coati* (*Viverra Nasua.*)
M. F. Cuvier complains of the poverty of the French
language in describing the multifarious colours which distinguish the animal tribes. This reproach, we apprehend,
is applicable to all languages, and to our own in a much
greater degree than to the French. Whatever advantages
the English language may possess, it certainly must resign
the palm here; for it possesses neither delicacy nor variety
of terms for the discrimination of colours. But the truth
is, that nature in this, as in most others of her wonderful
manifestations, is an overmatch for man and for his language. The magnificent splendour and the infinite variety
of hues in which she delights to clothe her wondrous works,
must ever set the ingenuity of man at defiance to describe
them. The attempt indeed to convey colours through any
medium, but that of vision, must ever prove, more or less,
a failure. No man ever formed a just idea of any animal
which he had not seen, from mere description. Hence the
indispensable necessity of figures to natural history, which
must very imperfectly attain its end without them. The
species which we are now about to describe, is sufficient to
demonstrate the unsatisfactoriness of *verbal* colouring, by
the errors of synonymy to which it has given rise; errors,

which could only be discovered by the direct and immediate comparison of the different species or varieties of the genus.

The Red Coati has been known a tolerably long time. Systematic writers have even expressed its character by the name which they bestowed upon it ; but the synonymy which they attached to that, destroyed on the one hand what they had established on the other. It is almost certain that this animal has never been described or represented faithfully; and if a contrary opinion has obtained, it is because what authors have said of another Coati, which is apparently but a fawn-coloured variety of the brown, has been always referred to this species. By a most singular chance, Schreber having copied the figure of the blackish Coati of Buffon, and having arbitrarily illuminated it, produced by accident a very near resemblance of the genuine Red Coati. But this only served to augment the error into which people had already fallen on the subject, and to confirm all the mistakes which were its necessary consequence. We shall say more on this subject when we come to the Brown Coati. We find no clear designation of the Red Coati among travellers, except in a note communicated by Laborde to Valmont de Bomaire, and inserted by the last in his Dictionary, at the end of the article *Quachi.* " We met," says the first, " in the woods of Guiana a large species of Quachi, the hair of which is of a bright red," &c. What is singular, most naturalists must have heard of this species, especially in modern times, notwithstanding the delusion under which they laboured respecting it.

The principal dimensions of an individual in the French Museum were as follows :

Length from the occiput to the origin of the tail, one foot, six lines.

Of the tail, one foot, four inches, four lines.

From the occiput to the end of the muzzle, five inches, nine lines.

Of the fore-foot, two inches.

Of the hind-foot, three inches, three lines.

Height to the shoulder; nine inches, nine lines.

Height in the hinder part, ten inches, six lines.

This Coati was a male. All the parts of its body, except the muzzle, the ears, and the spots of the tail, had a tint of bright and brilliant red, a little more sombre along the back, where the middle part of the hairs were black, but red everywhere else, and of a tint somewhat paler towards their extremities. The muzzle was blackish gray above, and gray on the sides. The ears were black, as also the lower part of the legs in front. The tail was covered with transversal spots of maroon on the upper part, which divided it uniformly into eight or ten parts. The lower jaw and the edge of the upper were white.

The fur, very thick and harsh, is composed of two sorts of hairs. The silky hairs are those from which the animal derives its colour, the woolly are gray, and in very small quantity. The eye is small and black, elongated transversely, but destitute of any accessary organs. The ear is small and rounded; the nose, which is prolonged considerably beyond the jaws, is terminated by a sort of glandulous protuberance, and the nostrils are oval, open in front, and are prolonged in a sort of cleft on the sides. The tongue is very soft and extensible. All the feet have five toes, armed with very long claws well adapted for digging. The three middle toes, equal in size, are the longest of all; the two external are shorter, and the thumb is shortest of all. The soles of all the feet are naked, and covered with a very soft skin. The animal in walking places only the extremity of the fore-feet on the ground, and does not even place the entire sole of the hind-feet on it, except when sitting down. In treating of the Brown Coati, we shall mention the remarkable tubercles which belong to this part, and which may yet perhaps be considered as characteristic. There are neither anal sacks nor glandular pouches The tail is tolerably thick at its base, but the animal does not make much use of it. He usually carries it raised, and contrary to the custom of many other animals, places it between his legs when he lies down, as if to repose himself upon it. In this situation it is folded round, pretty nearly in the manner we see dogs carry theirs.

The Red Coati has the same teeth as the rest. The cheek-teeth are six in number on each side of the two jaws. In the lower jaw are four false molars, a carnivorous, and a tuberculous tooth. In the upper, three false molars, one carnivorous, and two tuberculous teeth. But the carnivorous teeth take in these animals altogether the tuberculous character, in consequence of the development of those internal tubercles, which we have mentioned in our description of the Racoon. In each jaw are eight incisors and two canines, and these last are remarkable in their form ; they are depressed, and present on their front and back faces trenchant edges, which must constitute them very formidable arms.

Smelling is the predominant sense of the Coati. His nose is in perpetual motion, and he applies it strongly, as if trying to feel with it, to every object presented to his notice. He uses it also for digging and for pursuing worms, of which he seems extremely fond. In this labour he also assists himself with his fore-paws. His sight, hearing, and taste seem very obtuse. He exhales a strong and very disagreeable odour.

Sometimes this Coati which we have mentioned, took his food with his jaws. But most commonly he would carry it to his mouth with his paws, not in the way of handling, but by digging his nails into the provision which remained attached to them. In general these animals are very adroit in the use of their paws. They use them very cleverly in climbing and descending trees ; and they do not descend backwards like most other animals. They come down head foremost, hooking with their hinder paws, which they have an extraordinary faculty of reversing to great extent. The voice of this Coati was a gentle hissing when in good humour, and a very shrill and piercing cry when under the influence of pain or anger.

This same individual, without being precisely malicious, was never completely tamed; and though at times he would permit caresses, at others he would bite sharply. On this account it was necessary to keep him continually shut up. This confinement did not allow scope for the full impulse of his character, and complete exercise of his intelligence. But his disposition generally resembled so closely that of the Brown Coati, that we may conjecture that in such respects there is not much difference between the species, and the latter has been often completely tamed. Laborde says that the Red Coati lives retired in the largest woods, generally in companies of no more than three or four individuals of the species; while, on the contrary, the Brown Coatis live in large troops. But this information is not sufficiently important or authentic to illustrate the history of the species.

As the earth is subject to such numerous variations which exercise very opposite influences on animal life, and as each country in particular experiences the effects of many transitory and accidental causes, nature, by a necessary consequence, has bestowed upon her creatures the faculty of being modified in proportion to the extent and operation of such causes, and of being conformed to the various circumstances by which they are surrounded. This is one of the wisest regulations in the economy of the universe (if ndeed among so many proofs of omnipotent wisdom we may be permitted to single out any instance for eulogium in preference to others), and one indispensably necessary for the preservation of animal existence. In fact, in the very earliest steps which are made in the study of living beings, this grand truth must be felt and acknowledged. It is to this faculty that all individual, perhaps too, all specific varieties are owing, and most certainly all the races of our domestic animals. Without it, life itself would ere long be extinct from the surface of the earth. Were it withdrawn, all nature would speedily exhibit the most deplorable spectacle of disease, decay, decrepitude and death. This cannot be doubted for an instant, when we consider the effect which the most feeble of these modifying causes will produce upon animals not properly prepared to resist them. The natives of hot climates seldom pass with impunity into colder regions, unless the change be carried on by almost imperceptible gradations. And the case is the same with those which pass from Northern countries into torrid or even into temperate latitudes.

Notwithstanding the high importance of this law of nature, it has not yet been made the subject of any particular study. Every day we see doubts started on the characters of species and varieties, and we are in the most profound ignorance of the effects proper to each modifying cause, as we are also respecting the nature of these causes themselves. This is a branch of science totally neglected, although it might conduct to the most curious and useful discoveries. The transformation of the rough covering of the Argali (*Ovis Ammon*, Gm.) into the soft and beautiful fleece of the Merinos, the domesticity of the Dog, the subjection of the Horse, a thousand other phenomena of this description are incontestably verified, yet no attempt is made to investigate their producing-causes. But if on the one side it is important to make experimental researches on this point, it is no less so to ascertain the varieties which nature herself occasions, to establish the modifications which take place independently of our influence, the organs in which they occur, and the limits to which they arrive. With this view all the varieties exhibited by the Mammalia should, as far as possible, be represented and described, and on this occasion we shall present to the reader, along with our description of the Brown Coati, a notice of its fawn-coloured variety.

We have detailed at considerable length, in our description of the Red Coati, the principal organs of that species ; its size, its proportions, its teeth, its senses, its paws, its toes, the nature of the fur, and the principal uses which it makes of its limbs. All that we have said on this subject equally applies to the Brown Coati. M. F. Cuvier had both species under his inspection at once, and after the minutest comparison of their details, he declares that the only difference between them is in colour. We shall notice a few particulars omitted in our account of the Red Coati, and which the reader must consider as characteristic of both species. The tubercles of the paws exhibit very peculiar characters, which might of themselves suffice to distinguish the Coatis from the Racoons were there no other points of distinction between these sub-genera, such as the eyes, the elongation of the nose, the tail, and the general physiognomy. It is principally in the fore-feet that these tubercles are remarkable. In the first place, those with which the extremities of the toes are furnished, are very thick, and they are separated from those of the palm by folds of the skin of a very peculiar character. There the thumb communicates with a very large tubercle, divided into two parts, which itself communicates behind with another placed on the edge of the palm. The three middle digits rest upon one and the same tubercle, which is prolonged from the external side of the paw, and behind which another very strong one is found, which terminates the

palm on the side of the wrist. Finally, the little digit is connected with a very small tubercle, which communicates with
a part of the preceding. The sole of the hind-foot differs
less from that of the Racoons, the tubercles being similar
in number. The first, commencing from the side of the
first toe, furnishes its base, the next is in relation with the
two following toes, and the two others correspond to the
joining of the second toe with the third, and of this last
with the little toe. Lastly, a fifth tubercle is found behind
on the side of the heel. The external covering of all these
parts is a skin, extremely soft. It seldom or never happens
that two specimens of the Brown Coati are to be found
exactly alike. They are met with all variety of shades
between brown and fawn-coloured, depending on the greater
or less depth of the tint with which the extremities of the
hairs are coloured. Some have been found with the muzzle
entirely black, others had the tail without rings, others
were of a whitish gray, while the majority were a deep
orange yellow. In general on the upper parts of the body
the hairs were yellowish in their lower half, then came a
portion of black, and finally, the extremities were tipped
with fawn-colour more or less deep, which, according to its
depth, produces the sombre shade of the Brown Coati, or the
brighter tint of the fawn-coloured variety. None of these
differences happened to belong to sex. The lower parts and
the internal face of the limbs are of a yellowish gray, sometimes of an orange shade, and these colours at times rise
as high as the breast and over the sides of the neck and the
lower jaw; and behind these places is sometimes seen a
part of a white colour. The summit of the head is gray,
all the lower jaw white, the upper part of the muzzle
black, except that in the majority of specimens there is a
white line along the nose, and three other white spots
round the eye. One of these spots was above the eye, one
below, and the other at the side of the external angle. The

tail sometimes altogether black, more frequently covered with rings, alternately deep brown and fawn-coloured, is always black at the end. This is also the colour of the extremity of the paws.

An individual of the fawn-coloured variety was presented to the French Menagerie by General Cafarelli. Though very tame, it would never leave its cage, until it had tried to smell out every object around. When its distrust was abated, it would traverse the apartment, examining every corner with its nose, and putting aside with its paws every object that would be an obstacle in the way. At first it would not permit itself to be touched, but turned and threatened to bite when any one put his hand near it. But as soon as it was given something to eat, it became perfectly confident, and from that moment received all the caresses which were bestowed upon it, and returned them with eagerness, thrusting its long muzzle into one's sleeve, under the waistcoat, and uttering a little soft cry. It took a fancy to a Dog, and they both slept in the same cage but it would not suffer another to approach it. When it scratched itself with its fore-paws, it often made use of both at once ; and it had a singular custom of rubbing the base of its tail between the palms of its fore-paws, an action that appeared quite inexplicable. In drinking it lapped like Dogs, and it was fed with bread and soup. When meat was given to it, it would tear it with its nails, and not with its teeth, to reduce it to small pieces. It had six teats. Before it came to the menagerie it enjoyed complete liberty, and would run through haylofts and stables in pursuit of Mice and Rats, which it caught with great dexterity. It would proceed also into the gardens in search of worms and snails.

This species of the Coati is commonly sent into Europe from South America, where it appears to be found beyond the boundaries of Paraguay. They unite in small troops in woods near the habitations of the people, and cause much mischief in the sugar cane plantations. Buffon has represented two varieties, under the names of the Brown and Blackish Coati, which Schreber copied and disfigured by the colours which he employed to illuminate them. Linnæus published a tolerably good figure of the Coati, in the Acts of the Royal Academy of Sweden, in 1768. Pennant's figure of the Brown Coati is not good, nor is that of Margrave much better. We give the figure of one of the individuals before described.

It appears to us, far from improbable, that the Coati (at least, as far as discoveries have hitherto extended,) forms but one species. In all essential points, as we have seen, the red and brown exactly agree ; and, certainly, the mere difference of colour does seem a very insufficient ground of specific distinction. Even when that difference of colour is invariable, it seems to us, that it argues no more than the existence of a distinct race, and not that of a distinct species. If colour be once admitted as a criterion of species, there is no knowing where to stop, and specific distinctions must be multiplied ad infinitum. The white Horse will be a different species from the black, and the brown from the bay. From the preceding accounts, it appears that the Red Coati was more savage than the others ; but this, most probably, was the result of his education, or rather of his want of education. The disposition was accompanied with no corresponding traits of conformation, and cannot be admitted as a specific character.

The next animal to be noticed in the Baron's sub-division of the plantigrades is the KINKAJOU or POTO. It is a native of South America, has been often seen in Europe, and described and figured by skilful naturalists, and, certainly, is one of the most singular animals in the long list of the mammalia. Its correct classification is a matter of no small difficulty, according to all the systems hitherto received. It has, by turns, been attached to the Carnassiers, the Plantigrades, and the Quadrumana ; while, at the same time, it was acknowledged that, properly speaking, it belonged to none of them. It is not, perhaps, going too far to say, that it seems to be the type of a new grand division, equal in rank and importance to any of those now mentioned.

To judge by its general physiognomy, and, in some points, by its natural disposition, it might be taken for a Lemur. But its organs of mastication and of motion, are different. Its toes, claws, incisors, and canines, approach it to the Carnassiers ; but then its molar teeth entirely flat, and its prehensile tail, have no relation to these animals ; and if some of its traits connect it with the less carnivorous genera of the Plantigrades, such as the Bears, and Coatis, its physiognomy, and many points of its disposition, render it altogether different. The most probable conjecture seems to be that, under one point of view, its natural place would be immediately after the Quadrumana, between which animals and the Carnivora, it might establish a new link of connexion, as the Galagos do between the Quadrumana and the Insectivora. According to this sort of analogy, the Roussettes would no longer exhibit an anomaly among the Cheiroptera: they might be detached from them to form an intermediate step between them and the omnivorous Plantigrades.

Thus, we see, successively realized, every possible combination of organic life, whose infinite forms run as naturally into each other, as the threads which compose the tissue of an immense and ingenious piece of net-work. Indeed, there is no subject of more curious speculation than the existence of these connecting links between the various orders, genera, and species of living beings.

Examples of this description, will be found repeatedly

referred to in these pages, and they exist in sufficient number to convince us, that though, perhaps, the Platonic notion may not be correct, of a gradual and unbroken degradation in living beings, yet that there is a sort of circular chain which binds the numerous branches of the family of earth together, and indicates, perhaps, their universal descent from one common origin.

The Poto has been brought into Europe several times, and has been as often described. These descriptions, joined to the numerous and interesting observations made upon its habits, by the Baron de Humboldt, constitute the entire history of this singular animal. Vosmaër, we apprehend, was the first who published a description of it, and gave it the name of Poto, under which denomination, a Mr. Broker informed him that he had received a similar animal from the island of St. Christopher. The name, however, to all appearance, is not American : no traveller mentions it, and the Baron de Humboldt, never heard this name pronounced in America. It seems to be, originally, an African name. Bosmann assures us, positively, that the Negroes give it to an animal of their country, which, according to his description, would appear to belong to the Loris. Thus, in all probability, the Negro-slaves transplanted the name into America, and this would not be the first instance of a similar importation.

As to the name of Kinkajou, which this animal has received from Buffon, and other naturalists, that seems as little to belong to it as the other. We find it in Denis' description of North America, as the name of a carnivorous animal, which climbs trees, where it remains in ambuscade, flings itself suddenly on the deer, and fastens on their necks, with its paws and tail, until it has destroyed them by sucking their blood. Such habits do by no means accord with the character of the Poto, which, besides, is an animal not found in North America. The resemblance of name, and the resemblance of instinct and tastes would lead us rather to believe that the term Kinkajou, is the same as that of Kareajou or Careajou, and that both belong to the same animal, the Glutton of North America.

The Baron Humboldt has informed us of some of the denominations of this animal, which vary, no doubt, according to the language of the tribes within whose territories the Poto may be found. The Musica Indians, in the Mesa of Guandiaz, call it the *Cuchumbi*; and in the mission of Rio Negro, it bears the name of *Manaviri*. Either of those names would be preferable to Poto or Kinkajou, which naturalists will be obliged, sooner or later, to restore to their right owners. We do not presume, however, to depart in this point from the established usage, and change a name which has been consecrated by custom, however erroneously we may think it applied. It would be vain, indeed, to introduce a reformation, in a single instance, without carrying it farther, and the attempt at a general reformation of the nomenclature of Zoology, would, on our parts, be a piece of presumptuous temerity. We may presume, however, so far as to express a hope that the day will yet arrive, when this noble science shall be purged alike of popular appellations, which have no meaning, and of technical synonimes, which are only calculated to mislead—when the grand reform, so happily commenced by our illustrious author, in the higher divisions of the animal world, shall be carried with equal felicity into the lower, and when each genus and species shall receive a single, simple, scientific name, derived from its conformation, and expressive of its most influential characters. Either Zoology is entitled to the rank of a science, or it is not—if it be not, then have many illustrious men bestowed much pains upon it, to little purpose; if it be, (and, surely, there is no study which has a better claim,) such a reformation as we have ventured to allude to, will be indispensable to its perfection. We are

not ignorant of the difficulties attendant on such an undertaking, but we also know that there are men in existence,
capable of overcoming them ; men, marked out by destiny,
to enlarge the boundaries of science, to subvert the empire
of error, and to remove the heap of rubbish, with which
ignorance, prejudice, affectation of learning, or pains-taking
stupidity has clogged the portal of the temple of nature.

The Baron's brother gives a description of a specimen of
this genus, in the " Menagerie Royale." It was young,
and a female. Its size was nearly that of the domestic
Cat, and its physiognomy remarkably like a Lemur's. The
fur, also, in its smoothness, softness, and thickness, bore
a considerable resemblance to that of the animal just mentioned. It was not so high, but had very much of the gait of
the Makis, especially behind, and without being, like them,
quadrumanous ; it walked on the sole and palm altogether.
Its movements were slow, and apparently difficult, except
when it sprung forward ; its jumps were then extremely
rapid and energetic. Its large eyes, almost directed forward, seemed to complete the resemblance ; but a detailed
examination of its organs, soon proved that resemblance to
be merely superficial.

The Poto has five toes, without any thing like a distinct
thumb, on all the feet, armed with pointed claws. These
toes are united to the second phalanx by a membrane of
small extent, and their relation, in a decreasing proportion,
is as follows : the middle toe, the annular, the index, the
little toe, and thumb, and, in this respect, the fore and
hind feet exactly resemble. The sole and the palm are entirely naked, furnished with thick tubercles, especially at
the base of the claws, and covered with a very soft skin.
On the hinder-feet, the thumb and index remaining almost
close together, seem to be separated, habitually, from the
three other toes. The tail is prehensile, and covered every
where equally with hairs. Its extremity is not naked, as

in the Aletes, so that it may be considered as an organ
of motion, without being one of touch. The eyes are
simple, and the pupil is round, but it contracts to such a
degree, under the influence of light, that its diameter is reduced to scarcely a quarter of a line. The ears are rounded,
have no lobules, and are very simple in their internal tubercles. The nostrils are small, open on the sides of the
muzzle, and bear a very close resemblance to those of a
dog. The tongue is narrow, thin, very smooth, and of a
most disproportioned length. This has led to the supposition of some new organic peculiarities in its muscles. In
other particulars, the organs of taste appear unaccompanied
by any accessory part. The fur is composed of silky and
of woolly hairs, but both are of the same length, and very
difficult to be distinguished from each other. These hairs
are rather short, very numerous, and so pressed against
each other, that their direction cannot be distinguished by
the eye. To the touch, they appear, along the back and
sides, to lie from rear to front. On the tail and limbs,
they pursued the usual directions, and all the parts of the
body are equally covered with them, except the anterior
part of the muzzle, the external conch of the ear, the sole
of the feet, and palms of the hands, which are naked. This
animal had no mustachios. There are but two teats, which
are inguinal. The incisives are six in number, in each
jaw, and the canines two. Neither possess any peculiarity,
except a few longitudinal striæ or grooves. There are five
molars ; the two first which follow the canines, after a
little interval, are, especially in the lower jaw, small and
pointed, and have all the characters of false molars. The
three following are tuberculous ; in the upper jaw, their
coronal is nearly rounded, and a circle and border of enamel surrounds it. On their external edge, however, two
tubercles are observable, which seem to be unworn remains of the tooth. Of these teeth, the middle is the

largest; the two others are of equal size. In the lower jaw the three tuberculous molars are elliptical. The edges of the first exhibit two points, but the two others exhibit only a smooth surface, surrounded and bordered with enamel. These three teeth are opposed crown to crown, like all triturating teeth.

The colour of the Poto is, in general, of a yellowish-gray, and this colour assuming a more golden tint, prevails on the breast, the belly, and the sides of the cheeks. The eyes are black; the ears and muzzle, violet-colour; and the soles of the feet, and palms of the hands, flesh-colour. The nails are whitish. All the hairs are gray in the principal part of their length, and yellow at the points. The tail, at its extremity, is of a more sombre shade than the other parts of the body, though the colours are the same.

This animal was extremely mild, and very fond of being caressed. It passed the entire day in sleeping, lying on one side, the head reclining on the breast, and covered around by the arms. When wakened from its profound sleep, it at first complained, seemed to suffer from the light, and continually sought to conceal itself in some obscure corner, or to shade its eyes from the light. By caresses, however, in a little time, it would be induced to play, but the moment they ceased, the necessity of sleep and obscurity overcame it. As soon as the day declined, it would awake by slow degrees; at first it advanced a few paces in an irresolute manner, uttering a bleating sound, and putting forth its excessively long tongue. Presently, it would drink, lapping like a dog; and, at last, take its food, which consisted of fruit, bread, and biscuits. It sometimes ate meat, but it preferred vegetable nutriment. It took its provisions, at times, with its lips, but more usually carried them to its mouth with its fore paws. It climbed trees dexterously, and descended, catching with its hind legs, after the man ner of the Coatis, and completely turning back the foot. This action presupposes a very peculiar conformation in the bones of the leg. It often made use of its tail to prevent falls, and even to draw objects towards it, which it could not reach with its hands. Its voice, when calm, consisted in a little hissing, very soft; but it could utter stronger cries, like the barking of a young dog.

The celebrated and scientific traveller before mentioned, the Baron de Humboldt, tells us that the Poto makes use of its long tongue to suck honey, and that it is a great destroyer of the nests of wild bees. The missionaries, accordingly, have given it the name of the *Honey-Bear*. He adds, that this animal was formerly among the number of those reduced to a domestic state by the aborigines of the temperate parts of New Grenada.

Pennant calls this animal the Yellow Macauco.

The BADGER comes next; and though but a single species, yet constitutes a sub-genus in itself. It possesses a system of organization exclusively peculiar. No other species can, with propriety, be placed along side of it. We might imagine that it was withdrawn from all the ordinary influences which operate on animal life, by some particular and inexplicable power, impelling it beyond the common laws of nature, and we might be tempted, in this instance, even to accuse nature of impotence or irregularity, had we not learned rather to distrust our own conjectures, than to doubt of the power, the wisdom, and the infinite benevolence of the Creator.

But if the Badger be isolated, as a species, it yet enters very naturally as a genus or sub-genus of our author, into the series of those animals which are characterized by a tuberculous molar at the bottom of each jaw. These molars, in the Badger, are distinguished (the upper ones more especially) by their extent, the effect of which is to limit that of the carnivorous teeth, and, consequently, to diminish the animal's appetite of flesh, and his faculties of using it. In fact, the tuberculous molar of the upper jaw, occupies a space equal to that of the carnivorous molar, and of the two false molars which precede it: from whence it happens that the lower half of the under carnivorous tooth is augmented, that it may be properly opposed to the large tuberculous tooth above. This makes it, in fact, one-half tuberculous, and one-half carnivorous. The Badger, besides, has two false molars in the upper jaw, and four below. But the first of the latter, is merely a rudiment. The canines and incisives are similar to those of all the other genera of the Marten family, *i.e.*, the first resemble the canines of all the Carnassiers, and it is the same with the upper incisors. The middle incisor, in each lower maxillary, is not inserted on the same line as the two others, but much farther in. It is only parallel with them in the extremity of its crown, and on this account it projects more forward.

Although the Badger approaches the Martens by its system of dentition, it is far from resembling those fine-formed, light, and lively animals, in which particulars, probably, no other family of the Mammalia can equal them. It is on the contrary, heavy and gross; its body is thick, its movements slow, and its physiognomy announces neither promptitude of intelligence, nor vivacity of passion. Accordingly, we find that it leads a most gloomy and solitary life.

It is an animal entirely plantigrade, with five toes on each foot, united almost to their extremity by a thick membrane, not much susceptible of extension, and armed with digging claws, extremely strong. On the two feet, the second and third toes are equal, and the longest. The first and fourth come next, and the internal one is the shortest. The sole of the fore-feet is furnished with tubercles, very thick, and covered with a soft skin, one at the exremity of each toe. Three others, disposed in the form of a trefoil, are found on the middle of the sole, which is terminated behind, on the fore-feet, by a single tubercle, and on the hinder, by two contiguous ones. The tail is short, and rudimentary.

The organs of sense shew very little development. The eye is small, with a round pupil. The third lid is large enough to receive the cornea altogether. The ear has an external conch of small extent, and very simple within. The antitragus, alone, is remarkable for a thick tubercle, in the form of a semi-circle, which occupies its whole extent, transversely. The nostrils are surrounded by a very developed muzzle, composed of strong glands. They consist of two sinuses, large in front of the mouth, narrow on its sides, and the orifice of the olfactory conduit is found in their anterior part. The tongue is oblong, large, and altogether covered with very small papillæ, pointed, and even a little horned, but soft enough to make it appear smooth.

The fur is copious, and the haris very long on the body ; they are much less so on the head, the limbs, and the lower parts. They are both silky and woolly, yet they differ little from each other; the first are the longest, and the hardest. The mustachios are very small.

Immediately under the tail is a large transversal aperture, which conducts into a naked cavity terminating in a *cul-de-sac*. The sides of this cavity, though no glandular apparatus is visible, are yet covered with an unctuous matter. Underneath is a second and smaller pouch, in the midst of which the anus opens, and on each side of the anus is a tolerably large pore from which an unctuous matter escapes of a yellowish colour, and a most intolerable odour. The Badger has six teats.

Its colours are remarkable in their distribution. The head is of a slightly reddish white, divided on each side of the muzzle by a black band, which originates on the upper lip, takes in the eye, and terminates at the ear. The white of the head extends over the sides of the neck and terminates on the anterior half of the lower jaw. The top and sides of the body are of a dirty gray, which grows paler towards the flanks. But this colour is not uniformly spread. The black and white which compose it are disposed in spots or rather in confused and irregular masses. The throat, the under part of the chest, the belly, legs, and feet are of a deep black brown. The posterior part of the abdomen is of a reddish white, and the tail of a whitish gray. The ear is black, bordered with white above; and all the naked parts are of a tan-colour more or less deep.

On the parts whose colours are uniform, the hairs have but a single tint. On the gray parts the silken hairs are generally white with a black ring in the middle, and the woolly hairs white with a yellowish point.

The usual length of the Badger is about two feet and a half, and the tail six or seven inches.

The Badgers pass a great part of their time under ground in burrows which they dig with much dexterity. Two young Badgers were seen at their work by M. F. Cuvier; they were caught in the burrow of their mother, and placed in a fenced yard. They soon unpaved it, and made a burrow, where they passed an entire year, never quitting it, except by night, to take the food which was placed within their reach. From this, they were transferred into a moat, surrounded with walls, in the middle of which was a large mound of earth. These animals first sought all round the walls for a place in which they could dig. Having discovered an empty space between two stones, the upper of which was projecting, they tried to increase it; but as it was rather elevated, and they were obliged to stand on their hind-feet to reach it, it was with much difficulty that they tore away the plaster and stone which they wanted to get rid of. The male would then several times lie down at the foot of the wall, and the female mount upon his body to reach the hole more easily, which she was trying to augment. When they found that all their efforts were useless, they recommenced operations under another large stone, the only one in the place beside the former, which projected; but here they encountered a resistance which they could not overcome. Tired of their vain attempts on the side of the walls, under projecting stones, they turned their attention to the mound of earth, and worked, the female especially, with uncommon ardour and perseverance. At first they made little trenches or excavations all about this mound, and fixed themselves exactly opposite the place where they had made their second attempt against the wall. They commenced by removing the earth with their nose, then they made use of their fore-paws to dig and fling the earth backwards between their hind legs. When this was accumulated to a certain point, they threw it still farther with their hind-paws; and finally, when the most distant heap of earth impeded the clearance they were making from the hole, they would come walking backwards to remove it still farther, making use both of their hind and fore-paws in this operation, and they never returned to work at their burrow until they had completely removed this heap of mould out of their way. One of these animals would often lie down by the side of the other when it was digging, and seemed to annoy it as much in its labours as its own repose must have been disturbed by its coadjutor. During the night the burrow was finished.

According to the report of hunters, it appears that the Badgers furnish the bottom of their habitation with dry and soft substances of which they make a bed, and which they carry between their paws. This habitation is not a simple cavity, for excavations are found within its sides into which the animal also retires. Frequently many individuals are found in one burrow, and there is reason to believe that the male and female always inhabit together. These animals quit their retreats only in search of prey, and in winter they are many days without appearing, not that they fall into a lethargic sleep, but they are afraid of the cold, and being very fat, are not much pressed by hunger. Summer is the season of parturition, and the females bring forth two, three, and sometimes four little ones at a birth. They take great care of them, and sometimes fetch them to enjoy the sun on the edges of their burrows. From their second year the young Badgers can reproduce, and their life is probably extended to twelve or fifteen years. The Badger is carnivorous, but less so than the Dog. It will eat bread, fruits, &c., and is easily tamed. It lives and plays familiarly with Dogs, comes when called, follows the person who takes care of it, and soon learns to know him. They are found in all Europe as far as Norway, and in a great part of Asia. It is also probable that they exist in America. Buffon, in his Supplements, published one under the name of *Carcajou*, from that country; but that animal appears to belong to the genus Gulo.

Hunters distinguish two species of the Badger, one they call *Hog-Badger*, and the other *Dog-Badger*. But they only differ in some trifling particulars and slight shades of colour. One kind, say they, is deeper coloured than the other, and one digs more willingly in earthy soils, while sandy grounds are more to the taste of the other. These observations are far from being ascertained to be just, and even if they were, they could only give rise to two varieties of no importance.

The Badger does not appear to have been known to the Greeks; at least it has not been discovered that any author of that nation speaks of them. The Latins, on the other hand, have expressly designated them by two different names, *Meles* and *Taxus*. The last of these names has been adopted by naturalists for the genus, the first for the species. Thus the Badger is the *Taxus Meles* of methodical catalogues.

There is at present, in Mr. Cross's collection at Exeter Change, an albino variety of the Badger, of an uniform pale yellow colour, with red eyes.

Four species of the Linnæan families Viverra and Mustela have been separated by more modern methodical writers, and formed into a distinct genus. To these others have been added, which altogether compose the group or subgenus GULO of our author.

The teeth, and consequently the regimen and consequent habits of this group, are much more nearly allied to those of the Weasels than of the preceding subgenera ; but they bring the heel to the ground in walking. They may therefore be treated as intermediate between the digitigrades and plantigrades, possessing most of the physical powers and mental impulses of the former, combined with the mode of locomotion peculiar to the latter.

They have six incisors in each jaw, with a strong canine tooth on each side. There are five or four cheek-teeth on each side in the upper jaw, and six in the lower. The two first in the upper jaw when there are but four, and the three when there are five, are small unicuspidatous teeth, and may be called false carnivorous teeth, increasing successively in size ; the following or carnivorous tooth is large and strong, furnished with two points on the inner side, and a trenchant edge in front. The last is a small tuberculous or flattish tooth.

In the lower jaw the four first are false, presenting each but one point or edge ; the fifth is long and large, presenting two trenchant points ; the last is nearly flat. All the teeth touch each other successively. Here then we see a decided departure from the flattish surfaces of the cheek-teeth of the preceding divisions, a vestige of which only may be said to be found in the last cheek-tooth of this. The proportions of these flat to trenchant surfaces bespeak, almost with mathematical accuracy, the sort of food proper to the animal, and all those traits of character both physical and moral dependant upon that fact.

The head of these animals is long; the ears short and rounded, and the tongue smooth or rough in the different species; the legs are very short, and the body in general so elongated as almost to qualify them for the epithet vermiformed, which has been bestowed on some of the following divisions. They have no anal pouch, but a sort of vestige of it in some slight folliculi or folds of skin.

They have five toes, deeply divided, terminated by long bent nails, which rather approximate them to the foregoing than the succeeding divisions, as they seem better calculated for digging than offensive warfare.

The *Common Glutton* (*Mustela Gulo* of Linnæus, and *Ursus Gulo* of Gmelin) has the muzzle as far as the eyebrows black; the eyes small and black; the space between the eyebrows and the ears white mixed with brown; the lower jaw and the interior of the two feet are spotted with white; the legs, tail, back, and belly, are brownish black, but the sides of the body from the shoulders to the tail are maroon colour. There is no tubercle on the heel, which barely comes in contact with the ground when the animal runs. The body altogether is heavy. It is about the size of the Badger. It does not become torpid in winter.

The voracity of this animal, though excessive to a high degree, has been greatly exaggerated. To minister to this voracity it appears to have recourse to expedients in aid of ts physical lack of powers, which seem otherwise to be inferior to its wants. When a sufficient supply of small quadrupeds and birds cannot be procured, it is said to conceal itself on the horizontal branch of some tree, from which it will drop on Deer, even Horses, or other animals that may pass beneath, holding its situation and sucking their blood, till faintness and loss of blood sink them a complete captive to its voracity. When tamed, it has been said to have eaten thirteen pounds of flesh in a day.

The Common Glutton inhabits all the Arctic regions, Norway, Canada, and the uncultivated parts of the United States, where it is well known by its depredations on the magazines of provisions provided by the Indians.

Edwards describes the Quickhatch or Wolverene, which appears to be no other than the American variety of this species. His specimen had lost an eye, and from this trivial circumstance Linnæus applied to it the specific epithet of *Luscus*. The American Gluttons seem to be paler in colour those of the Old World.

The *Grison (Viverra Vittata*, Lin.) is one of the few animals that have the fur of a deeper colour underneath than on the back. The head, from between the eyes, top and sides of the neck, and back, crupper, flank, and tail, are of a pale gray, each hair being coloured alternately black and yellowish white ; the muzzle, lower jaw, under part of the neck, paws, and belly are black ; a pale gray or whitish line from each side of the head, which goes from between the eyes, passes over the ears to the sides of the neck.

The body is elongated, and its step is more decidedly plantigrade than in the common species. The toes are semi-webbed ; the ears are small ; the tongue rough ; the eye-pupils round ; and it has slight mustachios. The fur is of two sorts : woolly, of a pale gray colour, and silky black or annulated black and white, longest on the back, flanks, and tail. It measures about eighteen inches from nose to tail, which is about six inches long.

It is a very ferocious little animal, killing and devouring small quadrupeds, birds, reptiles, &c.; and although capable of domestication, to a certain degree, and apparently docile, when confined and well supplied with food, it will never fail to evince the sanguinary cruelty of its nature, whenever a less powerful creature falls within its reach.

It inhabits South América, especially Paraguay, where it is very common; Buenos Ayres, and the vicinity of Surinam, where, however, it is scarce.

An individual, possessed by M. F. Cuvier, had, notwithstanding its natural ferocity, been tamed to a very considerable degree. It appeared to recognise no person in particular, but it was fond of play, and, for that purpose, all, comers were alike to it. It seemed to derive pleasure from being stroked down the back with the hand. When invited to play, it would turn over, return with its paws the caresses addressed to it, bite gently the fingers it could seize, but never so as to hurt or wound them. One might almost have imagined that it felt the degree of resistance which the skin was capable of making, and proportioned the force of its bite accordingly, when it meant only to express its joy. It knew the fingers of a person without seeing them. Nevertheless, this animal preserved its ferocity for all those living beings that could become its prey. Even when satiated with food, it testified, in a lively manner, the desire of getting possession of such animals. One day, it broke the bars of its cage to attack a Lemur that was within reach, which it mortally wounded. When it could catch a bird, it killed it directly, and laid it by for provision, as was its custom to do with the meat it received, when it had eaten sufficiently.

The *Taira* or *Galera* of Brown's Jamaica, (*Mustela Barbara*) is about the size of the Common Marten, to which, also, it bears a similarity, in general form. It is this species which is said to have but four cheek-teeth in the upper jaw, on each side. The fur is uniformly black, except that there is a large white patch covering the under part of the throat.

C. Hamilton Smith Esqr. del.
Mas Paris.

London Published by G. B. Whittaker Decr. 1824.

Brown sc.

THE TAIRA.

MUSTELA BARBARA L.

D'Azzara, who describes this animal, at considerable length, under the name of the Great Weazle, gives us no account of its habits. It is said, however, to dig a retreat for itself in the earth, and, as to its general habits, to resemble the Grison. It is an inhabitant of South America.

There is a specimen, in the French Museum, drawn by Major Hamilton Smith, and which, by his kindness, we are enabled to present here.

The *Rattel* is another species of Glutton, with a thick and heavy body. The head is moderately large, devoid of external ears; the tongue aculeated; the fur composed of rough long hairs, ash-coloured on the forehead, upper part of the head, neck, shoulders, back, and tail. It is black on the muzzle, round the eyes, on the lower jaw, the ears, under part of the neck, belly, breast, thighs, and legs, the gray colour being separated from the black by a brighter gray line, about an inch in width.

This species seems to have been considered as exclusively African, and is designated, by Linnæus, under the addition Capensis; but General Hardwicke gives us an account, in the ninth volume of the Linnæan Transactions, of an animal *proper to India*, which we shall, in part, subjoin, presuming it to be the same as the species in question.

He says, " The claws are unequal, those of the fore-feet very long, and awl-shaped; the three middle ones much longer than the two lateral; the interior toe very remote from the rest; the claws of the hind toes remarkably short, nearly equal, and bearing no comparison to the strength o the fore-feet.

" This animal is found in several parts of India, along the courses of the Ganges and Jumna, in the high banks which, in many parts, border these rivers. It is rarely seen by day, but at night visits neighbouring towns and villages, inhabited by Mahomedans, and scratches up the recently-buried bodies of the dead, unless they are quickly covered with thorny bushes.

" The natives, when encouraged by the expectation of purchasers, dig the animals out of their subterraneous retreats, and take them alive. The full-grown ones are with difficulty secured, and seldom bear confinement long, but roll and beat themselves about till they die. When taken young, they are very manageable, docile, and playful. It is a bold animal ; its hide remarkably thick, and its strength too much for most dogs of common size. Its general food is flesh, in any state; but it is remarkably eager after birds, and living rats seem almost equally acceptable. It has an inclination to climb upon walls, hedges, and trees : this, however, it seems to execute clumsily ; but seldom falls, and will ramble securely upon every arm of a branching tree, that proves strong enough to bear its weight, without much motion. This species burrows with great facility, scratching the earth like a Dog, with the fore-feet, and expelling the loosened soil to the distance of two or three yards backward. In ten minutes, it will work itself under cover, in the hardest ground, and is restless till it can form such a retreat to sleep in. It sleeps much by day, is watchful during the night, discovering inquietude, by a hoarse call or bark proceeding from the throat. The hair is short and wiry, nor has it any of the softness of fur. It is known to the natives of Hindostan by the name of *Beejoo*."

The *Nyentek* of the Javanese, the *Gulo Orientalis* of Horsfield, has been described by that eminent naturalist, in his Zoological Researches, and beautifully engraved by Taylor. It seems to be confined to some of the mountainous parts of Java, and is very rare, the Doctor never having been able to see it alive. It is somewhat smaller than the English Polecat, and is rather more slender in the body than the Gluttons in general. The fur is thick, consisting of long hairs, closely arranged, silky at the base, of a brown colour, and somewhat glossy, with a slight tint of reddish-brown. In certain lights, it appears diversified with grayish and tawny. This fur covers the greatest part of the body and head, and

T. Bradley. sc

THE MASKED GLUTTON.

GULO LARVATUS — Tem.

London. Published by G.B. Whittaker. June 1825.

C. H. Smith Esqʳ Delᵗ

Mus. Amstrodam.

the whole of the tail and extremities. The sides of the head, the neck, the throat, breast, and a broad patch on the top of the head, which passes gradually, decreasing in breadth, to the middle of the back, are white, with an obscure tint of Isabella-yellow, of different degrees of intensity. This colour, also, exists less distinctly in a longitudinal band, along the lowest part of the abdomen. The tail is nearly half the length of the body, is somewhat bushy, and terminated with long bristly hairs.

To these described, admitted species, we must add some notice of two drawings, in our possession, of animals which have been referred to this sub-division. In doing so, however, here, and in many other places, in which we may think it necessary to insert figures from our collection, without having had the opportunity of inspecting the subjects whence they were taken, or examining the character of their dentition, by which alone the species may be ascertained, we cannot but express a hope that our motives will be properly appreciated. Possessed of an extensive collection of many figures, which cannot, by mere superficial detail, be referred with certainty to particular groups, it seems, nevertheless, an unnecessary fastidiousness, injurious, perhaps, in some measure, to the cause of science, to withhold them. Whatever, therefore, is not said of them positively, must be taken conditionally, and their location in particular, in the Cuvierian system, is merely presumptive, and subject to investigation.

The first of these is from an animal in M. Temminck's celebrated museum, and is named by him, *Gulo Larvatus*, the *Masked Glutton*. It is larger and longer than the Polecat. Its colour is a mixture of olive-brown and gray, but the end of the tail and the feet are black; the ground-colour of the head is black, but a white streak passes down the forehead to the nose; there is also a whitish circle round each eye, and a pale band passes round the throat from ear to ear.

The second was in Mr. Bullock's late museum, and was referred to this sub-genus. It may be called, conditionally,

the *ferruginous Glutton, Gulo Castaneus.*   It measured
nearly four feet, from the nose to the end of the tail, which
was two-thirds the length of the body.   It was long, slender,
and vermiformed like the Weasels ; but the limbs were extremely robust.   The head was broad, and-depressed ; the
eyes were very near the nostrils ; the ears were far back,
and the whole appearance of the animal strongly indicated
a predacious and savage nature.   The fur was long and
rough, of a dark brown and- chestnut-colour mixed ; the
tail was nearly black, and the feet sepia.   The habitat and
manners of this animal were entirely unknown.

THE DIGITIGRADES.

WE are now arrived at the second tribe of the Carnivorous
family, distinguished by their quick and light mode of locomotion on the extremities of the toes instead of the whole
sole of the foot, from heel to toe, in the manner of the
Plantigrades.

This mode of arrangement, like every other human invention, is imperfect, inasmuch as we observe certain species, which by their general analogies, must be placed with
the Digitrade Carnivora, still to approximate the Plantigrade mode of walking.

Activity, and, consequently, the digitrade step, is a necessary ingredient to the perfection of carnivorous regimen.
All the species, therefore, in this tribe are more exclusively
flesh-eaters than the preceding tribe of the third, or Carnivorous family, of the order in question.

The four first sub-divisions of the Digitigrades of our
author, distinguished by a single tuberculous tooth at the
back of the upper jaw, form a group which may be conveniently contemplated by the English reader, under the name
of Weasels, from which the true Viverræ are removed.

It is extremely interesting to trace the progress of Nature,
in all her works, as she inclines from one state of things,
through various and almost imperceptible gradations, to

Griffith sc

THE FERRUGINOUS GLUTTON;

GULO CASTANEUS.

London Published by G.B.Whittaker March 1827.

another. The first dawn of animal life is so nearly allied to vegetable existence, that we are puzzled in concluding which to call it: organization improves, and the semivegetable zoophites are exchanged for others, in which animal life assumes a more decided form: we then pass, imperceptibly, by an infinite number of species, linked, as it were, in some one or more particulars, one with another, through the insects and worms, mollusca and crustacea, to the osseous animals. Here again, as with the rest, nothing is constant but inconstancy; no two species are alike; and, although many may be found corresponding almost altogether in construction, faculties, and pursuits, yet they will differ from each other relatively to the means bestowed on each.

Among the flesh-eating animals, the Felinæ and the Hyænas (to be treated of shortly) may be considered as purely or perfectly carnivorous. Their powers are more or less calculated for offensive warfare, and their teeth are not adapted to the mastication of any other than animal food.

The various species hitherto known by the name of Weasels, with the exception of a few, stand, in this respect, next in order among the carnivorous quadrupeds, since the physical character of the teeth shows, that they are destined to seek in flesh their principal aliment; though a slight departure from the carnivorous form indicates a corresponding approach to the substitution of a vegetable diet. Their disposition, nevertheless, is extremely cruel; but from inferiority in size and powers, they are capable only of an inferior degree of mischief.

They have a large, perfect molar tooth, placed behind the carnivorous teeth, in the upper jaw. The other cheek-teeth also, although they have cutting or carnivorous lobes on the outer side, are more or less tuberculated on the inner; a character, which indicates a slight approach to the use of a vegetable diet, as it enables them, though in a small degree, and very clumsily, to masticate this sort of food.

The last or molar tooth takes a direction inwards with

the other cheek-teeth, and exposes a very large and flat surface. The reversed figure of the upper jaw, in the opposite
plate, is intended to exhibit it. The corresponding tooth
to this, in the Cat tribe, which we shall take occasion to
call the auxiliary carnivorous or cheek-tooth, is much
smaller, is placed more on an inclined plane in the mouth,
and seems destined to receive the cutting edge of that op
posite to it; whence it cannot act as a molar or grinding
tooth, but merely as facilitating the cutting operation. But
the large flat tooth of the Weasels is met by a corresponding flat surface, in the opposite teeth of the lower jaw, the
last of which is small, and perfectly flat; the third lobe,
also, of the last but one, or largest, is flat, and both these
flat surfaces are brought into contact with the opposite flat
tooth before described; so that if any substance be placed
between them when the mouth is about to close, it will be
squeezed or pounded; while any thing placed on the flat
tooth of the Felinæ would be exposed to the action of the
cutting edge of its opposite, and consequently be divided,
and not pounded.

The Weasels are very slender and long, and possess a peculiar pliability of body, which enables them to pass through
very narrow and winding apertures; whence they are called
vermiform animals, and a *verminium genus* by Ray.

The head is small and oval, and the forehead flattish;
the jaws are rather short; the external ears are short, and
rounded; the tongue is nearly smooth; their legs are very
short in proportion to the length of their bodies, having
five toes before and behind, armed with strong, curved,
acute claws, which, in many of the species, are very slightly
retractile. The tail is of a moderate length. They have no
glandular pouch near the anus, for, we must recollect the
true viverræ are not included, but they have some small
glands placed there, that secrete a fatty substance, which
has a strong, and, to many, a very disagreeable odour,
although it is highly prized by others.

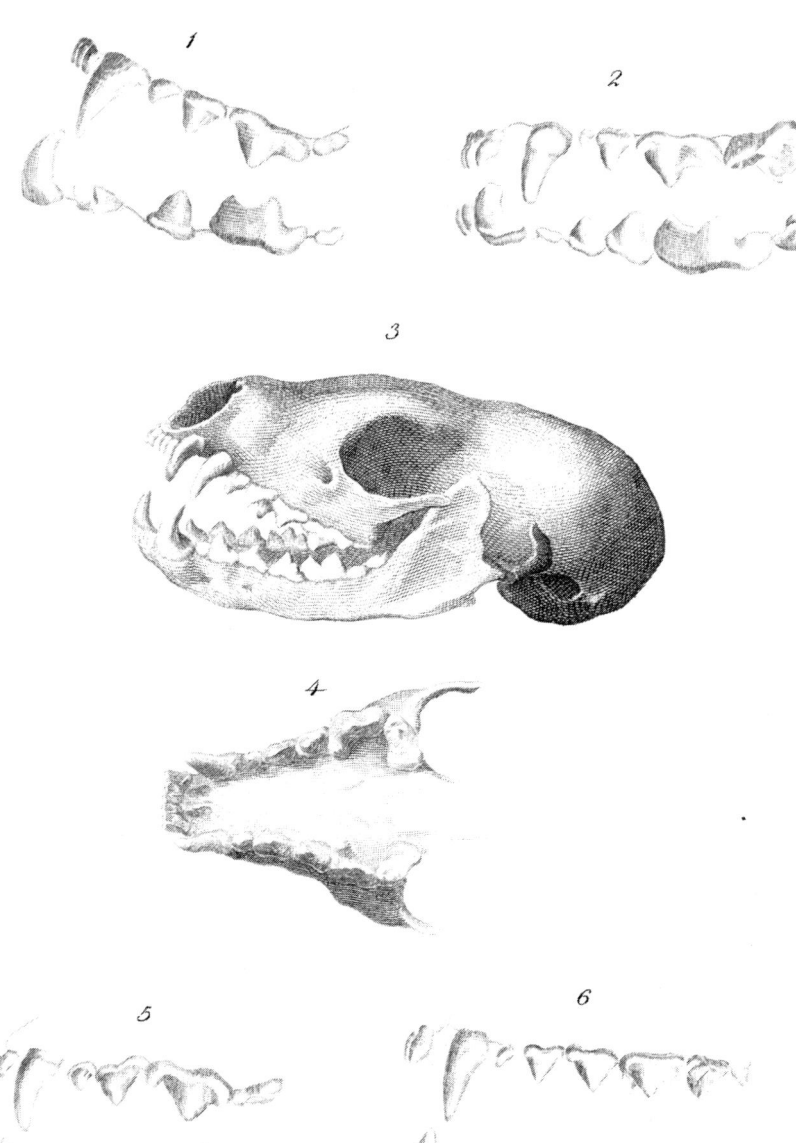

1 Polecat.

2 Mephitic Weasel.

3 Martin.

4 Martin Upper jaw reversed.

5 Otter.

6 Civet.

London. Published by G.B. Whittaker March 7.1827.

Their teeth, however, differ amongst the several species, and seem to indicate that they are not all equally carnivorous; some groups being found to vary from others in this particular. Hence, as we have seen, they are divided by the number and conformation of the cheek-teeth, into the sub-genera of the *Putorii*, or Polecats; the *Mustelæ*, or Martens; the *Mephites*, or Mephitic Weasels; and the *Lutræ*, or Otters. The viverræ are classed with the Dogs, on account, also, of a similarity of the teeth of these animals with those of the canine genus.

The first of these, the PUTORII or POLECATS, have no tubercle on the inner side of the carnivorous tooth in the lower jaw; the tuberculous tooth in the upper jaw, very long; two false cheek-teeth in the upper, and three in the lower; and the muzzle shorter and thicker than that of the next sub-division, or Martens.

The fur of the *Common Polecat* is of two sorts, the one long and shining, of a brown-black colour; the other, silky, short, and yellowish, or fulvous-white. Hence the animal is brown on those parts most furnished with long hairs, as on the back, &c.; and yellow, where the other sort most prevail, as on the belly; the legs and tail are black-brown; round the mouth, at the corner of the ears, and on the forehead, there is some white; the body is about eighteen inches long; the tail, six or eight inches; the head is rather shorter than that of the Marten, and displays additional powers of jaw; the ears are small and round; the eye is small, and the eye-pupil is elongated transversely; the tongue is aculeated; the feet are pentadactylous, and the toes are semipalmate; the thumb is very short, and the third and fourth finger, of equal length, are the longest. The noxious vapour of the Polecat proceeds from a yellowish viscous secretion, produced by glands on each side of the anus.

From the habits of this species, it is extremely destructive, as every animal it can conquer falls a victim to its appetite. It is strong and active; and, by bringing all the feet near together, and drawing the back into an arch, springs with great force on its intended victim, which it generally kills expeditiously, and with a single bite on the head, making a wound scarcely perceptible. Its facility of passing through a small hole, enables it to get admission to outhouses and barns ; and if no sufficient aperture be found below, it is in general able to find and reach one on or under the roof, to enable it to proceed to its cruel office of devastation within, or to lie concealed till a fit opportunity offer for its predatory operations. Its work of destruction is also frequently more extensive, from its habit of sucking the blood, and leaving the carcass of its prey until it can find a convenient time for dragging it to its hiding-place. Even the finny race is not secure from the attack of the Polecat; fish-ponds are exposed to its depredations, as well as poultry-yards, dairies, warrens, preserves, and beehives; the hole of a Polecat has been found to contain the mutilated remains of a number of eels.

It either takes possession of a rabbit-hole, or prepares for itself a subterraneous retreat, which is in general found to be protected by the ramifications of the roots of a tree ; a practice apparently originating in that degree of intellect and foresight, which the Creator has so remarkably bestowed upon all the races of animals, when necessary for their preservation or propagation. Here, or in some secure hiding-place, under a hay-stack, or in a barn, or outhouse, the female produces her young, generally five or six in number at a time, which she accustoms, when very young, to suck blood and eggs.

When disabled, irritated, or dying, the fetid smell from this animal is almost insupportable ; and the place where it is destroyed will not lose the scent for a considerable length of time. It is very tenacious of life; and a scuffle with a Polecat should be conducted with caution. If not able to escape from a man, it will, in desperation, attack him;

and when no longer capable of either, will seem to show malevolence, even in death, by emitting its offensive vapour.

The name by which the Romans designated this animal does not seem to be certainly known. Its modern scientific epithet, Putorius, is very applicable and descriptive, though others of its American congeners seem more eminently entitled to the distinction.

It should seem that the *Ferret* is a southern or mere albinose variety of the Polecat, at least if the generative faculty be made a test of variety, for they will produce an offspring partaking of the appearance of both. The Ferret is smaller than the Polecat, and is of a uniform lighter colour. The eyes are red.

It does not appear to be indigenous either in France or England, but to have been imported from the northern parts of Africa, as reported by Strabo; for when a Ferret is lost here, as is very common in the chase in summer, it is generally understood that it does not survive the following winter. As we are enabled to turn its sanguinary inclinations and predacious habits (which are, perhaps, not much inferior to those of the Polecat) to our advantage, the Ferret is fostered and preserved by art in our climate, which would soon destroy it if left, to nature.

It is bred in this country in casks or boxes, and fed on bread and milk, with flesh occasionally, to encourage its carnivorous appetite. In this state it is trained to enter the burrows of Rabbits and Rats, being previously muzzled, both to prevent its destroying the game, and to anticipate its habit of lying down to sleep in the burrow after being saturated with the blood of its victim, from which even smoke will not always rouse and remove it. Its appearance alone is sufficient to drive out the terrified tenants of these retreats, which are caught in purse-nets fixed before their holes; or if they escape this snare, become exposed to the attack of men and dogs without.

This animal seems to be the least domesticated of all those

that have submitted to the power of man. It seems not even to know, much less to have any affection for its master; if he calls it, the animal neither answers nor pays any attention; if he caresses, it exhibits no gratification; and if it once succeeds in escaping from its confinement, it never will, as many others do, return to it again. Time and habit, which act generally on the individual or on the race of other domestic animals, never soften the natural unbending character of its disposition, nor has it ever been brought to any other use than that of destroying Rabbits and vermin.

A breed is produced between this animal and the Polecat, which is much prized for the chase of Rabbits, Rats, &c.

The females are smaller than the males, breed twice a year, and generally produce five at a time.

The *Javanese Ferret*, named conditionally by M. Cuvier *Nudipes*, has the same system of dentition and of senses as the Polecats, and differs only in the sole of the feet and colour of the fur. The former is more naked; the latter is bright golden yellow, except on the head and at the end of the tail, which are yellowish white. These perhaps may not amount to specific differences.

The *Sarmatian Weasel* is shaped like the Polecat, but differs materially from that species in the colours of the fur, being yellow brown spotted with yellow and white patches It measures about fifteen inches from nose to tail, which is about six or seven more.

Pallas states that this animal is never seen to drink unless it be of the blood of its prey. He gives an ample description of its manners, &c., which, as they accord very much with those of its congener, the Common Polecat, we shall, for brevity's sake, not notice further than by observing, that it is the relentless enemy of almost all animals it can conquer, and that it lies in a hole during the day of which it has either dispossessed another animal or dug for itself. When irritated, it emits a scent and erects the fur in the manner of a Cat. Its beautifully spotted skin is at present much in request among the furriers.

The *Common Weasel* is one of the smallest of this numerous race, but is the most extensively diffused over the earth's surface. Its general length is about seven inches, with a tail measuring two and a half. It is of a pale red or yellow brown colour, whiter beneath. It emits an offensive odour, in common with many of the tribe, but it is an elegant little animal. It feeds on Field-mice, Birds, &c., and will attack animals larger than itself. It lives under roots of trees or in banks, and will run up a wall in pursuit of its prey. Its body, altogether, is extremely flexible. Though a ferocious little creature, there are instances of its being perfectly tamed.

There is a variety, which has been described by Linnæus, in his Fauna Suecica, as a distinct species, under the name of *Mustela Nivalis*, which has since been classed sometimes as a variety of this, and sometimes of the Ermine; but although it has the colour of the latter, it seems to be more nearly related to this, because of its being of the same size; and the black hairs which terminate the tail are much fewer than those of the Ermine, the black also is of a different tinge. It breeds twice or thrice a year, generally brings forth four or five at a birth, and deposits them on a bed of dry leaves in a hollow tree.

The Stoat or Ermine, like the common Weasel, is widely spread over the earth, and is found in America, as well as in Europe and Asia. It is in general about ten inches long, and the tail is half the length of the body. This animal exhibits, in a remarkable degree, a peculiarity which is proper also to a few others: the whole upper part of the body is of a red brown colour, during summer, but this vanishes in winter, when the upper part becomes perfectly white, and the belly yellowish. The tip of the tail is at all times black. In the latter state it is called the Ermine, when the fur is greatly esteemed and in much request, particularly for ornamenting habiliments of office and dignity.

This extraordinary mutation of colour, however, is nearly confined to those individuals that are met with in high latitudes, as in Norway and Siberia : but the Stoat, which is very common in England, is seldom found white in this country, though in the winter it is occasionally seen here, and more frequently in Scotland, in a sort of intermediate condition, appearing to be assuming the pure white dress, yet as if the force of the cause, whatever it is, were insufficient to do its office completely. Buffon had one of these animals brought to him in its white dress. He kept it confined, and observed the change of colour, which commenced early in March, and was completed by the 17th of that month. The animal died afterwards, in consequence, as is stated, of substituting milk for its diet instead of flesh.

It is surely a task well worthy the attention of the physiologist, to ascertain the exciting cause, as well as the mode, by which this strange operation of nature on an animal body is produced, We have data enough to conclude, perhaps, that it is the result of climate, since it is observed in those animals chiefly, if not wholly, which are found in the Polar regions ; we know also that the new colour is produced together with a new coat or fur, but why the new fur should reflect rays differing from the old is not explained ; in short, we seem as yet to be quite in the dark as to the excitement, the mode of operation, and the object intended.

The habits of the Stoat correspond in general with those of the Polecat, though, being smaller, it is less capable of mischief. It will eagerly attack a rat, and soon overcomes and kills it by an almost imperceptible wound in the head or throat.

Captain Ross, in his voyage of discovery, found an Ermine in lat. 73° 37', where it must necessarily have been exposed to an intense degree of cold.

The *Siberian Weasel,* or *Chorok,* is described by Dr. Pallas. It resembles the Polecat in size, form, and proportions, but differs in its colour, which is of an uniform bright yellow, with the face brown and the nostrils white. Its name indicates its country, where it is an inhabitant of the mountainous forests. It is said to eat vegetables as well as flesh.

The *Water Polecat,* called also the *Smaller Otter,* the *Mink* or *Norek,* is found on the banks of rivers, in the north and east of Europe, from the Arctic Ocean to the Black Sea, as well as in America. In autumn it frequents the rivers and lakes, and in spring the torrents. It feeds principally on Fish, Frogs, Crawfish, and Aquatic Insects. It has the feet semipalmate; but its teeth correspond with those of the Polecat tribe rather than with the Otters, and its tail is round; whence it may, with propriety, be named the Water Polecat. Its general colour is brown, but the jaws are white; it emits a scent of musk.

Pallas thinks this is the Minx of America. It is named *Tutucuri* by the Zealanders; *Nœrs* by the Prussians; and the skin is called *mœnk* by the furriers. It is very like the Sable, and is sometimes fraudulently sold for it, although it is by some highly prized, and very much esteemed.

The *Cape Polecat,* or *Zorillo.* (The *Zorille* of Buffon. *Viverra Zorilla,* Gm.) Buffon, Gmelin, and others, describe this animal under the name of Viverra Zorilla, or Zorille, a name by which the Spanish Americans distinguish the Mephitic Weasels, and which means a little Fox. The errors of so useful and fascinating a writer as Buffon must necessarily be difficult of eradication; and this transferring of the proper name of one species to another has induced much confusion on the subject of the Mephitic Weasels. The specific character of the teeth, in which this animal agrees with the Polecat tribe, and the absence of that superlatively noxious stink, which is emitted by the Mephitic Weasels, properly speaking, separate it from the latter

U 2

animals; though it possesses, in common with them, claws calculated for digging, which consequently indicate, to a certain degree, its mode of life, and distinguish it from the other Polecats ; whence it is separated into a distinct genus by Iliger : it is also assimilated to the American Mephitic Weasels in appearance. Its colour is black, with three dorsal white stripes, extending from the occiput to the tail, which is spread, and generally carried erect. Notwithstanding these similarities, it must not now be confounded with the transatlantic Mephitic race, from which it differs in the teeth in approaching the Putorii of the old world.

Major Smith suspects, that the Zorillo of the Cape forms a family of several species ; at least the stripes indicate great diversity of disposition in the colours of the several individuals, which have come under his observation. The opposite figure is from one of his drawings.

However interesting the habits and manners of animals may be, it would soon, perhaps, be deemed an unnecessary prolixity, to say much on the subject in relation to every species. The accordance of their characters and pursuits with their physical description may be said to be mathematically correct ; and if a group be once formed, corresponding in material conformation, more especially as it regards the leading characters of the teeth, whatever be their relative disproportion of size and strength, and wherever they may be found, either at the pole, or under a vertical sun, their characters will be similar, and their pursuits the same.

The differences in the teeth which distinguish the MARTENS properly speaking, from the Polecat family, have been already noticed in the text ; we shall merely add that they have four carnivorous teeth, instead of three, in addition to the molar tooth, in the upper jaw, and four below. The last lower carnivorous tooth has a rounded lobe on the inner side, which fits them something less perfectly for a carnivorous regimen, and enables them, in the same ratio, to masticate vegetable matter, and consequently indicates a slight diminution in the cruelty and ferocity of their nature. The muzzle of the former is rather large, and their claws sharp, and slightly retractile.

Every remove in the works of nature is by a gentle gradation; nothing is abrupt: but although these gradations are observable in different groups of the Weasels, it is hardly to be expected, that the character of each will differ from its preceding subdivision very apparently. They are only to be observed by minute inspection, which is equally necessary to discover their consequences; and the means of the latter investigation are much less in our power than the former. If the cruel experiment were to be tried of keeping the Polecat and Marten without animal food, and on vegetables only, probability indicates that the latter would be the survivor.

The *Beech Marten* (*Mustela Fagorum*, Ray,) was first described by Gesner and Aldrovandus, under the epithet *domestica*, which has more reference to the conduct of the animal, in secreting itself in outhouses and buildings, than to any peculiar disposition it evinces to become tame and the associate of mankind, which it will not do unless it is taken quite young, and brought up in confinement. It was afterwards mentioned by Ray, who calls it Martes Fagorum; whence the French name of it, *la Fouine;* although it does not appear to be particularly fond of the beech-tree.

The body, from the nose to the anus, is about sixteen inches long, and the tail eight inches; the fur is of two sorts; the first is long and close, and the lower half of it of an ash-colour, but the upper half is brown-black, having a reddish tinge in some lights; the second is short, close, very soft and fine, and of a whitish or pale ash-colour: it is visible through the long hairs. The legs and tail are blackish; the throat and neck of a clear white.

This animal is frequently found near rural habitations; and the female generally brings forth her young in barns, or holes in rocks, which she takes care first to line with moss. She generally produces from three to seven young ones at a time.

The *Common Marten*, (*Mustela Abietum*, Ray.) This and the preceding species were confounded together by Linnæus, although it was distinguished by Ray, in his excellent Synopsis of Quadrupeds, by the name of *Martes Abietum*, with reference to its being found generally in pine-forests.

It is rather larger than the Beech Marten, the body, from the nose to the anus, being about eighteen inches long, and the tail about nine or ten inches. The fur, like that of the preceding, is formed of two sorts of hair ; the first is long and close, the base of which is ash-colour, but the middle yellow, and the tips dark brown. The second sort of fur is very fine, rather downy, of a yellowish ash-colour, and not entirely hidden by the long hairs ; on the chest, fore legs, and tail, it is brown black ; on the throat and neck clear yellow.

This animal is very wild, and hardly ever forsakes the thick forest, where it climbs trees with the greatest facility, by the aid of its sharp claws, in search of birds and their nests. It also attacks squirrels and other small quadrupeds. In the spring the female brings forth two or three young ones, which she generally places in the nest of a squirrel that she has killed ; or in that of a buzzard, owl, or some other bird of prey. It is found in the whole of the north of Europe, and also in North America, near Hudson's Bay.

The skin of this species is three times as valuable as that of the Beech or House Marten.

The *Sable* (*Mustela Zibellina*) is universally known by its rich fur. It is brown, with white spots about the head;

and gray on the neck; and differs from the preceding Martens in having fur to the extremities of the toes; a natural indication that it is an inhabitant of the cold and frozen regions of the earth, as we have seen in the text.

The Sable, which is so remarkable for the beauty of its skin, is also inferior to none of its kind in what we call instinct. It is capable of being rendered very docile; a remarkable instance of which is related by Steller, in one that was domesticated in the palace of the Archbishop of Tobolsk, which used to wander about the city, and visit the neighbours. It will attack and destroy a hare, though larger in size than itself; and it is said also to kill the Ermines and Siberian Weasels.

It is principally an inhabitant of woody countries; lives in holes in the trees, and not under ground; and hunts during the night, particularly if it be clear and fine; but if otherwise, retires to sleep.

If pressed by hunger, it follows Bears, Gluttons, and Wolves, as the Chacal does the Lion, to partake of the overplus of their meals. It will also then eat fruit, particularly that of the service-tree. It is about the size of the Common Marten. The hairs of the fur will lie any way in which they may be placed. A single skin, of the best quality, is said to fetch twelve or fourteen pounds.

The females, towards the end of March or the beginning of April, produce from three to five young.

The *Vison* is a native of South America. Buffon says, that although its skin was well known in the fur trade, the animal to which it belonged was not strictly ascertained; that the name has been variously applied; and that no description, to be relied on, has been given of it. But he adds, that he has inspected the animal, and found it to belong to the family of Martens; and Cuvier confirms his classification. The Vison, like the Pekan, is partly aquatic in its habits. It is larger than the Polecat, and of a beautiful chestnut colour, except the point of the chin, which is white. In size and shape it corresponds with the Common Martens. Its paws are covered with hair to the nails; and are semipalmated, not altogether palmated, as stated by Gmelin, which probably induced Dr. Shaw to place it with the Otters (*Lutra Vison*). It is probably the Minx of Lawson.

The American Martens are still in some obscurity, and some species have probably been confounded. This, by the disposition of its colours, approaches very nearly to the *Martes Lutreola* of the north of Europe, and has been often confounded with it, although Cuvier has placed the Martes Lutreola along with the Polecats, and this with the true Martens. The most striking distinction between the two is in the brown-black tail, and the point of the lower jaws only being white, not like the Martes Lutreola, where both the upper lip, and the chin, and neck, are white all through. They appear to agree very much in habits, and in the character of the semipalmated feet covered with hair.

They are generally found on the edge of rivers, and burrow under the ground. They feed principally on Fish, Water-birds, Rats, and the eggs of Tortoises, &c. ; but they sometimes approach the houses. The female brings forth three or six young at a time.

The *Pekan Weasel*, (*Mustela Canadenis.* L.) is an American species of the Marten family of the old world, with which it corresponds in all the specific characters. It is, in general, about eighteen inches long, and the tail measures about a foot.

This animal inhabits holes in the banks of rivers ; and feeds more especially on such small quadrupeds as live near the water, and on fish.

The head, neck, shoulders, and upper part of the back are varied with gray and brown hairs; the nose, back of

C. Hamilton Smith Esq.ʳ del.ᵗ

THE WHITE-EARED WEASEL.

MUSTELA LEUCOTIS. Temminck

London. Published by G. B. Whittaker. Sep.ʳ 1825.

the neck, tail, and legs are black-brown.   It has mostly a white spot under the throat.

The *Fisher-Weasel* (*Mustela Pennanti*) is decidedly distinct, and is so named by the fur-hunters of America. It is considerably larger than the Common Weasel; is of a glossy silvery black colour, which is paler towards the fore quarters, and slightly rufous about the nose; the tail and legs are velvet black; the hair is silky, and the fur beautiful. The head is small, the ears short, and the claws are very much crooked.   It inhabits the banks of rivers and lakes, and pursues the fish, which are its principal prey, with ease and effect.

The *Zorra*, (*Mustela Sinuensis*. Humboldt.)   The Baron Humboldt describes this species of American Marten as having the body less vermi-formed than the race in general; of a blackish gray colour, with the under parts and insides of the ears white; the tail half the length of the body, and but little covered with hair.

The *White-eared Weasel*, (*Mustela Leucotis*. Temminck.) In the museum of that celebrated naturalist, M. Temminck, is preserved a Weasel, which he has named Leucotis, apparently because of the whiteness of the ears.   It is twenty inches long.   The fur is of a deep glossy sepia brown colour, like the beaver, but the insides of the ears are white.

It seems not improbable that this and the preceding may be the same as the White-Cheeked Weasel of Pennant.

THE MEPHITIC WEASELS.   Most surprising accounts have been given, by almost all writers on the animals of America, of certain Weasels, found in various parts of that continent, which are provided by nature with a very singular but effectual mode of self-defence, in the power they possess of emitting, at will, a most insupportable and disgusting stench, which seems equally noxious to every animal, those of their own species only excepted. Such extraordinary powers of defence seem the more unaccountable, when it is considered, that the predacious habits of these animals, in common with the Weasels in general, seem rather to demand means and weapons for offensive operations, with which, indeed, they are otherwise well provided, than so strange a protection against the attacks of others. Timidity of disposition, accompanied with celerity of motion, afford a frequently availing defence to many of the herbivorous animals against their natural enemies ; but it is not apparent why extraordinary powers, for mere self-preservation, should be granted to animals, whose existence depends on their capability of overcoming and destroying others ; and it does not appear that they actually capture or destroy their prey by means of their vapour, but merely call it into action when irritated, or attacked, simply in self defence.

It is scarcely possible to discover, from books, how many species of the mephitic Weasels exist. A probability is assumed, that the same species has been described by different writers, under different names, and the numbers of them, consequently, erroneously increased. The principal external differences in them appear to consist in the number of white stripes, which pass down the back and sides of the animals, on a black ground. The descriptions of the mephitic stench of all seem nearly to accord.

Buffon collected together several accounts, from various voyages and travels, from which, and from the observation of a few skins, he established four species. These he called, Coase, Conepate, Chinche, and Zorillo; and the supplement to his work contains a fifth, under the name of Mouffette de Chili.

The first of these, the Coase, does not, certainly, correspond with any known animal ; and, as has been surmised,

THE CHINCHE OF BUFFON.

VIVERRA MEPHITIS_Gm.

C. Hamilton Smith Esq.r del.t   J.t Basire sc

London Published by G. B. Whittaker. Dec.r 7.t 1825.

may be described from a mutilated skin of the Coatimondi,
a plantigrade animal, whence it has been dismissed as supposititious.

The Conepate is thought to be the animal described by
Catesby, and not the Yagouaré of Azara, as this traveller
conjectured. It is the Viverra Putorius of Gmelin.

The Chinche, the Viverra Mephitis of Linnæus, appears
also to be distinct.

The Zorillo was certainly described by Buffon, from a
specimen of the Cape Marten before-mentioned ; and the
name he attributed to it belongs properly to the animal described in his supplement, under the name of Mouffette de
Chili. It is, probably, the same as the Viverra Conepatl of
Hernandez, and the Mapurito of Mutis, which Gmelin
adopted as a distinct species.

The Baron Cuvier, in his " Ossemens Fossils," in particular, seems to have bestowed very great pains to clear up
the existing difficulties on the subject of these animals. He
quotes, from various writers, the description given by each
of the Mephitic Weasels : and we shall subjoin the result
of his inquiries : premising, that their cheek-teeth correspond in number with those of the Polecat sub-division ; but
the molar, or flat tooth, is larger ; and the opposite tooth to
it, in the lower jaw, has two tubercles. The claws on the
fore feet are also very long, calculated for digging, and indicating subterraneous habits ; but their most distinctive
peculiarity consists in the pre-eminently offensive vapour
they emit, which exceeds any thing of the kind other Weasels are capable of, and separates them, in a remarkable
manner, from all other animals.

Azara (Animaux de Paraguay, t. 1. p. 211) describes the
Yagouaré, which has two white bands, extending to the tail ;
but which bands, he says, are altogether wanting in certain
individuals, and are but slightly indicated in others.

Kalm (Voyage, p. 452) describes the Skunk of the Americans, which has one dorsal white band, and another on each side.

Gemelli-Carreri (Voyage, t. 6. p. 212) mentions the Zorille merely as being black and white, with a very fine tail.

Gumilla (Hist. Nat. de l'Orénoque, t. 3. p. 240) describes the Mafutiliqui of the Indians, having the body spotted with black and white.

Lepage-Dupratz (Hist. de la Louisiane, t. 2. p. 86) describes the Puant, the male of which is of a fine black colour: and the female black, bordered with white.

Fernandez (Hist. Nov. Hisp., c. 16. p. 6) describes the Ortohula of Mexico, which is black and white, with yellow in some parts. Fernandez also mentions the tépémaxtla, which is without any yellow.

Humboldt (Vag. Partie Zoologique) describes the Zorra of Quito, which is a plantigrade, with two white bands.

Hernandez describes the Ysquiepatl, having several white stripes.

Catesby (Carol. ii., p. 62. tab. 62) figures the Mephitic Polecat, marked with nine white stripes.

Buffon (t. 13. pl. 40) has the Conepate, with six white rays.

Hernandez (Mexico, p. 332) describes the Conepati, having but two white rays.

Mutis (Act. Holmiens., 1769, p. 68) named from him the Mutis (*Viverra Mapurito*, Gm.), with a single white stripe, extending only half-way down the back. This has been recognised by the Baron Humboldt.

Buffon (Suppl. t. 7. pl. 57) describes the Mouffette de Chili with two white stripes on the back, which unite behind the head, and form a crescent.

Buffon gives also another, (t. 13. pl. 39) under the name of Chinche (*Viverra Mephitis*, Gm.), with two very large white stripes behind.

Feuillée (Journal du P. Feuillée, p. 272) also describes the

C. Hamilton Smith. Esq.r del.t

THE MEPHITIS OF CHILI.

BUFF. SUP. TOM. VII.

London. Published by G. B. Whittaker. Dec.r 12.th 1825.

G. Whittle sc.

Chinche, having two white rays, which go off, and are dispersed on the sides.

Raffinesque (Ann. of Nat.) describes the Mephitis Interrupta, which is brown, with two short white parallel rays on the head; eight on the back, the four anterior of which are equal and parallel, and the four posterior rectangular.

Molina describes the Chinga black, with a band of round white spots along the back.

And Shaw (Gen. Zool. v. i. p. 2) presents us with the Mephitic Weasel of Bengal, with spots on the head, four white dorsal stripes, and a furry tail.

We shall now proceed with the description of the Yagouaré of Azara.

This animal is generally identified with the Mouffette de Chili of Buffon, and the Viverra Conepatl of Gmelin. It is described at length, by Azara, as an inhabitant of South America, and generally found in the open country rather than in the forests. It lives on insects, eggs, and such birds as it can seize by surprise. Its motion is gentle and gliding, and it carries its tail horizontally. It will not run from a man; and, indeed, exhibits no signs of fear at the sight of any animal, however powerful ; but if it perceive itself about to be attacked, it curves its back, raises its hairy tail into a vertical position, and then ejects, with considerable force, its urine, which is mixed with such an insupportably fetid liquid, produced by certain glands for the purpose, that neither man, dog, nor any animal, however fierce, will venture to touch it. If a single drop of this most powerful liquid fall on a garment, it is rendered absolutely useless; for washing it twenty times over will not destroy its horrible stench, which it will even diffuse throughout the whole house in which it is kept. Azara declares he was not able to endure the disgusting stink which a Dog, that had received it from the Yagouaré a week before, communicated to some furniture, although the Dog had been washed and scrubbed with sand above twenty times.

This animal is comparatively slow in its motions; for although it gallops occasionally, it does not then go faster than a man. It digs holes in the ground for retreat, and deposits its young in them. Its fetid urine, when ejected in the dark, is said to emit a phosphoric light.

When they are hunted, it appears the natives irritate them first with a long cane, in order to make them void their urine, and exhaust their means of defence. They will also approach by surprise, and, seizing them by the tail, will quickly suspend them by it, in which situation they are incapable of emitting their offensive liquor; and the hunters are enabled to destroy the pouch in which it is secreted, before they kill and skin them. When taken by these means, and deprived of their strange mode of annoyance, they are said to be sometimes domesticated.

Azara observed a considerable tendency in his species to variety; and he found also that their skins became subject to change their colours, when kept any time, which seems to strengthen the probability, that although animals in a wild state, in general, are much less subject to vary than those that are domesticated, yet this tribe of Weasels may be more particularly liable to this influence, whence individuals of the same species may have been described so differently, and treated as distinct.

The above observations are intended to convey, as nearly as possible, the present state of knowledge, or, rather, the real state of ignorance existing in regard to the distinctness of the species of these animals. Individuals of the Mephitic family of Weasels are very seldom brought to this country; and the Zorillo of the Cape, which has been occasionally to be seen in our menageries and collections, has been very generally confounded with the transatlantic animals; a confusion which results from the similarity of the African Zorillo to the Mephitic Weasels, as well as from the former animal being improperly called by an American name; but little additional information can be expected on this subject, except from those who have for some time resided in America, and are urged to observation by an energetic and inquisitive mind.

Major Hamilton Smith, though he admits the confusion of writers, or the varieties of the species, is not prepared to discard any that have been described and figured as different. He inclines to think the Coase of Buffon, though not corresponding with any animal elsewhere described, is distinct, principally because the character of the figure has something positive, which a mutilated skin could hardly have produced. He thinks it may be one of a race, also, but little known, namely, the smaller Gluttons or Grisons, of at least three undescribed species of which the Major has made drawings; and he would, consequently, not dismiss it as factitious, but leave the matter open for future investigation.

The Major acknowledges, also, the Conepatl of Buffon, as well as the Chinche, which he has frequently seen, and is well figured in Buffon's work. Catesby's animal he also has little doubt is distinct. It is longer, more slender, has the nose more pointed, and the markings very different from the Conepatl, or any other. The Mouffette de Chili is very nearly allied to a drawing in the Major's collection, though it differs in some degree, resulting, possibly, from the type being found in a different part of the country from that of Buffon's animal.

The opposite figure is of the male, of the animal commonly known in America by the name of Skunk. The white marks differ in shape in the female. The hairs are long in

the tail: two-thirds, from their root upwards, they are
white ; the remaining third of each hair is black.

These Mephitic animals are very clumsy, and not nearly
so active as their congeners ; whence a certain awkward-
ness, resulting from their make, which may be the cause
of their being provided with their singular mode of defence;
and thus, as their means of flight are limited, nature has
supplied them with powers the most effectual, not merely
for self-defence and preservation, but also for actual annoyance. It is a known fact, that young and sporting dogs,
unacquainted with their quality, sometimes pounce upon
them; but the dash of fetid liquid in their nose instantly
forces them to quit the animal; they then dig, with miserable
whinings, in the earth, rub their noses into it, and scratch
themselves so violently at the same time, as to produce considerable bleeding. They are seldom appeased till exhausted'
with fatigue, and never will pursue a second of the same
species. Washing and baking clothes is insufficient ; and
Mr. Skidder, the owner of the New York Museum (as Major
Smith states), had a set of clothes spoilt, which, after wash
ing, were hung upon the roof of his house, full fifty fee.
high, and yet could be very distinctly smelt some distance
off in the streets, or the square near the house. On one
occasion, as the Major was travelling by the coach, the
vehicle gained upon a Skunk, which was attempting to get
through a fence, which any other species would have passed
in a moment ; not succeeding, however, in its endeavours
before the coach came up with it, it emitted the Mephitic
vapour, and, by a wisk of the tail, sent it on the seat of the
driver, next to whom sat a young buxom American girl, all
of whose clothes were completely ruined by a few drops.

The residence of Major Smith, in South America, was
principally in the east and north east low-lands of that continent, which are parts but little frequented by the Mephitic

J. Landseer. sc.

THE TELEDU.

MYDAUS MELICEPS.

London. Published by G. B. Whittaker Dec.r 1824.

Major C. Hamilton. Smith. del.

Paris Museum.

Weasels. His observations were principally made on such as are met with in North America.

M. F. Cuvier, in his lithographic work, gives a figure and description of the Chinche, but as he seems to think it the same animal as the Yagouare of d'Azara, we shall avail ourselves but little of it. His individual sat habitually on its paws, with the tail elevated.

The *Teledu* (*Mydaus Meliceps* of Dr. Horsfield) is very nearly allied to the Mephitic Weasels of America, but differs from them in its truncated pig-like muzzle, and shortened tail: it is also more decidedly plantigrade than those animals.

The figure of this animal, from a drawing by Major Hamilton Smith, of a specimen in Paris, differs from the beautiful engraving by Taylor, in Dr. Horsfield's work, principally, in having the transverse white stripe lower down on the shoulders.

" The Teledu," says Dr. Horsfield, " has a peculiar external character and physiognomy. Although it generally agrees in size with the Polecats of Europe and America, the circumstances which influence its appearance are entirely different. The heavy form of the body, as well as the head gradually narrowed to an obtuse point, call to mind the figure of a Hog. The shortness and strength of the neck, and the manner of walking, by placing the entire sole of the foot on the ground, contribute further to give to the animal a sluggish appearance. The eyes are placed high in the head, and in their size and disposition have considerable resemblance to those of a Hog; the eye-lids are rigid, and well provided with eye-brows, consisting of minute bristles; the irides are of a dark colour, and the pupil is circular. The ears are nearly concealed by the hairy covering of the body; but these organs are provided externally with an oblong concha, which surrounds the posterior part, and passing the lower extremity of the meatus auditorius, forms a small curve inward. No whiskers are perceptible, but a few long straggling hairs arise from the upper lip. The covering of the Teledu is adapted to the elevated and cold regions which it inhabits. The fur is composed of long delicate hairs, silky at the base, which are closely arranged, and afford a very warm coat to the body. On the sides of the neck the hairs are lengthened, and have a curved direction upward and backward; on the top of the head, meeting from before and behind, they form a small transverse crest, and on the abdomen they are thinly disposed, and afford, in some parts, a view of the naked skin. The colour of the hairs is blackish-brown, more or less intense on every part of the body, except the crown of the head, a streak along the back, and the extremity of the tail. These parts are white, with a slight tint of yellow. The mark on the head has a rhomboidal form, obtuse and rounded anteriorly, but gradually attenuated as it passes to the shoulders, where it unites with the streak on the back : in some individuals this streak is interrupted. On the abdomen, the brown is of a lighter hue, inclining to grayish or rufous. The covering is subject to several variations. The tail is scarcely half an inch long, but the hairs covering and surrounding it project above an inch from the body. The limbs are short and stout, and the feet agree in structure with those of the allied genera, being formed for the plantigrade manner of walking. The claws are united at the base by a thick membrane, which envelopes this part as a sheath. Those of the fore feet are nearly double the size of those of the hind feet. In place of the pouches and reservoirs of fetid fluids with which several genera of this family are provided, the Mydaus has two glands of an oblong form, about one inch long, and half an inch wide, near the extremity of the rectum : they are placed opposite to each other, and are individually furnished with an excretory duct nearly half an inch long, which communicates with this intestine.

In the middle of each duct is a very minute aperture, surrounded by a muscular ring, somewhat swelled, which enables the animal at pleasure to discharge or to retain the fetid fluid secreted by the glands. The ducts enter the rectum about half an inch within the external aperture. The internal surface of these glands is covered with numerous wrinkles disposed transversely. The fluid secreted by them is perfectly analogous, in its odour, to that secreted by several species of Mephitis in America, particularly to that of the Mephitis Striata of Fischer. Having experienced that of the latter, which is known in most parts of North America by the name of Skunk, I readily recognised it in Java."

The dentition of the mammiferous genera is the main foundation of our author's arrangement of that branch of the Animal Kingdom : in this particular the species in question agrees, except in a slight variation of the incisive and canine teeth with the Mephitic Weasels of America. In other characters, however, it differs from them, and M. F. Cuvier, who describes the Teledu under the name Telagon, refers to these discrepancies, as follows:—The Chinche-Mephitis dimidiata, has a rounded head, a short pointed, not very broad muzzle, which calls to mind the head of the Fitchet, or rather of the Cat, if the muzzle of this were less obtuse. On the contrary, the head of the Telagon calls to mind the elongate muzzle and snout of the Badger, with a face still narrower. The Chinche further has a large tail, furnished with long bushy hairs, which it elevates as a plume on its back, in the same manner as Squirrels. The Telagon, on the contrary, is almost deprived of this organ, its tail being scarcely an inch long, and very scantily provided with hairs. The examination of the bony parts further confirms the propriety of separating these animals into distinct genera. The elongation of the head of the Telagon, and the narrowness of its muzzle, are the cause that the grinders are individually more separated from each other, and that the front teeth, instead of being placed nearly in a straight line, are disposed in form of a very small arch or curve. There is also a difference in the relative arrangement of the grinders in the jaw-bones, which affects the communication of the nostrils and the posterior parts of the mouth.

The minute details of Dr. Horsfield are highly interesting to the professed zoologist, but in a general point of view it must be admitted, that the multiplication of divisions tends to hinder the progress of science, and we should, therefore, wish to see the Mydaus Meliceps treated as a species of the Mephitic Weasel, differing in certain specific characters.

" The Mydaus Meliceps" (we again quote Dr. Horsfield,) " presents a singular fact in its geographical distribution. It is confined exclusively to those mountains which have an elevation of more than 7000 feet above the level of the ocean ; on these it occurs with the same regularity as many plants. The long extended surface of Java, abounding with conical points which exceed this elevation, affords many places favourable for its resort. On ascending these mountains, the traveller scarcely fails to meet with our animal, which, from its peculiarities, is universally known to the inhabitants of these elevated tracts ; while to those of the plains, it is as strange as an animal from a foreign country. A traveller would inquire in vain for the Teledu at Batavia, Semarang, or Surabaya. In my visits to the mountainous districts I uniformly met with it, and as far as the information of the natives can be relied on, it is found on all the mountains. It is, however, more abundant on those which, after reaching a certain elevation, consist of numerous connected horizontal ridges, than on those which terminate in a defined conical peak. Of the former description, are the Mountain Prahu and the Tengger Hills, which are both distinctly indicated in Sir Stamford Raffles's Map of Java; here I observed it in great abundance. It was the less common on Mountain Gede, South of Batavia ; on the Mountain Ungarang, South of Semarang; and on the Mountain Ijen, at the farthest eastern extremity; but I traced its range through the whole Island.

" Most of these mountains and ridges furnish tracts of considerable extent, fitted for the cultivation of wheat, and other European grains. Certain extra-tropical fruits are likewise raised with success: Peaches and Strawberries grow in considerable abundance, and the common culinary vegetables of Europe are cultivated to great extent. To most Europeans and Chinese, a residence in these elevated regions is extremely desirable ; and even the natives, who in general dislike its cold atmosphere, are attracted by the fertility of the soil, and find it an advantage to establish villages, and to clear grounds for culture. Potatoes, Cabbages, and many other culinary vegetables are extensively raised, as the entire supply of the plains in these articles depends on these elevated districts. Extensive plantations of Wheat, and of other European grains, as well as of Tobacco, are here found, where Rice, the universal product of the plains, refuses to grow. These grounds and plantations are laid out in the deep vegetable mould, where the Teledu holds its range as the most ancient inhabitant of the soil. In its rambles in search of food, this animal frequently enters the plantations, and destroys the roots of young plants; in this manner it causes extensive injury, and on the Tengger Hills particularly, where these plantations are more extensive than in other elevated tracts, its visits are much dreaded by the inhabitants : it burrnws in the earth with its nose in the same manner as Hogs, and in traversing the hills, its nocturnal toils are observed in the morning in small ridges of mould recently turned up.

" The Mydaus forms its dwelling at a slight depth beneath

the surface, in the black mould, with considerable ingenuity. Having selected a spot, defended above by the roots
of a large tree, it constructs a cell or chamber, of a globular
form, having a diameter of several feet, the sides of which
it makes perfectly smooth and regular ; this it provides with
a subterraneous conduit or avenue, about six feet in length,
he external entrance to which it conceals with twigs and
dry leaves. During the day it remains concealed, like a
Badger in its hole; at night it proceeds in search of its
food, which consists of insects and their larvæ, and of
Worms of every kind : it is particularly fond of the Common
Lumbrici, or Earth Worms, which abound in the fertile
mould. These animals, agreeably to the information of the
natives, live in pairs, and the female produces two or three
young at a birth.

" The motions of the Mydaus are slow, and it is easily
taken by the natives, who by no means fear it. During my
abode on the Mountain Prahu, I engaged them to procure
me individuals for preparation ; and as they received a
desirable reward, they brought them to me daily in greater
numbers than I could employ. Whenever the natives surprise them suddenly, they prepare them for food ; the flesh
is then scarcely impregnated with the offensive odour, and
is described as very delicious. The animals are generally
in excellent condition, as their food abounds in fertile
mould.

" The structure of the teeth affords to the Mydaus but feeble
means of defence ; the front teeth in the lower jaw have nearly
a horizontal position, and the canine teeth are comparatively
small and weak. The animal being slow in its motions, its
manner of defence is of a negative nature, and, as in the
American Mephitis, consists in preventing the approach of an
enemy by an intolerably offensive odour: hence these animals
have received the names of Mephitis, Mydaus, Stifling Wea-

sel, Bete-puante, &c. The effort by which the fetid matter is projected, is described by the natives as a crepitus ventris : the muscular coat of the glands, as far as I have ascertained, serves only to propel the fluid into the rectum, at the pleasure of the animal: its discharge, as a means of annoyance to its enemies, is effected by a general effort of the abdominal muscles. On the Mountain Prahu, the natives who were most active in supplying me with specimens of the Mydaus, assured me that it could only propel the fluid to the distance of about two feet : the fetid matter itself is of a viscid nature ; its effects depend on its great volatility, and they spread through a great extent ; the entire neighbourhood of a village is infected by the odour of an irritated Teledu, and in the immediate vicinity of the discharge it is so violent as in some persons to produce syncope. The various species of Mephitis in America differ from the Mydaus in the capacity of projecting the fetid matter to a greater distance.

" The Mydaus is not ferocious in its manners, and taken young, like the Badger, it might easily be tamed. An individual, which I kept some time in confinement, afforded me an opportunity of observing its disposition ; it soon became gentle, and reconciled to its situation, and did not, at any time, emit the offensive fluid. I carried it with me from Mountain Prahu to Bladeran, a village on the declivity of that mountain, where the temperature was more moderate. While a drawing was made, the animal was tied to a small stake ; it moved about quietly, burrowing the ground with its snout and feet, as if in search of food, without taking notice of the bystanders, or making violent efforts to disengage itself : on earth worms (*lumbrici*) being brought, it ate voraciously ; holding one extremity of a worm with its claws, its teeth were employed in tearing the other : having consumed about ten or twelve, it became drowsy, and making a small groove in the earth, in which it placed its snout, it composed itself deliberately, and was soon sound asleep."

THE OTTERS, in the system of dentition, at least, present but little deviation from the sub-division of Polecats and Martens, except that their teeth are more developed in certain parts. We shall, therefore, add nothing on this head, to the Baron's observations in the text, but merely present a figure of the teeth, in illustration of his description.

The large flat head and short ears of the Otters are so far singular in their kind, as to remain indelibly impressed on the memory, by a single inspection. Their palmated feet, and compressed tail, have been adverted to in the text. They have a small gland near the anus, which secretes a fetid liquor. Their fur is of two sorts, the one short and thick ; the other, long, shining, and close. Their natural regimen is piscivorous, but they will also eat small quadrupeds, and gnaw the bark of trees, and, in a state of domestication, in particular, may be kept on bread and milk, and vegetables.

The *Common Otter, (Mustela Lutra)* is a very destructive and ferocious water animal; it destroys its prey by biting off the head, and leaving the remainder; thus killing many more than are necessary for its sustenance.

> . . . . Rapine and spoil
> Haunt e en the lowest deeps: seas have their sharks ;
> Rivers and ponds enclose the rav'nous pike ;
> He, in his turn, becomes a prey—on him
> Th' amphibious otter feasts . . . .
> . . . . nor spears
> That bristle on his back, defend the perch
> From his wide greedy jaws ; nor burnish'd mall
> The yellow carp ; nor all his arts can save
> Th' insinuating eel, that hides his head
> Beneath the slimy mud ; nor yet escapes
> The crimson-spotted trout, the river's pride,
> And beauty of the stream.

It is found in all parts of Europe, and in the north of Asia, but the limits of its location are not defined. It averages about two feet in the length of its body, and the tail is about sixteen inches. It is very fierce; and, when hunted, will often turn on the dogs, and bite them severely. Notwithstanding the natural ferocity of its character, which, however, is principally directed against fish, there are many instances of its having been tamed, and rendered of considerable service in fishing. Buffon, in his original edition, expressed his doubts of this though, in the supplement, he retracted them.

It is not properly amphibious, or capable of living either on land or in the water. It is true that it is an excellent diver, and can remain a considerable time under water; but it has been known to have been drowned when entangled among weeds in the pursuit of fish, an instance of which the editor lately witnessed, in one that got into an eel-trap, and could not return.

As it cannot live long without respiring, the shore is its natural residence; but as it pursues its prey in the water, we find its organization adapted for its double destination. Its short limbs, flattened head, and compressed body, are well adapted for swimming. On shore, nevertheless, it moves with facility, and may be said, indeed, to run rapidly. It prepares its retreat either under a rock or the root of a tree, where it generally passes the day on a bed of dry grass. At night, it sallies forth in pursuit of its prey. We are ignorant of all the circumstances of its reproduction, as well as of the first state and appearance of the young. It is towards the beginning of April that young Otters are first seen. The mother does not appear to bestow her maternal cares upon them for any great length of time, for, in May, they are observed to take the water in pursuit of food. At about two years of age they are adult.

M. F. Cuvier has had several Otters, which were very

familiar, and were kept on bread and milk. I am not, therefore surprised, says he, that they were so far trained, as Gesner relates, as to catch and bring fish to their master. They
are animals not wanting in intelligence, and which have
scarcely any other instincts than those of choosing a fit retreat, and furnishing it with dry grass. Instinct is ever
less predominant, in proportion as intelligence is more so.
Their flesh is not palatable, but their fur is much esteemed.

The nails of the Otter are very short, and grooved;
the fingers have, underneath, toward their extremity, a
round tubercle; a large tubercle, in four lobes, is found in
the middle of the palm, and there is a third at the base of
the carpus. To the hind feet there are only the simple
rounded tubercle of the toe, and that of the middle of the
palm, which is divided into three lobes. The eye is furnished with a third lid, which appears entirely to cover the
cornea. The ear has, on its lower part, a lobe, which may
be considered analogous to the tragus, and a slight swelling
on the opposite side, answering to the antitragus. In the
upper part of the cavity of the conque, another slight protuberance, which may be considered to correspond with a
branch of the helix or anthelix, although these branches do
not, in fact, exist. The nostrils are surrounded with thick
glands, and open in the lower part of the angular furrow,
which forms their orifice. The upper lips have long, stiff
mustachios. It is uniformly brown above, and whitish underneath.

A variety of the Common Otter is found in the neighbourhood of Paris, covered with a number of small round spots,
irregularly placed.

The *Brazilian*, or *American Otter*, *(Lutra Brasiliensis)*
is brown, or yellowish, with a throat generally white, but
sometimes yellow. It is something larger than our Otter,
and is found in the rivers both of North and South America.

This Otter appears to be gregarious, inhabiting the

THE CANADIAN OTTER.

C. Hamilton Smith Esq.ʳᵈ del.ᵗ

Published by G.B. Whittaker. Sep.ʳ 1825. London

Griffith

rivers in small troops. Occasionally, but not always, it swims with the head above the water ; and by its manners, as well as the noise it makes, seems to be menacing : but it is never known to do harm, even to bathers, in the water inhabited by it. It either digs or takes possession of a hole in the banks. Several females inhabit a single burrow, and bring forth their young together.

Azara mentions one which was domesticated. It ate fish, meat, bread, cassavas, and other things, though it preferred fish to all other food. It went about the streets, and returned of itself to the house, knew its owner's family, and followed them like a dog, though it soon became fatigued by exercise. It knew and answered to its name, and sported with the dogs and cats of the house; but as it bit severely in playing, no one was often willing to sport with it. It was never known to attack the poultry, or any other animal, except a very young pig, which it would have killed, had it not been rescued.

The American, or Brazilian Otter, is occasionally met with at a considerable distance from water, when it is supposed to be in pursuit of a new domicile. On land, it moves very slowly, and almost on the belly ; runs or gallops very clumsily, and may be easily caught, and held by the skin of the back, in which situation it is not able to offer much resistance. The tail, though very flexible, is generally carried straight. It measures about five feet, from nose to tail.

The Otter of Canada appears to be distinct from that of Brazil. The latter is much longer, though it does not exceed the girth of the former. The neck of this species is also shorter ; and the tail is compressed from near the base to the end, while that of the Brazilian animal is compressed only near the tip.

Dr. Horsfield treats the Otter of Java as distinct from the common species, under the name of *L. Leptonyx.* The head, he says, is narrower and more lengthened, and the ears have, comparatively, a posterior situation. The neck is considerably shorter, and the tail is shorter in proportion to the body, and more acutely terminated. The anterior extremities are longer, and the entire habit is more slender and extended ; the claws, also, are weaker and shorter than in the common Otter.

The *Sea Otter (Mustela Lutris)* is full twice the size of the Common Otter: the body is very long, and the tail about one-third the length of the body. Its skin, shining like velvet, is the most esteemed of all furs, and, consequently, the most expensive. It is black, with a shade of brown ; but, about the head, there are, in general, more or less of white hairs. The hinder legs, in particular, are very short, and placed nearer the anus than in quadrupeds in general, which assimilates it to the Seal, to which it bears a considerable general affinity ; it sometimes weighs as much as seventy or even eighty pounds. It is found, perhaps exclusively, in the northern parts of the Pacific Ocean, where the Asiatic and American continents nearly approach each other, and in the intervening islands. It is said, that a single skin is sometimes sold, in the Chinese or Japanese markets, for upwards of twenty pounds sterling.

During winter, the Sea Otter confines itself to the ice near the sea-shore, or to the shore itself ; in summer, it ascends the rivers, as far as the fresh-water lakes, in company with its single female. The latter is gravid eight or nine months, and brings, generally, but one young at a birth.

They are said to feed on fuci, as well as fish and crustaceous animals, but the teeth do not appear to indicate it.

Individuals seem to differ, in having more or less white about them ; and a white-headed variety is known, of which we are enabled to present a figure.

THE SEA OTTER.

WHITE HEADED VAR.

Griffith sc.

London. Published by G.B. Whittaker March 1.1827.

The head, throat, chest, and fore paws of this are white, but sprinkled, as it were, with individual brown hairs.

We now proceed to the Baron's subdivision of Carnivora, with two tuberculous teeth on each side, at the back of the upper jaw.

The first subgenus of this division is that of the Dogs (*Canis.*)

It is not without considerable hesitation that we venture to append any observations on the canine race. Had we temerity to go to the task with the view of even self satisfaction in the execution of it, at least a volume in the stead of a few pages must be devoted to the subject.

Without disturbing the integrity of the entire species, its countless varieties all well deserve an accurate physical demonstration. Without pretending to attribute to the mere brute, even *in specie,* the powers of mind which distinguish mankind, the intellectual sagacity of the race, with its endless modifications, challenge a strict investigation.

Almost every nation of the earth, intertropical, temperate, and polar, possesses its own peculiar variety of the Dog; the theatre of observation, therefore, is the world itself, and anything like a description of the whole race would require a much more intimate knowledge of the surface of the earth than is at present obtained.

The intelligence and moral qualities of the Dog, which it possesses in common with the rest of the class, though much greater in degree, form a subject perfectly impenetrable by our limited faculties; we contemplate the effect but can by no means arrive at the cause; we may speculate and conjecture, but can never demonstrate.

It would, indeed, be no difficult task to collect a great number of anecdotes of the Dog, highly amusing in themselves, and not altogether devoid of instruction. It is in these that the real character of the animal is evinced, but our limits rather oblige us to presume that the reader is acquainted with many of them, and to devote the little space we have in attempting to illustrate the present state of zoological knowledge in regard to the race, which may not perhaps be quite so familiar to the English reader.

There is indeed ample room for a separate and extensive work on the Caninæ, replete too with original and instructive matter ; for all that has hitherto been done is but partial and unsatisfactory, at least when the subject is considered in a comprehensive point of view. Almost all the original observations we have upon the subject come from France, and partake perhaps too much of an exclusive and national character. Thus Buffon, in his canine pedigree, would trace almost all the varieties to the French Mâtin, a Dog not known out of France, and even there apparently but ill defined.

Buffon's canine history also abounds with hypotheses, which, if false, are so much the more difficult of eradication, in consequence of the fascinating language in which they are conveyed.

M. F. Cuvier, in the present age of zoological improvement, has done by far the most on the animals in question. We propose, therefore, in this part of our undertaking to notice briefly most of the more important varieties, arranged according to the plan of that eminent zoologist ; in the synoptical table to refer more particularly to all of them he has mentioned, and in the graphic department, with the exception of two or three plates, to confine ourselves to figures of British Dogs, under their popular appellatives. We shall, however, premise a few general observations.

It is said that the Shepherd's Dog, transported into the temperate climates, and among people entirely civilized, such as England, France, and Germany, will be divested of its savage air, its pricked ears, its rough, long, and thick hair, and, from the single influence of climate and food alone, will become either a French Matin, an English Mastiff, or a Hound. The last, whether Staghound, Foxhound, or Beagle, transported into Spain or Barbary, where the hair of quadrupeds in general becomes soft and long, will be converted into the Land-Spaniel and the Water-Spaniel, and these of different sizes. The gray Matin-Hound, transported into the north, becomes the great Danish Dog; and this, sent into the south, becomes the Greyhound of different sizes. The same, transported into Ireland, the Ukraine, Tartary, Epirus, and Albania, becomes the great wild Dog, known generally by the name of the Irish Wolf Dog. If these premises be corrects, it follows, that these varieties of the Dog are not of original creation, but result from climate, or other unknown causes, acting on the first species.

In pursuing this observation in regard to animals which appear still more foreign from each other, we find that the common Dog breeds together with the Wolf and the Fox; and, although zoologists and comparative anatomists have ascertained, that there is a certain similarity of physical structure in all these animals, whence they have classed them in one genus, yet the Wolf, the Fox, and the Dog, are very distinct animals when not viewed scientifically. One or two species of the Cats and the genus of the Hyænas, are nearly as much like Dogs as Cats, or, in other words, are really very like both these genera; and yet an animal is found in the *Canis Venatica* of Burchel, connecting the Dog with the Hyæna almost without an interval; certain species of the Cats approach nearly if not completely to the Weasels; certain Weasels again to the Seals and Bears; and these last to the herbivorous and granivorous races.

With facts like these constantly occurring, and they are almost endless in zoology, there is great room for wonder that the genera and species have been kept so distinct as we actually find them, and as fossil osteology seems to evince they have ever been.

A question connected with the subject before us has often been discussed by naturalists, traces of which are to be found in the most remote antiquity. It has been conjectured that the Creator produced only the germs of existing beings, and that these have been conformed by surrounding influences, so as to produce the result we see before us. The development of these germs, it is said, is proportioned to the more or less favourable state of these influences, and animals of the most simple organization, as the Polypes, are in fact nearer the immediate work of the Creator than those less imperfect, as Man and the Mammalia. The latter are the production of secondary causes, and could not have arrived at their present state of perfection and complication, but by having passed through the intermediate conditions between them and the most simple beings.

Such an hypothesis may at least have one word in its favour. Instead of limiting the power of Omnipotence, it seems rather to place it in that point of view in which we ought to regard it, that is, as simple in its means, immense in its views, and infinite in its results. The principal, indeed the only rational support of this hypothesis, is to be found in the variations we observe, especially in the canine race, and very generally in all others, which we can only attribute to secondary and accidental causes.

It is not our business to state an hypothesis merely for the purpose of refuting it, but as this has been suggested to account for the phenomenon under consideration, it may not be improper to advert to it. Such a system of creation, however, loses much of its probability, when it is considered that the tendency to variety seems almost exclusively confined to the more perfect animals, or at least is observed to prevail less as we descend in the scale of organization.

The intellectual, moral, or invisible works of the Creator

in his various creatures must ever be the subject of hypothesis and conjecture; and, however convenient in grouping the animal kingdom may be the proportions of cranium and face, the strictest analysis of the cerebral masses will never detect the mental faculties of the animal to which they belonged.

The most eminent writers, poetical and prosaic, have exercised their oratory in describing and eulogizing these highly useful and interesting animals. The subjugation and domestication of them by Man may be called reason's conquest of nature; and, as our author observes, it is the most complete, singular, and useful conquest man has ever made. It is true that, in the refined state of society in which we live, this is not so apparent; but a little observation on the state of such of our fellow-creatures as are yet beneath us in intellectual improvement, will probably satisfy us that we owe originally much of our advance to, and progress in civilization, to the powers of the Dog.

In illustration of the services of the Dog in the earliest stages of civilization, we cannot refrain from quoting the facts and reflections in relation to the genus to be found in Mr. Burchell's Travels in Africa:

"Our *pack of Dogs*;" says he, "consisted of about five-and-twenty of various sorts and sizes. This variety, though not altogether intentional, as I was obliged to take any that could be procured, was of the greatest service on such an expedition, as I observed that some gave notice of danger in one way, and others in another. Some were more disposed to watch against men, and others against wild beasts; some discovered an enemy by their quickness of hearing, others by that of scent: some for speed in pursuing game; some were useful only for their vigilance and barking; and others for their courage in holding ferocious animals at bay. So large a pack was not, indeed, maintained without adding greatly to our care and trouble, in supplying them

with meat and water ; for it was sometimes difficult to procure for them enough of the latter ; but their services were
invaluable, often contributing to our safety, and always to
our ease, by their constant vigilance ; as we felt a confidence that no danger could approach us at night without
being announced by their barking.　No circumstances
could render the value and fidelity of these animals so conspicuous and sensible, as a journey through regions which,
abounding in wild beasts of almost every class, gave continual opportunities of witnessing the strong contrast in
their habits, between the ferocious beasts of prey which
fly at the approach of man, and these kind, but too often
injured, companions of the human race.　Many times when
we have been travelling over plains where those have fled
the moment we appeared in sight, have I turned my eyes
towards my Dogs to admire their attachment, and have felt
a grateful affection towards them for preferring our society
to the wild liberty of other quadrupeds.　Often, in the
middle of the night, when all my people have been fast
asleep around the fire, have I stood to contemplate these
faithful animals lying by their side, and have learnt to
esteem them for their social inclination to mankind.　When
wandering over pathless deserts, oppressed with vexation
and distress at the conduct of my own men, I have turned
to these as my only friends, and felt how much inferior to
them was Man when actuated only by selfish views.

　" The familiarity which subsists between this animal and
our own race, is so common to almost every country of the
globe, that any remark upon it must seem superfluous ; but
I cannot avoid believing that it is the universality of the
fact which prevents the greater part of mankind from reflecting duly on the subject.　While almost every other
quadruped fears Man as its most formidable enemy, here
is one which regards him as his companion, and follows
him as his friend.　We must not mistake the nature of the

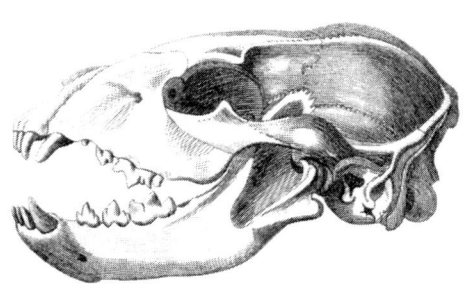

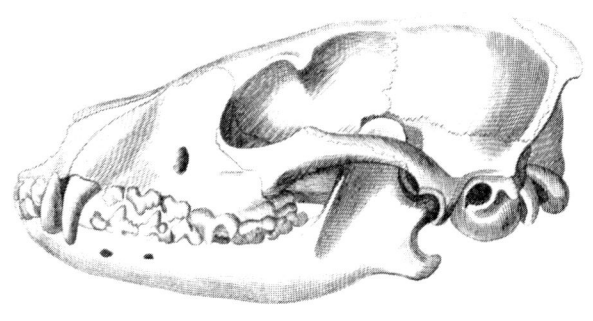

1. GENUS HYÆNA.   2. GENUS CANIS

SP. H. CROCUTA.   SP. C. AUREUS.

London. Published by G.B. Whittaker. June. 1825.

case : it is not because we train him to our use, and have made choice of him in preference to other animals, but because this particular species feels a natural desire to be useful to man, and from spontaneous impulse attaches itself to him. Were it not so, we should see in various countries an equal familiarity with various other quadrupeds ; according to the habits, the taste, or the caprice of different nations. But everywhere it is *the Dog* only takes delight in associating with us, in sharing our abode, and is even jealous that our attention should be bestowed on him alone : it is he who knows us personally, watches for us, and warns us of danger. It is impossible for the naturalist, when taking a survey of the whole animal creation, not to feel a conviction that this friendship between two creatures so different from each other, must be the result of the laws of nature ; nor can the humane and feeling mind avoid the belief that kindness to those animals from which he derives continued and essential assistance, is part of his moral duty."

The upper cheek-teeth of the Dog are six on each side, the three first are sharp trenchant, called by our author false molars ; the following, a carnivorous tooth, has two cutting lobes, beyond which on each side are two flat teeth. In the lower jaw there are seven, four false molars, a carnivorous tooth, with the posterior part flat, and two tuberculous teeth behind it. The length of jaws and muzzle vary greatly. The tongue is smooth. The ears are extremely variable. There are five toes on the fore-feet, and four on those behind, furnished with longish nails, obtuse, and not retractile, and the mammæ are ventral. The eye-pupils are circular in all the species except the Foxes.

The females are pregnant sixty-three days, and produce generally three, four, or five at a time ; but some of the more fertile species bring from six to ten ; and instances do occur of thirteen whelps at a litter. They are born with the eyes closed, which do not open for ten days or a fortnight. They live ordinarily fourteen or fifteen years; but frequently suffer much from age and decay in their latter days. Their size is indefinite; in the museum at Dresden is a perfect specimen only five inches long.

Notwithstanding the endless varieties of the Dog, and the near relationship which numerous instances of the mixtures of the breeds evince with the Wolf and Fox, the Dog, properly so called, is always distinguishable by its tail, which in all cases takes an arched direction, more or less perfect in the different varieties; and it is observed, that all Dogs which have any white about them, always have the tail tipped with this colour.

Mr. Pennant cites Galen, Hippocrates, and Pliny, to prove that the ancients were fond of the flesh of Dogs as food. He states also that the New Zealanders and inhabitants of the Society Islands eat them at the present day. The Chinese are said also to be fond of this sort of food, which is commonly sold in their markets; and the celebrated Captain Cook's recovery from a serious illness at sea was much accelerated by the broth and flesh of a Dog.

It would be desirable to arrange the varieties of this species in successive groups, as they diverge from the original stock; and much research has been made for this purpose, from which different opinions have resulted as to the original type in a state of nature, which has been so remarkably excited to variety by domestication or other causes. Some have considered the Dog as a domesticated Wolf; others think that it is a Chacal; and many, observing that Wild Dogs are found always to have the ears erect, have, from this circumstance principally, concluded, that the Shepherd's or Wolf-Dog is the original root. Since, however, the shape of the head has so much excited the attention of naturalists, it has been found that some Dogs correspond more in this particular with the Wild Dogs than with any domesticated variety; and the Dingo, or New-Holland Dog, a half reclaimed animal, and its like, are placed at the head of the list, as being supposed to be nearest to the wild and original stock. Thus M. Frederic Cuvier has arranged the varieties of the Dog, upon this principle, into three groups, each differing materially in the shape of the head, and the length of the jaws and muzzle.

Without determining which of the known varieties is the most ancient, or deciding upon the claim of pureness of blood and descent to which each may pretend, we shall merely refer to the anatomical principles, which form the ground work of this arrangement.

The first of these, which includes the Greyhounds and their consimilars, have the head more or less elongated; the parietal bones insensibly approaching each other; and the condyles of the lower jaw placed in a horizontal line with the upper cheek-teeth.

The next group of Dogs includes much the most intelligent, interesting, and useful varieties. Their head and jaws are shorter than those proper to the first division, but they are not so completely truncated as in those of the third. To speak anatomically, the parietal bones do not approach each other above the temporal fossæ, but, on the contrary, they widen so as to enlarge the cerebral cavity and the forehead. The Spaniels, Hounds, Shepherd's, and Wolf-Dogs, and the still more useful Siberian and Esquimaux races of this genus, are included under this description.

The third subdivision of the Dogs has the muzzle more or less shortened; the frontal sinuses considerable; and the condyle of the lower jaw extending above the line of the upper cheek-teeth. The construction of the heads of these animals renders the capacity of the cranium smaller, when compared with the jaws and face, than in the preceding divisions.

The first division includes among others,

The *Dingo,* or *New Holland Dog,* the head and elon gated snout of which half-wild variety are like those of a Fox. In its other proportions it agrees with the Shepherd's Dog. It is about two feet six inches long, and about two feet high. The fur, composed both of silky and woolly hairs, is of a deep yellowish brown colour, lighter on the lower parts of the body.

It is very voracious and fierce; and Mr. Pennant mentions one that was brought to this country, which leaped on the back of an ass, and would have destroyed it in a short time, had not the animal been rescued. It is very active, and runs with the tail stretched horizontally, the head elevated, and the ears erect.

The *Dhole,* or Wild Dog of the East Indies, is made like the Dingo, but the hairs of the tail are not bushy. It is of a uniform bright red colour, and is found in South Africa, and in various parts of the East, where it is named Dhole.

The *South American* half-reclaimed variety is about the size of a Spaniel. The head has much of the character of the last, but the hairs are longer, particularly on the tail. The back is brown gray; the spots on the flanks and legs are ochrey; and the ground colour is gray, lightest on the belly.

This animal is very much like a Wolf; and probably the same as is noticed by the early voyagers to America, who assert that the Indians tamed Wolves.

The *North American Dog* of the Indians is also a half-tamed breed, which differs materially from the South American race, though it corresponds, apparently to identity, with the Dogs found in the Falkland Islands. It is said, indeed, that the Spaniards landed this breed of animals on these islands after the Falkland Island dispute with England, in order to make any attempt of our countrymen to settle there difficult or impossible.

We present portraits in outline of these half-reclaimed

1. The Dingo of New Holland
2. The Dhole of India
3. South American
4 North American

London Published by G.B.Whittaker March 1827.

varieties of the Dog, in order to show their similarity in
the length and shape of the muzzle and head.   They are
from drawings of the respective animals from specimens
indigenous in Asia, Africa, and North and South America.
And it is worthy of observation, that, although the endless
domesticated varieties of this genus differ so materially
from each other, from the pointed nose of the Greyhound
to the truncated muzzle of the Bull Dog, the Wild Dog,
wherever it may be found, has the elongated jaws of the
Dingo, Dhole, and North and South American semi-barba-
rous breeds here portrayed, as well as that of the Wolf and
Fox.   These may, therefore, be said to exhibit a sort of
average representation of the Wild Dog all over the world.

The following brief sketches are of the most prominent
domesticated races proper to the division with an elongated muzzle.

The *Albanian Dog* has been noticed by historians, naturalists, and poets, ever since Europe first began to
be raised into consequence and importance.   A super-
natural origin, and infallible powers, have been attributed to it.   Diana is said to have presented Procris with
a Dog, which was always sure of its prey ; together with a
dart, which never missed its aim, and always returned to
its owner.   To the former the canine genealogists of antiquity attributed the origin of the celebrated race of the
south-east of Europe, particularly Molossus and Sparta.
The very fine breed of Dogs, now found very plentifully in
this corner of Europe, particularly in Albania, accords
with the descriptions existing of its progenitors, indigenous
in the same countries, and does not seem to have degenerated.

They are as big as a Mastiff ; their thick fur is very long
and silky, generally of different shades of brown; their
tail is long and bushy ; the legs seem more calculated for
strength than excessive speed, being stouter and shorter

than those of the Greyhounds; their head and jaws are elongated, and the nose is pointed.

The *French Mâtin, (Canis Laniarius.* L.*)* The French writers seem to consider this variety or breed as the most important of the race, and as the progenitor of many others; the reason for which is not very apparent, unless it is, that a venial patriotism is apt to decide in favour of our own country, when certainty and truth are unattainable. Mr. Pennant identifies it with the Irish Greyhound (*Canis Graius Hibernicus* of Ray), and there certainly seems every reason to conclude, that the Molossian or Albanian breed, the French Mâtin, and the Irish Greyhound, possibly, also, the Danish Dog, and the Greyhound, and its varieties, are ramifications from each other.

This variety has the head elongated, and the forehead flat; the ears are partly erect, but pendulous towards the tips. It is about three feet long, and two feet high; very muscular, but active. The colour is ordinarily a yellowish-fawn, with blackish, oblique, and parallel, but indistinct rays. It will attack the Wolf or Wild Boar eagerly, but is more commonly used in France as a House or Sheep-Dog.

The *Irish Greyhound* is much like the last, if not the same animal; but is said to attain a larger size, and is sometimes seen four feet in height. It is to this breed that the Irish owe the extirpation of Wolves from their island, since which time the race has gradually disappeared, and is now become extremely rare.

The *Great Danish Dog* is presumed, by Buffon, to be the Mâtin transported to a northern latitude. It is commonly white, marked all over with small round black spots; and is generally used as a Stable-Dog, and to accompany a carriage.

The *Common Greyhound (Canis Grajus.* L.*)* is familiar to every one, and is very remarkable for its elongated jaws and compressed head, as well as for its speed, which exceeds that of all other Dogs.

When we compare the Greyhounds with other varieties, in reference to the form and proportion of the head, we perceive that it terminates the series of those whose forehead is flat, and muzzle elongated.

This flatness of the forehead is produced by the obliteration of the frontal sinuses from those cavities which are formed at the base of the nose, and which, being immediately connected with the nasal cavities, and covered with the same membranes as they are, increase the sense of smell. This is ordinarily accompanied with an extraordinary slenderness and length of the legs, as well as a great contraction of the abdomen; phenomena, which, although not explained, are without exception.

This obliteration of the frontal sinuses, in weakening the powers of smell of the Greyhound, contribute, probably, to the development of their other senses, by the necessity induced of exercising them more exclusively. The sight and hearing of this variety are excellent, and although they are as domestic as any of the race, the conque of their ears is but semipendent; notwithstanding which, they have the faculty of elevating and moving them with as much facility as the unreclaimed races. They are destitute of the fifth toe found in the other varieties.

The Greyhound is but little susceptible of education; his intelligence is limited, and he seems to conceive with slowness and difficulty, while other varieties do so with facility. His sentiments, however, are very strong, and he is, more than any other, alive to caresses; indeed, his emotions, on being noticed, are so strong, if we may judge, at least, by the violent and irregular movements of the heart, that it seems difficult to believe how they can be borne. This want of intelligence, joined to high sensibility, however, seem to divest the Greyhound of any exclusive affection; he has no personal attachment, but is alike delighted with all who notice him.

The *Scotch Greyhound* has long, curling, stiffish hair, generally white, inclining to a reddish-brown tinge. It is also called the Wiry-haired Greyhound.

The *Russian Greyhound* has also long and bushy hair. The tail forms a spiral curl.

The *Italian Greyhound*. The *Turkish Greyhound*. These are small varieties of this group, which are very timid, and seem to suffer much from the cold of this part of Europe. The former is either white or sable-coloured. The latter has the skin nearly naked.

These which next follow, are included in the second subdivision of the Dogs of Mr. F. Cuvier, before alluded to.

The *Shepherd's Dog*. (*Canis Domesticus.* L.) This well-known animal is covered with long shaggy hair, and has little personal beauty to recommend it. The colour is, in general, varied black and gray. The ears, unlike those of most of the domesticated varieties, are short and erect; and the tail, which is bushy, is sometimes found directed horizontally, or even pendent, but more generally a little curved.

The peculiar and eminently-useful services of this variety to the shepherd, appear almost to arise from an intuitive disposition in the animal, rather than from laboured training; at least, there is an astonishing aptness exhibited by it in acquiring its lesson; with an apparent interest, patient perseverance, and courageous fidelity, accompanied by a discriminating sagacity in the performance of its task, when acquired, as notorious as it is surprising.

This breed is confined to the temperate and southern parts of Europe; and in England there are two varieties of it: first, the Shepherd's Dog, properly speaking, or that which is the usual attendant on the flocks while in their pastures; and, secondly, that which may be called the Drover's Dog, which is larger than the former, and more usually employed to assist in driving Sheep to the London market.

The *Terrier.* Two distinct varieties are used for the purpose of entering the burrows of Foxes, Badgers, &c., in hunting, both of which are thence called Terriers.

The first is generally black on the back, sides, head, and tail; but has the belly, neck, paws, and tip of the tail, a bright or reddish-brown, with a spot of the like colour over each eye. The hair is short; the tail is carried slightly curved upwards; the ears are short and erect; and the snout is moderately elongated. Though small, it is a very resolute Dog, and a determined enemy of Rats, Rabbits, and many other animals, in the pursuit of which it evinces an extraordinary and untaught alacrity. Some of them will draw a Badger from his hole.

The other species of the Terrier alluded to, is generally of a dirty white colour, except about the eyes and ears, which are brown. It stands higher before than behind; has the muzzle more truncated than the other, and beset with stiff bristles; the hair, all over, is rather long and curly; and the ears are partly erect, and partly pendulous.

This is, perhaps, in general, more powerful than the other. It is equally courageous, and quite as well fitted for the purposes from which they both take their name. It is sometimes called the Scotch Terrier.

The *Wolf, or Pomeranian Dog,* (*Canis Pomeranus,* L.) has the hair short on the head, feet, and ears, but long and silky on the body and tail. It is white, black, gray, or yellowish in colour; and has almost all the sagacity of the Shepherd's Dog, accompanied with much more strength. It is also used as a guard for the flocks, particularly in countries pestered with the Wolf, which it never fails to attack with success, while the former can only frighten that animal.

The *Siberian Dog,* (*Canis Sibiricus.* L.) app---- to be

nearly related to the last, and very like it, except that it is covered with long hair, even on the head and paws. Mr. Pennant adds, that the other varieties in the inland parts of Russia and Siberia, are chiefly from the Shepherd's Dog; and there is a high-limbed, taper-bodied kind, the Common Dog of the Calmuc and independent Tartars, excellent for the chase, and all other uses.

This breed is trained to the most important services in its cheerless native country, which appear to be very ill repaid, if the accounts we have of their treatment be correct. During the short Siberian summer, they are said to be turned adrift, to seek their own sustenance ; and, at the commencement of winter, they are taken home for a series of fatiguing labour. Four of these Dogs are attached, by pairs, to a sledge, and before them is placed a leader, on the good training of which much of the utility of the set seems to depend. These sledges carry but one person, who guides them principally by his voice, with the assistance of a stick, and of reins fastened to the collars of the Dogs. It is said, they will thus draw a sledge between seventy and eighty miles in a day ; and when the falling snow hides the beaten track from the sight of their master, they will keep or regain it by the power of their scent.

The *Esquimaux Dog.* This highly useful breed is described by Mr. Desmarest, as having the head shaped like that of the Wolf-Dog; the tail spreading and curved ; and the ears erect. The hair is of two sorts ; one silky, which is thinly scattered ; the other woolly, which is extremely thick, very fine, and curly, and may be pulled off in flocks from the animal. The colour is black, or reddish-gray, with large marks of white.

The *Spaniel.* (*Canis Extrarius.* L.) The Spaniel has the hair very long, in parts. It is generally white, with large brown, liver-coloured, or black spots, of irregular shape and size. The nose is sometimes cleft. The ears are very long and pendulous, and covered with long hair This race came originally from Spain, whence its name.

The *Setter.* The Setter is sometimes called the English Spaniel. It corresponds, in every point, with the true Spaniel ; but is trained more immediately for field-sports.

The *Alpine Spaniel.* The Alpine, or St. Bernard's variety of the Spaniel breed, exceeds all others in size and beauty. It generally reaches two feet in height at the shoulders, and full six feet from the nose to the end of the tail. There is a peculiarity about the corners of the eyes of this animal, which is attributed to the snow, and to the high windy regions it inhabits.

Two of these Dogs are sent out to scour the mountain, in search of lost or wearied travellers ; one with a warm cloak fastened on his back, the other with a basket tied round his neck, containing a bottle of cordial. They are frequently of the most eminent use in meeting the traveller, in these snowy and dangerous regions, in time to lead him to the convent. It is said, that, in cases where a man has been found by them in an exhausted state, perishing with cold and fatigue, they will lie close to him, and afford warmth from their own bodies, to assist his resuscitation.

The *Newfoundland Dog.* This admired species is also highly useful in its native country and climate, where it is employed for many purposes of labour, particularly drawing wood on sledges to the sea-coast, which they do without a driver, and return by themselves for more. Four of them are said to draw three hundred weight, on these sledges, a considerable distance.

They are fitted by nature and inclination for the water, being semiwebbed between the toes, which greatly facilitates their swimming; and many instances are to be found of their saving persons from drowning. Their disposition is extremely docile, though their powers are great.

The *Smaller Spaniel, King Charles s Dog, (Canis Brevi-*

pilis. L.) is a small variety of the Spaniel, prized as a fancy lap-dog in proportion to its diminutiveness. It is sometimes found entirely black, and is then called, in England, King Charles's Dog, from the liking evinced by our second Charles for this variety.

The *Maltese Dog.* The *Lion Dog*, (*Canis Leoninus.* L.) These, also, are small species of the Spaniel. The first is supposed to have sprung from the intercourse of the little Spaniel with the smaller Water-Dog. It has the hair, all over the body, extremely long and silky, and generally pure white. The other has long silky hair about the head, neck, shoulders, and extremity of the tail ; but on the other parts it is short, giving the little animal a leonine appearance. It is probably bred between the Little Spaniel and one of the naked varieties.

The *Great Water-Spaniel* (*Canis Aquaticus*, L.), has long curly hair, and is, in other respects, much like the large Land-Spaniel ; but the head is larger and rounder.

The small Water-Spaniel is presumed to be the offspring of the great Water-Dog and the Little Spaniel. It is very much like the former animal ; but the curly hair is more silky, and like that of the Land-Spaniel.

There is also a useful variety of this breed between the Water-Spaniel and Shepherd's Dog.

These animals are used as finders in shooting Water-Fowl, which their great fondness for water, and consequent aquatic habits, enable them to bring to the sportsman when the birds are shot, and have fallen into this element.

The *Hound,* (*Canis Sagax.* L.) The Hounds have the muzzle nearly as long as that of the Dogs included in the first division, but much larger ; their head is large and round ; the ears are large, long, and pendulous ; the limbs long and strong ; the body is thick and long ; the tail elevated ; the hair uniformly short, and the colour is white, with large irregular, black, brown, or yellow patches.

The largest variety of the Hound used for stag-hunting, is also sometimes trained to follow the scent of blood, and is thence called the Blood-Hound. This variety was formerly much fostered in Great Britain; and was probably of particular use during the existence of the severe forest-laws.

The King of Saxony kept a breed of Hounds of immense size and powers, for Boar-hunting. They were larger and taller than our largest Mastiff, and had the transverse dark shades on the body which characterize this animal in general rather than the Hound. The ground-colour was white, and the markings of a reddish or brownish yellow, in the different individuals. There is, as before stated, in the Museum of Dresden, a Dwarf Dog, which attained two years of age. Major·Smith observed, that this diminutive animal measured only five inches and a half in length, which was just the length, from the corner of the eye, to the tip of the nose, of a specimen of the Saxon Boar-Hounds he saw.

The Fox-Hound is a smaller variety of the Stag or Blood Hound, used in Fox-hunting. It is extremely persevering in the chase.

The Harrier is a still smaller variety of this species, used in Hare-hunting. There are, again, particular breeds of the Harrier, as the Beagles and Southern Hounds, which rather interest the sportsman than the zoologist.

The name of *Talbot* appears to have been applied to the several varieties of the Hound.

The *Pointer*, (*Canis Avicularius*, L) The muzzle of this variety is rather shorter and smaller than that of the Hounds in general; the head is shorter; and the ears, which are smaller, are partly erect and partly pendulous. There is a large breed, called the Spanish Pointer, which is considered as having greater acuteness of scent than the smaller or English Pointer. The Dalmatian Pointer is a beautiful spotted kind, which is white, with very small black or yellow spots. It is sometimes erroneously called the Danish Dog.

The *Turnspit*, (*Canis Vertagus.* L.) There are two varieties of the Turnspit; one with the fore-legs crooked, the other with the legs straight. The head is like that of the Pointer and Hound.

The third subdivision of M. Frederic Cuvier includes the following varieties.

The *Bull-Dog*, (*Canis Molossus.* L.) The round, thick head, turned-up nose, and thick pendulous lips of this formidable Dog, are familiar to all. The nostrils of this variety are frequently cleft.

The want of that degree of discernment which is found in so many of the canine varieties, added to the ferocity of the Bull-Dog, make it extremely dangerous, when its courage and strength are employed to protect the person or property of its owner, or for any domestic purpose; since, unlike many of the more sagacious, though less powerful Dogs, which seem rather more anxious to give the alarm, when danger threatens, by their barking, than to proceed immediately to action, the Bull-Dog, in general, makes a silent but furious attack; and the persisting powers of its teeth and jaws enable it to keep its hold against any but the greatest efforts, so that the utmost mischief is likely to ensue, as well to the innocent visitor of its domicile, as to the felonious intruder.

The savage barbarity, which, in various shapes, is so apt to show itself in the human mind, particularly when unchecked by education and refinement, has encouraged the breed of this variety of the Dog, in order that gratification may be derived from the madness and torture of the Bull and other animals, when exposed to the attacks of these furious beasts; and it is observed, that, since the decline of such sports, Bull Dogs have diminished in number; an instance whence we may learn how much the efforts of mankind operate on the domesticated genera of the animal kingdom.

The internal changes which determine the external characters of this Dog, consist in a great development of the frontal sinuses, a development which elevates the bones of the forehead above the nose, and which leads in the same direction the cerebral cavity.

But the most important change, and that, perhaps, which causes all the others, although we cannot perceive the connexion, is the diminution of the brain. The cerebral capacity of the Bull-Dog is sensibly smaller than in any other race, and it is, doubtless, to the decrease of the encephalon that we must attribute its inferiority to all others in every thing relating to intelligence. The Bull-Dog is scarcely capable of any education, and is fitted for nothing but combat and ferocity.

A fifth toe is occasionally found more or less developed on the hind feet of this race.

This, like all other races far removed from the primitive type, is difficult of reproduction; the males are seldom amorous, and the females frequently miscarry. Their life, also, is short, though their development is slow: they scarcely acquire maturity under eighteen months, and at five or six years show signs of decrepitude.

There is said to be a variety of the Bull-Dog found in Thibet, which is of a black colour.

The *Pug-Dog* may almost be called a diminutive variety of the Bull-Dog, to which it is nearly assimilated in appearance, though its tail is more curled. But this animal differs altogether in disposition from the Bull Dog, being as timid as the other is courageous.

The Mastiff, (*Canis Anglicus*. L.) This powerful breed is considered as English; it is said, however, to be bred between the Irish Wolf-Dog and the Bull-Dog. The ground-colour is generally a dirty white, with numerous dark hairs all over the body, and transverse stripes of a darker hue It is a very large and powerful dog, and being much more capable of training, and not less courageous than the Bull Dog, it is much fitter for domestic purposes. It is frequently known to protect its master's house and property by menaces only, even when a stranger is completely within its power; and will not be excited to violence, unless an imprudent perseverance should render it necessary for the protection of its charge ; and, in such cases even, it has been known to pull a man down, and stand over without hurting him a considerable time, till its master appeared.

This breed was assiduously fostered by the Romans, while they had possession of this Island ; and many of them were exported to Rome, to combat other animals in the amphitheatre.

There is a degree of generosity about this animal, which commonly attends true courage ; and, as if conscious of its superiority, the Mastiff has been known to chastise with great dignity the impertinence of an inferior. An instance is recorded of one, which, being frequently molested by a Mongrel, and teased by its continual barking, at last took it up in its mouth by the back, and, with great composure, dropped it over the quay into the river, without doing it any further injury.

Buffon's work, with Daubenton's additions, contains figures of the *German Dog*, or *Mopse ;* the *Iceland Dog ·* the *Little Danish Dog*, which is said to be improperly named, as there is no similarity of make or size between this and the Great Danish Dog; the *Bastard Pug ;* the *Artois Dog*, which is supposed now to be extinct; the *Naked*, or *Turkish Dog ;* and a variety of it with a sort of mane.

To the foregoing profiles of Wild Dogs from distant parts of the earth, resembling each other almost to identity,

Fig. 1.

Fig. 2

Fig 3

Fig. 4

F. 1   The Greyhound

F 2   The Alpine Spaniel

F 3   The Mastiff

F 4   The Bull-Dog

London. Published by G. B. Whittaker. Feb? 1825.

specimens are selected for the opposite plate, which exhibit the disparity in the shape of the head and jaws, incident to the domesticated varieties, that have induced the modern subdivisions of this variable race.

We must now quit these humble companions and faithful friends of Man, and proceed to a review of their rougher and more intractable congeners.

The docile character of the greater part of these animals must not induce us to forget that we are now treating on genera, decidedly of the true carnivorous type, properly to be termed beasts of prey. Those whom we have hitherto surveyed, were either, in part, frugivorous, and many of them proportionally gentle in their disposition ; or if carnivorous and cruel, yet wanting strength completely to enforce the demands of their sanguinary appetite, or to allow it any extensive range of annoyance. But the animals now before us, have both the power and the will for devastation and carnage. They may be termed, without impropriety, the aristrocratic order of the carnivorous tribes, and, like similar orders among men, at certain periods of history, they maintain their pre-eminence, by remorseless rapine, unsated thirst of blood, and inextinguishable ferocity.

Let us not, however, run into the popular error, that the ferocious disposition of the carnivora is unconquerable. It is a common opinion, that these animals, ever thirsting for blood, and stimulated into fury by the mere sight of their prey, are alike insensible to the voice of kindness and the rod of correction, and will resist, by the mere force of their native instinct, every means successfully employed in the taming of other species. Buffon, with his usual eloquence, in speaking of the Tiger, says, " His only instinct is a perpetual rage, a blind fury, which knows nothing, which distinguishes nothing. His disposition is,

Z 2

perhaps, at alone, among all animals, which is utterly inflexible. He is to be tamed neither by force, by restraint, nor by violence. He is equally irritated by kind and by severe treatment, the omnipotence of habit has no influence over his iron nature, and he will tear the hand which is daily extended to present him sustenance, without compunction or discrimination."

This error arises from a consideration of the habits acquired by these animals in their native forests, in a state of uncontrolled freedom, abandoned to themselves, and thrown entirely on their own resources for the support of their existence. Exclusively of their sanguinary appetites, and the sentiment of self-preservation, surrounded, as they are, by victims or by enemies, their actions must perpetually tend to the acquisition of the first, and the removal of the second, and must consequently be violent and cruel. Place them in different relations, and under other influences, and the case will be widely altered.—Commit them, betimes, to the care of man, and they will assume other manners, their destructive impulses will be weakened, more sociable feelings will be developed, and these terrible Carnivora, whose very name spreads terror and dismay, will manifest a capacity for the kindest affections, and submit with confidence to the voice of their benefactors.

It may be remarked as a curious fact, that the larger Carnivora are more easily tamed than the small. The truth is, that gifted, on the one hand, with superior strength, they are also possessed of superior intelligence on the other. They have more of that faculty which approximates to human reason, and less of blind instinct, than their weaker and more diminutive brethren. Instinct is a fatal enemy to education, and those animals, in which its manifestations are most frequent and surprising, are precisely those which, where it is not exerted, are the most unintelligent and unsusceptible of culture. The smallest Carnivora have been tamed, but they retain, in their domesticated state, characters exclusively peculiar to themselves, and derived, unquestionably, from the peculiarities of their cerebral structure.

The *Wolf*, which forms the subject of our present essay, is a striking proof of the truth on which we have been just insisting, and shows how much the character of Carnivorous animals varies, according to the circumstances under which it is developed. Submitted by the inscrutable fiat of nature to the domination of sanguinary appetites, intelligent to discover, and powerful to enforce the means of their gratification, we behold them in a state of nature, attacking every thing which has life, and spreading hatred and consternation around. But as the animals which are their destined prey are provided with activity to fly, sagacity to elude, and not unfrequently with strengh and courage to resist, so they, in their turn, must have the power of appreciating various circumstances, and of accommodating themselves to different situations. They must know when to employ force, and when to make use of stratagem; at what periods audacity will best serve their purpose, and when to assume the semblance of timidity. Thus, different faculties are summoned into play, and as the necessity of love exists in all beings endowed with sensibility, and which re-produce by sexual union, it is sufficient to place the most ferocious animals where they shall have no appetites to satisfy by violence, no enemies to combat or fear, where they shall have benefits to receive, and security to enjoy, almost, in appearance, to change their very nature, and to produce within them the kindliest sentiments of gratitude, of confidence, and of affection.

Experience confirms what reasoning would have led us to conclude. There is no Carnivorous animal which cannot be tamed by proper treatment, and which will not, to a certain degree, become affectionate and familiar, to those who attend and feed it. But this disposition is evinced in very different proportions by different species, and even by different individuals.

The Wolf is one of those ferocious animals in which attachment may be carried to the greatest extent, and which presents us with one of the most singular examples of the developement to which the desire of affection may attain—a desire so extraordinary, that it has been known to prevail, in this animal, over every other necessity of his nature

The individual, instanced by M. F. Cuvier, must undoubtedly have been, naturally, of a very excellent disposition. Brought up like a young Dog, he became familiar with every person whom he was in the habit of seeing. He would follow his master every where, seemed to suffer much from his absence, was obedient to his voice, evinced, invariably, the most entire submission, and differed, in fact, in nothing, from the tamest of domestic Dogs. His master being obliged to travel, made a present of him to the Royal Menagerie, at Paris. Here, shut up in his compartment, the animal remained for many weeks, without exhibiting the least gaiety, and almost without eating. He gradually, however, recovered; he attached himself to his keepers; and seemed to have forgotten all his past affections, when his master returned, after an absence of eighteen months. At the very first word which he pronounced, the Wolf, who did not see him in the crowd, instantly recognised him, and testified his joy by his motions and his cries. Being set at liberty, he overwhelmed his old friend with caresses, just as the most attached Dog would have done after a separation of a few days. Unhappily, his master was obliged to quit him a second time, and this absence was again, to the poor Wolf, the cause of most profound regret. But time allayed his grief. Three years elapsed, and the Wolf was living very comfortably with a young Dog, which had been given to him as a companion. After this space of time, which would have been sufficient to make any Dog, except that of Ulysses, forget his master, the gentleman again returned. It was evening, all was shut up, and the eyes of the animal could be of no use to him ; but the voice of his beloved master was not effaced from his memory ; the moment he heard it, he knew it ; he answered, by cries, indicative of the most impatient desire ; and when the obstacle, which separated them, was removed, his cries redoubled. The animal rushed forward, placed his two fore-feet on the shoulders of his friend, licked every part of his face, and threatened, with his teeth, his very keepers, who approached, and to whom, an instant before, he had been testifying the warmest affection. Such an enjoyment, as was to be expected, was succeeded by the most cruel pain to the poor animal. Separation again was necessary, and from that instant the Wolf became sad and immoveable ; he refused all sustenance ; pined away ; his hairs bristled up, as is usual with all sick animals ; at the end of eight days, he was not to be known, and there was every reason to apprehend his death. His health, however, became re-established, he resumed his good condition of body, and brilliant coat ; his keepers could again approach him, but he would not endure the caresses of any other person ; and he answered strangers by nothing but menaces.

Such is the recital of a scientific naturalist, himself an eye-witness of the facts which he relates, and who, we may well believe, as he himself asserts, has exaggerated nothing in his account of them. It is the narrative, not of an ignorant exhibitor, or an ambitious traveller, but of a philosopher, not less distinguished for his patient habits of obser_vation and comparison, than for the soundness and calmness of his general deductions. We dare not, therefore, refuse it a particle of credit, however little it may agree with the popular notions concerning the disposition of the Wolf, and the reports of travellers concerning it. But this species has hitherto been known only in its wild state, surrounded with enemies and dangers, among which no feelings could be developed, but those of fear, hatred, and distrust. Certain it is, that Dogs suffered to run wild in the woods, from birth, become just as savage and ferocious as Wolves, and yet we cannot suppose that they are so essentially. So true it is, that to acquire a complete knowledge of the character of a species, of its fundamental intellectual qualities,it must be seen under every circumstance adapted for their manifestation. On this subject, as on most others, the world, we trust, is fast outgrowing the prejudices of its childhood. The love of the marvellous and the terrible is rapidly losing its hold upon the public mind. Tales of wonder are either banished to the nursery, or reserved for the purposes of temporary amusement; and, even to those, they often fail to contribute, from their want of veri-sim litude, and the superior attractions of truth. A traveller, who should attempt to entertain us now, like Mandeville or Raleigh, would meet but an indifferent reception. The wide dissemination of knowledge has given birth to better taste, and sounder judgment; and the present era, we trust, will be characterized, in history, as the true age of reason, philosophy, and science, alike removed from the follies of superstition, and the rashness of impiety *

To return from this digression:—Extraordinary as those feelings, which we have been just describing, may appear in this species, we find the germ of them in the attachment which the young Wolves exhibit for each other, the tenderness of the She-Wolves for their young, and that affection

* It is, perhaps, needless to remark, that the above observations apply only to England, France, and America, and, with some limitations, to Germany. In the latter country, though illumination is widely spread, the marvellous still predominates in literature, and the mystic, in some branches of philosophy. As to the rest of the world, the less we say about it, perhaps, the better.—E. P.

which universally accompanies physical love. The author just cited gives us another instance of this affectionate disposition in one of this species, the degree of which is utterly unaccountable. A She-Wolf, in this particular, evinced more sensibility than the most attached and faithful Dog could possibly do. At the least word, expressed with kindness, the slightest pat of encouragement, she would press against you, turn in all manner of ways, as if to touch you better, and send forth a soft and plaintive cry, expressive of the pleasure which she felt ; nay, her emotion was so powerful as, *solvere vesicam, et facere ut copiosè urinam redderet.* But it was not merely to her master that she testified this extraordinary feeling ; it was produced by the caresses of every person who approached her. It would appear that it was merely the caresses which produced this effect, and that (unlike the last example) there was no discriminating sentiment of regard.

These were not the only examples of Wolves completely tamed in the Royal Menagerie. In 1800, there was a She-Wolf there, which had been caught in a snare, and which, though taken when adult, became so thoroughly tame, that she lived familiarly among the Dogs, with which she reproduced several times. She would bark like them, whenever she perceived a stranger, and she was so completely cured of her taste for poultry, that she might be suffered, with impunity, to enjoy the utmost freedom.

In his wild state, the Wolf exhibits none of the characteristics we have been detailing. Surrounded by enemies, and living always in fear and distrust, he is gloomy and brutal. In the gray of the morning, or at the approach of evening twilight, during the night in summer, or in the most sombre days of winter, he stalks forth in search of food, which, in cultivated countries, he rarely finds in abundance. It consists for the most part of the dead remains of domestic animals ; and in thinly-wooded tracts, of

Frogs, Field-mice, and other of the smaller animals. In large forests, where game is more abundant and the neighbouring population thinner, the Wolf becomes stronger, and bolder, and his frame exhibits more energy and elasticity. During the winter he retires to the recesses of lofty woods, in the neighbourhood of inhabited places; in the summer he keeps the open fields, concealed amid the ears of corn. The females are in heat in the month of January. They are immediately followed by all the Wolves in the neighbourhood, who settle their pretensions by the most sanguinary combats. The strongest, having driven away the rest, attaches himself to the female, and never quits her until the young are educated. Gestation continues a little more than sixty days, during which period the mother is busy in preparing a nest for her young, in some situation best adapted for shelter and concealment. She furnishes it with moss and with her own hairs, which she easily plucks out for the purpose, as it is the moulting season. She brings forth from three or four to eight or nine, according to her age, and for the first days she never quits them. The He-Wolf supplies her with food, and the suckling lasts about two months, but the young Wolves begin to eat at a month old. At first the parents only give them half-digested meat, which they themselves disgorge; and during this time one of the two always remains to guard the family. By degrees they feed them with fresh meat, and lastly bring them small living animals. After this, they make them join in the chase. About November or December, the young ones occasionally remove from their parents, and begin to live without them; but they still remain in habits of connexion from six to eight months. In fact, it is only sexual necessities that finally divide them altogether. Then they form another link, so that the Wolves cannot with strict propriety be called solitary animals, for though they live not in troops like Dogs, yet they

END OF PART IV.

live in families. This circumstance, as it presupposes the moral qualities necessary for such a situation, is sufficient to explain the examples of affection which we have instanced.

In the early days of winter, before the sexes couple, many of these animals are met pursuing their prey in concert. It has been remarked, that while one of them follows the game step by step, the others follow on the right and left waiting the moment when their victim shall make a turn aside to take the diagonal line and intercept his path.

The size, the proportions, and the physiognomy of the Wolf, bear no indistinct resemblance to those of the larger Mastiffs. The colour is a grayish fawn, irregularly distributed, which renders a detailed description of it somewhat difficult. It is moreover very rare to meet two Wolves exactly similar. It would appear that as they grow old, the gray predominates, while the fawn-colour is more general with those of younger age. In general, the head, neck, shoulders and back, sides, and crupper, are black mingled with fawn. This last colour predominates on the thighs, and the extremities, where it is paler. The under part of the neck and the breast are of the same tint. The tail is the colour of the thighs, and is terminated by some black. The internal face of the limbs, and especially of the thighs, is of a dirty grayish fawn. The belly, the lower jaw, internal face of the ear, and edge of the upper lip, as far as the lower part of the jaws, are white. A black longitudinal spot is found on the fore-legs on their anterior part above the carpus, and the head between the two ears is gray.

As to the organs of sense, motion, generation, and the teeth, the Wolf entirely resembles our common Dogs. Should we deem any further detail on this subject necessary, we shall give it when we come to the Chacal.

Of the synonymy of this animal it is unnecessary to speak.

It has been known in Europe in all ages, and should we cite the authors who have spoken of the species as existing in other countries, we should hazard the committal of grievous errors. It is certainly to be found in a large portion of Asia. We may also believe that it exists in Barbary, and Major Smith has seen and examined many specimens in North America, as far as the Isthmus of Panama. But the limits within which it may be found are by no means precisely established.

We should have been disposed to consider the *Black Wolf* in no other light than that of a black variety, particularly as many species of other genera of the large mammalia are found to have black consimilars, had not the Baron, in the text, treated it as distinct. It very common ; more so than the ordinary species or variety south of the Pyrenees.

Independently, however, of the deep and uniform black colour, in which it differs from the common species, there is a deviation also in the relative position of the eyes, and still more in the character of the fur, at least in the specimen from which our figure was taken, which is now in the menagerie of the Tower; the eyes appear to be rather nearer the ears than in the common species, and the hair seems to be much more erect and plentiful ; indeed, about the neck and throat, it is so bushy, as to shorten, considerably, the apparent length of that part. The tail, also, is more villose. In other respects, we observe no difference between the two species.

The above-mentioned individual has a companion in the same den, where, at least, they do not indulge in *ennui* from confinement, being almost constantly at play with each other.

As they are full of life and vigour, it would be no safe experiment to procure their dimensions, but they appear to be little or nothing short of those of the Common Wolf.

THE BLACK WOLF *London. Bay. Var.*

C. LYCAON. L.

London, Published by G.B Whittaker, Feb. 7 1825

T. Landseer del. et sc. Tower

Hearne and Mackenzie speak of the White Wolf, a variety, originating not in the supposed morbid excitement to ordinary albinism, observed so generally in the various species, but rather from the action, whatever be the *modus operandi* of a high latitude. Mr. Warder, also, informs us, that the Wolves found in the United States vary considerably both in dimensions and in colour. In the northern states, they are yellowish or reddish-brown, with a black dorsal line; while more to the south, they are found entirely black. Great caution, therefore, seems necessary, that these mere varieties be not admitted into systematic catalogues as distinct species.

To the specific characters of the Red Wolf or Agouara—Gouazou of d'Azara, as given in the text, we shall add nothing. Azara states that Agouara signifies the Fox, and Gouazou merely great, consequently, that the compound word implies that the animal is the largest of the Foxes.

This naturalist possessed one at about three months old, which was kept tied up. When any one approached, he would growl like the Dog, but in a more confused and louder manner. He lapped in drinking, and pressed his food under his fore-paws, while he tore it with his teeth; he was particularly fond of rats and small birds, and ate also of sugar-canes and oranges.

The figure is so like that of a Dog, that any one seeing it in the fields, without knowing it, would naturally take it for that species, for he has no other difference than in his mane and large erect ears, with their concavities turned in front; his legs, also, and figure are rather more slim. The mane, however, is a strong specific distinction, whence its scientific epithet jubatus.

He inhabits the low swampy lands, is a good swimmer, nocturnal in his habits, hunting solitarily, with much courage and agility, almost all other quadrupeds, even the stag.

This species seems much troubled with intestine worms, which has induced some exaggerated and ridiculous notions among the vulgar.

The *Mexican Wolf*, of a reddish-gray, mixed here and there with blackish, appears to be distinct, but it is not sufficiently described.

The *Chacal* is one of those species of the mammalia most widely extended throughout the warmer regions of the ancient world. It is found in Africa, from Barbary to the Cape of Good Hope ; in Syria, in Persia, and throughout the entire of southern Asia. Intense cold alone seems to present a bar to its multiplication. Humid or dry climates, sheltered countries, or exposed and arid plains, appear to suit its constitution equally well, provided there be sufficient warmth. It is not less common on the frontiers of Sahara, than on the confines of Senegal, in the mountains of Abyssinia, than on the shores of the Persian Gulf. It would seem that this species had received from nature the faculty of modifying and conforming itself to circumstances in a more eminent degree than others, that it might perform a more extensive part in its destined occupation. The Felinæ, and some other Carnivora, disdain to touch any thing, except living prey, unless, indeed, while suffering the extremity of hunger ; but the Chacals will feed on carcasses with avidity, and seem to partake, with the Hyænas and Vultures, the supposed office of ridding those countries, where life is most abundantly re-produced, of the remains of those organized bodies which, otherwise, would poison the atmosphere by their spontaneous decomposition.

All travellers, who have been in those countries where the Chacal is found, agree in mentioning the ravages occasioned by his voracity, and his dreadful nocturnal cries, which re-echoed by all the Chacals in the neighbourhood, produce the most discordant and lugubrious of all possible concerts, utterly depriving all hearers of repose, who have

1. THE CHACAL.   2. THE CHACAL OF SENEGAL

C. AUREUS. GM.   *C. ANTHUS. F. CUVIER.*

1

2

F. Bradley sc.

London. Published by G.B. Whittaker. Feb.? 1825.

not been long accustomed to it*. They add, that these
animals live in troops, inhabit burrows, which they them-

* " The Chacal's shriek bursts on mine ear
When mirth and music wont to charm."

LEYDEN.

As we have quoted this passage, we shall not refuse to embel-
lish our pages with the entire of the beautiful ode from which it is
taken.   It was written in the East Indies, at a time when the author was in momentary expectation of his dissolution, which soon
followed, from the fatal effects of a " *coup de soleil*."   It is addressed to an Indian gold coin:

" Slave of the dark and dirty mine!
What vanity has brought thee here ?
How can I love to see thee shine
So bright, whom I have bought so dear !
The tent-ropes flapping, lone I hear,
For twilight converse arm in arm ;
The Chacal's shriek bursts on mine ear,
When mirth and music wont to charm.
By Cherical's dark wandering streams,
Where cane-tufts shadow all the wild ;
Sweet visions haunt my waking dreams
Of Teviot loved while yet a child,
Of castled rocks stupendous pil'd,
By Esk or Eden's classic wave,
Where loves of youth and friendships smiled,
Uncurst by thee, vile yellow slave!
Slave of the mine ! thy yellow light,
Is baleful as the tomb-fire drear ;—
A gentle vision comes by night,
My lone deserted heart to cheer ;
Dim are those eyes with many a tear,
That once were guiding stars to mine ;
That fond heart beats with many a fear,—
I cannot bear to see thee shine !
For thee, for thee, vile yellow slave !
I left a heart that loved me true,
I crossed the tedious ocean-wave,
To roam in climes unkind and new.

The

selves excavate ; disinter dead bodies, and, when impelled by hunger, may become dangerous even to men.

The Chacal can be tamed with tolerable facility, but he always preserves an extreme timidity, which he manifests by concealing himself on hearing the slightest unusual sound, or on seeing any person whom he does not know. His fear, too, has a character different from that of other wild animals. Among the latter, it is nothing but the sentiment of self-preservation, the result of some apparent danger, and is as powerful a stimulus to resistance as to flight, when the latter has become impossible. The Chacal, on the contrary, like a Dog, which fears the chastisement of his master, flies when he is approached, but the moment you reach him, you may touch him in all manner of ways without any attempt, on his part, to resist or injure you. This apparent contradiction seems the result of this natural instinct, which impels him to distrust every strange species, and of his acquired knowledge, which has taught him that there is no real danger. This, perhaps, is the state which is nearest to the most perfect tameness. There are many animals which will not fly the presence of man, but which, at the same time, will not suffer themselves to be touched. Others will not fly, but will not receive caresses, except from those whom they are accustomed to

> The bleak wind of the stranger blew
> Chill on this withered heart :—the grave
> Dark and untimely met my view,
> And all for thee, vile yellow slave !
> And com'st thou now so late to mock
> A banished wand'rer's hopes forlorn,
> Now that his frame the lightning shock
> Of sun-rays tipt with death has borne
> From friends, from home, from country torn,
> To memory's fond regrets the prey.
> Vile slave ! thy yellow dross I scorn,
> Go ! mix thee with thy kindred clay."

see, and who are in the habit of ministering to their wants. But it is rare to see an instance of an animal who will fly, and yet suffer itself to be touched with impunity. As soon, however, as the Chacal knows the persons who approach him, he will fly no longer; he will even come and yield himself to their caresses.

This great facility of being tamed, and proneness to submission, remarked in some Chacals, would tend to confirm the idea of certain naturalists, who have deemed this species to be the original source of our domestic Dogs. In fact, the organization of the Chacals is entirely similar to that of the Dogs, and when these last re-enter the savage state, they assume, in all respects, the mode of existence of the Chacal. They form numerous families, dig burrows for themselves, feed on carcasses, and pursue their prey in concert. One essential difference, however, exists between them. The Chacals exhale an odour so strong and disagreeable, as must ever have prevented men from suffering them to approach too closely, or from making them the companions of their house and table. There is no reason for supposing that, in a domestic state, they would have lost this offensive peculiarity*. This of itself may be sufficient to refute the notion of the Chacal being the original root from which our common Dogs have sprung, though some have not noticed it, while others have. The fact is, that the presence of a single Chacal would be sufficient to poison a whole habitation.

The Chacals have always been compared to the Foxes, but they cannot, with any propriety, be said to appertain to a class of animals so generally considered nocturnal. They are, in fact, (with the exception of smell just mentioned,) genuine Dogs. Like these, the pupil of their eyes is round, the eye itself is simple, that is, without any accessory organ.

* It must not be forgotten, however, that all the Mephites or Skunks lose, in a great degree, their offensive smell, and the power of producing it, when in captivity, as Major Smith assures us.

The nostrils extend to the end of the muzzle, and open on its middle and sides. The ears are pointed, with a tubercle on the external edge. The tongue is extremely soft, and there are mustachios on the upper lip, above the eyes, and on the sides of the cheeks. The feet have four complete toes, but the anterior have the rudiment of a fifth toe on the internal side, and on the same feet there is a horny production behind the articulation of the wrist. The claws are short and thick. Six incisors and two canines are in each jaw. But it is unnecessary, in this place, to dwell longer on such characteristics. The coat is well furnished with hair, especially the tail, which resembles that of Foxes. In short, the Chacals, as to habit of body, movements, use of the senses, intelligence, instinct of concealing their food, &c., exactly resemble the Dogs.

The general colour of this animal is dirty fawn-colour above, and whitish underneath. The tail is a mixture of fawn-coloured and black hairs.

The *Chacal of Senegal* appears to belong to a species essentially distinct from that of the Chacal properly so called, that is, from the animal found in the central regions of Asia, and, perhaps, through the entire extent of Africa; which lives in troops, and feeds on carcasses.

The denomination of the Chacal of Senegal may be, in some degree, improper, as the true Chacal is also found in this country, but still there is no great inconvenience attending the use of it. We insert the figures of both.

This branch of the family of the Dogs of the old continent is liable to some obscurity, from the general uniformity of its organization. It seems to consist of the Common Chacal, the present species, the Adive, the Corsac, and Mesomelas. We may presume that the Europeans who inhabit Senegal, do not distinguish the common species from the other Chacals, and it may be as well, for the purpose of marking its differences, that we should begin by a rigorous determination of the characters of this species, that we may clearly prove that it appertains to none of these. It should be observed, however, that the Mesomelas is said to belong to the vulpine sub-division.

The Corsac and Adive (as the Baron conjectures) appear to be the same, if the Adive be the little Indian Dog called, in Malabar, *Nougi-Hari*. In fact, there are many of these Dogs in the Cabinet of the French Museum, sent by M. Lechenault. And when we compare the description given, by Guldenstadt, of the Corsac, to that given by Buffon of the Isatis, we shall find that there is no difference.

The Corsac is not larger than the Common Weasel, and the tail, which is very long in proportion to the body, descends three inches lower than the feet, when it is completely pendulous. All the upper parts of the body, and also the tail, are of an uniform grayish-fawn, of a very soft tint. The limbs are entirely fawn. The end of the tail is black, and, about three inches from its root, on the upper part, there is a white spot. All the lower parts of the body are of a yellowish-white. Individuals have been observed, which seem to blend this species with the Chacal of Pondicherry and the Common Fox without an interval.

The Common Chacal we have already described.

The Mesomelas is gray and fawn-coloured. Its size about the same as that of the Chacal, being doubly larger than the Corsac. The tail descends nearly to the earth. The hairs of the back have fawn-coloured, black and white rings, but as these are generally very large, the tint resulting from them is not so uniform as that of the upper parts of the Chacal. The black and white is irregularly mixed, contrasting strongly with each other, and with the brilliant fawn-colour of the other parts The tail is fawn-colour, with the extremity black. The colour of the back, which is broad towards the point of the animal, and descends over the shoulders, grows narrow behind, and is not two inches broad over the crupper. The ears of the Mesomelas are twice the size of those of the Chacal.

Having thus characterized these three species, we shall be better enabled to perform the same office for the Chacal of Senegal. It is not possible to confound it except with the Chacal. It is much larger than the Corsac, and wants the triangular dorsal patch of the Mesomelas. But we shall soon see that it differs essentially from the Chacal.

Its forms and proportions are more light and elegant than those of the last-mentioned animal. It is about fifteen inches high to the middle part of the back. Its body from tail to occiput is about fourteen inches long. The head, from occiput to the tip of the nose, seven inches, and the tail is ten inches. Back and sides are covered with a deep gray fur, sullied by a few yellowish tints. The gray is not uniformly spread, and, occasionally, the white and black, with which the hairs are tinged, become visible. The neck is grayish-fawn, still more gray upon the head, especially on the cheeks and below the ears. The upper part of the muzzle, the limbs, hinder part of the ears and tail, are of a pure fawn colour. The rest of the body is whitish. On the back and tail the hairs are long, short and smooth elsewhere.

The gait and motions of this animal are the same as those of the Dogs. When afraid, it claps its tail between its legs, and shows its teeth. This is not, however, a menace of anger, for the moment it is re-assured, by a few kind words, it will approach and lick one's hand. Its voice is soft; it is a prolonged sound, and not loud barking like a Dog, or Common Chacal. The cry, by which it evinces desire, is like that of a young Dog, and it always cries at hearing other animals do so. It exhales a tolerably strong odour, but much less so than that of the Chacal.

All the other parts of its organization are like those of Dogs in general. It is, therefore, unnecessary to say more on this subject.

Travellers have, unquestionably, mentioned this animal, and confounded it with the Chacal. This may account for what we mentioned in our essay on the last animal, namely.

that some travellers speak of the disagreeable odour it exhales, while others affirm the reverse. The affirmation refers to the Common Chacal, the negation to this species. M. Fred. Cuvier proposes to call this new Chacal, *Anthus*, the name of a family in Arcadia, an individual of which was every year metamorphosed into a Wolf, according to Pliny.

We shall now, as an appendage to the Chacals, speak of some mules, the result of an intercourse between the two last-mentioned species.

The assemblage of individuals, whose copulation is prolific, constitutes a species, and the exact knowledge of species is the basis of Natural History. As individuals themselves are the work of nature, so likewise are the relations in which they stand towards each other. Indeed nature appears to attach more importance to those relations, than to the individuals which are placed in them. The continuance of a species is evidently an object of greater consequence with her, than the preservation of each or any of the animals which compose it; and, in truth, the principal cause of the formation of the latter seems to be the conservation of the former. In some of the living tribes, individuals are called into existence obviously for no other purpose. The entire of their ephemeral life is comprised in birth, reproduction, and death. They propagate their kind, and then return to nothing. Nature demands nothing but the preservation of the species, and in contributing to that, their destiny is completely fulfilled.

There are few phenomena, therefore, more worthy of the attention of the philosopher, than the reproduction of species, and that independently of the mystery of fecundation This truth was felt by the illustrious Buffon, and if he has not been invariably consistent in the application of the general rule which he himself laid down, it was from the want of facts which might lead him to appreciate the exceptions presented by the production of mules. Those which we are about to notice, are said, by the French naturalists, to be the first ever produced from the intercourse of two species entirely wild, but this seems very doubtful, as similar instances have occurred here several years ago. Mules have often been born from the intercourse of two domestic species, such as the Horse and Ass, Sheep and Goat, or from that of a domestic and wild species, as the Wolf and Dog, Ass and Zebra, &c. From this it has been inferred, that the deviation of instinct, exhibited in such unions, was the result of the influence of man over these animals. It appears, however, that this is by no means a necessary condition to the production of this phenomenon. A domestic state, or even a long captivity, are not requisite for this purpose. In the case before us, the animals were very young, and had not been together in the same cell more than six months before intercourse ensued. It was odd enough too, that these animals were far from having lived in a good understanding with each other. They never played together. Each retained its own corner, and the female being the strongest, often made the other feel its superiority.

These animals we have already described as the Common Chacal and the Chacal of Senegal. The latter was the female. At the end of sixty-two days after the beginning of their intercourse five young ones were born. The mother at first evinced some inquietude, but at last received her offspring to her maternal cares. This adoption of the young by the mother is a fact worthy of attention, under the various circumstances in which it may occur. The sen_timent from which it springs, and which is instantaneously produced in all females the moment of birth, when they find themselves in a state of security, seems to be developed very imperfectly, and sometimes not at all when they are under the influence of captivity and restraint. The feelings

which arise in such a state seem opposed to those of maternal love, to domineer over the animal, and to be the
cause of the monstrous action we sometimes witness of the
parent devouring its offspring. The first manifestation of
the internal sentiment consists in the care which the mother takes to clean the young immediately after birth ; but
it never exists in its full force until she has permitted them
to suckle. Previously to that time, she may be mistaken,
destroy, or abandon them. These are accidents more especially liable to occur at the first birth. They are less to be
feared afterwards, probably because the organization of
the animal is more developed and perfected, and the creature can better resist the influence of those moral and physical causes which would otherwise pervert its natural
instincts.

These little Mules at their birth were seven inches long
from muzzle to tail, and the latter two inches and a half.
The ears and eyes were closed. The conch indeed was free,
but its tubercles obstructed the entrance of the auditory
canals. In about ten days, as in young dogs, these organs
were opened. These young animals were covered with a
soft and thick coat, woolly on the body, silky on the head
and paws. It was generally of a fine slate-coloured gray,
mingled in some parts with a tint of fawn. There was a
white transverse line on the breast between the two fore-
legs. This colour, in about forty-nine days, changed to a
dirty fawn

During the suckling three of these mules died. The two
surviving ones evinced almost from the moment of birth a
remarkable difference of character. One shewed no symp-
toms of fear, while the other constantly manifested the most
lively terror. The first became familiar, and even gave
tokens of affection, the other remained wild, and it seemed
very improbable that any attention would succeed in taming
him. Yet these animals were brought up exactly alike. Such

examples only shew, that though education may modify, it cannot change the natural character; and that there are cases in which the latter will completely resist its influence.

The FOXES are separated by zoologists into a distinct group among the Caninæ, distinguished by the eye-pupil of an elongated shape.

The *Common Fox* is one of those animals whose habitat is most widely extended over the surface of the globe. It is found in all the middle and northern regions of the old and of the new world. The faculty of rapid multiplication and diversified extension, which it possesses in so eminent a degree above the other carnivorous tribes, must in a great measure be attributed to its instinctive choice of such places of concealment as are accessible to none of its enemies except man.

The Fox is not a little particular in the choice of his quarters. When he purposes to establish himself in a neighbourhood, he visits every part of it, fathoms the extent of every excavation, and carefully examines every spot that promises a convenient place of refuge in the hour of danger. As soon as he appropriates an habitation suitable to his wants, he instantly commences to scour the country, reconnoitres every post around, ascertains the resources placed within his power, and the nature and degree of the dangers with which he may be threatened. Constantly under the guidance of the most extreme and cautious prudence, and never leaving any thing to the result of chance, he lays himself down with tranquillity to taste the pleasures of repose. A repose thus guarded and secured is the only one that his natural timidity will permit him to enjoy. The excessive suspicion of his character renders every new object a source of distrust and inquietude. He is uneasy until he has discovered what it is, and approaches for the purpose of observation with slow and hesitating steps, and by indirect and circuitous paths. Accordingly whenever he is agitated by a permanent source of fear, he betakes himself to flight, and proceeds to seek in some other retreat that security which he can no longer enjoy in his present abode. He passes the live-long day at the bottom of his hiding-place, and sallies forth in search of prey, only during the obscurity of twilight and the darkness of night. Guided with equal certainty by the sense of smelling as of sight, he glides along the trenches of the field to surprise the Partridge on her nest, or the Hare within her form. Sometimes he will lie in ambush near the burrows of Rabbits, into which he even occasionally penetrates, and sometimes with the cry of a Dog, he gives chase to those animals in the open plain. When game of this description fails, he will subsist on Field-Mice, on Frogs, on Snails, and on Grasshoppers. In cultivated and well-inhabited countries, the Fox finds new resources. He approaches the habitations to collect the refuse of provisions thrown out of kitchens, &c. He penetrates into poultry-yards, where he makes terrible devastation ; and in autumn he will enter the vineyards, and feed upon the grapes, which fatten him, and diminish in some degree the disagreeable odour of his flesh. But he does not limit himself to the quantity of food necessary to appease the hunger of the moment. Instinct leads him, where there is abundance of prey, to lay up provision for the future. When he invades a poultry-yard, he kills all he can, and carries away successively every piece, which he conceals in the neighbourhood to retake them at a more convenient opportunity.

This character of extreme prudence in the Fox is a main cause of his preservation. It renders him extremely difficult to be destroyed or taken. As soon as he has acquired a little experience, he is not to be deceived by the snares which are laid for him, and from the moment in which he recognises them, nothing, not even the severest pangs of hunger, can induce him to approach them. Le Roi, in his letters upon animals, informs us that he has known a Fox to remain fifteen days in his subterraneous hole, that he might not fall into the snares with which he had been environed.

This timid prudence, however, completely disappears in the female Fox when she has young ones to nurse and to defend. The maternal instinct which in all species, the human not excepted, is probably the strongest of all feelings, effaces in the instance before us the specific character of the animal. There is no sentiment so completely disinterested as this, none in which the sacrifice of self is so instantaneous and so complete. The mother will not hesitate a moment to endure the utmost privation, to brave the most appalling danger, nay, to encounter the certainty of death for the preservation of her infant offspring. She that but a little before was all gentleness, shrinking timidity, and fastidious delicacy, who could not bear " the winds of heaven to visit her face too roughly," becomes on the sudden bold, fierce, and resolute, unshaken by all that is trying, and unrevolted by all that is disgusting. The female Fox watches incessantly over her young, provides for all their wants with unwearied assiduity, and exhibits an audacity very foreign to her general disposition against their most formidable adversaries.

If we might presume to conjecture at the proximate cause of this maternal instinct, we should be inclined to trace it, like many other powerful sentiments in animal nature, to some sensation of physical pleasure, by which its exercise is accompanied. Even in man, those feelings which assume, for a time, the completest domination over his constitution, have sensual pleasure as their origin and object, however remote their apparent distance from such a source may be, however they may be glossed over by high-sounding names, or to whatever degree of refinement they may be spun by those mighty casuists, vanity and self-love. All our feelings and ideas, however refined and abstracted, are resolvable in their last analysis into physical sensation, and the closer their connexion is with this primal source, the more impetuous and commanding is their influence. If this be the case with man, it is much more strikingly so with the brute creation.

About the month of February, the Foxes are in heat. They are then heard to utter very sharp yelpings, which commence like the barking of a Dog, and end in a sound resembling the cry of a Peacock. Gestation continues for from sixty to sixty-five days. When the female is ready for parturition, she prepares a bed for her young with leaves and hay. The cubs are generally from five to eight in number, and born like Dogs, covered with hair, and having the eyes shut.

As the vicinity of the Fox is productive of nothing but inconvenience to Man, and as its intelligence augments its resources against danger, the Fox-chase has always afforded a subject of occupation and amusement to great landed proprietors. Many crowned heads, both in our own and foreign countries, have been passionately devoted to this sport. Among others, Louis XIII. of France gave to this species of hunting the preference over all others, and even brought to perfection the employing the Hound instead of the Terrier, which last, previously to his time, had been constantly used for this purpose. This piece of information we derive from Robert de Salnove, lieutenant of the chase to that royal lump of imbecility.

At about three or four months old, the young Foxes quit their burrow. They abandon their parents with all convenient speed, and at two years of age their growth is completed.

The Fox averages about two feet and a half, or a little

more, from rump to muzzle. Its medium height is about one foot. A fawn colour intermixed with black and white constitutes its characteristic hue. The fawn predominates on the head, along the spinal column, the flanks, the posterior part of the limbs, and the sides of the tail. Grayish-fawn sprinkled with white prevails on the thighs and shoulders. The under part of the neck and breast anteriorly, a kind of half collar at the bottom of the neck, and a narrow spot commencing at the internal angle of the eye, and descending towards the throat, are black. But it is superfluous to dilate on the colours of so well-known an animal.

The coat is thick, especially on the tail and back. In winter the woolly hairs are more abundant than the silky, and at that season the fur is more valuable. In summer the silky hairs predominate, and their number is not great.

The physiognomy of the Common Fox, its slender muzzle, large head, and shortness of limbs, in comparison of the body, are well known. With one exception, the organization of the Fox and Dog are precisely similar. This exception is the eye, which in the Fox resembles that of our domestic Cat, and not that of the Dog. The pupil contracts in a strong light, and appears only a narrow and longitudinal section. It opens and assumes a circular form only during twilight or night. This animal consequently, like the Cat, avoids the light, and prefers obscurity and darkness.

The Fox has been always known. The Greeks named it *Alopex*, and the Latins *Vulpes*. This last name has been most usually given to it by authors since the restoration of letters. Gesner and Johnston have given very good figures of it, and those of Buffon and Schreber are very exact. Its scientific name is *Canis Vulpes*.

A variety of the Fox has been found, principally in Burgundy and Alsace, and described as distinct under the name of *Alopex*. Its colour is somewhat of a deeper red, and its fur thicker than that of the ordinary Fox, and from this last-mentioned peculiarity it derives a thicker and more squat appearance. Some are also found which have more black hair than common along the dorsal line, and across the shoulders, and to those the name of the European Cross Fox has been applied. The cross disposition of the black stripes is met with in three or four species or varieties of the Fox ; but the distinctive epithet is applicable only to a South American species, in which this character is peculiarly remarkable.

The Egyptian Fox is treated as distinct by Geoffroy, under the name of *Canis Niloticus.* It differs very slightly from the common species.

Of the *Tri-coloured Fox (Canis Cinereo-Argenteus,)* it may be observed that the name might in point of fact be applied to the other species, as the white, the fawn-colour, and the black are combined in the fur of almost all the Foxes. But improper as it may be, it must be retained, as it has been so long received, and as we are ignorant of the name which this animal bears in the middle and southern regions of North America, which are its native countries. It is astonishing how little care is taken by travellers to ascertain the proper names of the animals of those countries which they traverse, even when the means of such information are completely within their reach. The influence which such information must exercise on the progress of natural history would give a double value to their researches. The history of any species can evidently be the result only of a very long series of observations, which it is utterly impossible for any single individual to make. To the first observations of this description the second should be naturally attached, for the purpose of giving them their full portion of utility, the third to the second, and so on, until all the necessary information is acquired. Without this plan, we are liable to endless repetitions, which can produce nothing but regret for the labour which has been expended on them. In fact, the knowledge of the native name is essentially necessary to enable us to know of what animal any traveller speaks, and of which, in all probability, he cites but a few characteristics, very insufficient for the purposes of a clear distinction. On this point the ancients appear to have been much more careful than the moderns. The generality of the latter can bear no comparison in this respect with Marcereau, Hernandez, Pison, &c. Such, however, as the Baron Humboldt, Dr. Horsfield, our respected friend, Major Hamilton Smith, Peron, D'Azzara, and a few more, are honourable exceptions to this remark, and stand at an immeasurable distance above the generality of our modern travellers.

Such reflections as these are sure to suggest themselves to the mind, when we come to consider the Foxes of North America. There are few animals of which travellers have spoken more, but there are few whose history has been treated of with less detail, and with less attention to any thing like method. It is difficult to know what use to make of the numerous notes upon the Foxes, which we find scattered over the works of Hearne, Mackenzie, Bartram, &c.

Without doubt they had seen, as many others had seen before them, this species, the tri-coloured Fox. But notwithstanding, until very lately, the animal has been known only from the account of Schreber, who has given a very imperfect figure, drawn in all appearance from a stuffed specimen, and a bare description of colours.

M. F. Cuvier describes an individual sent from New York to the French menagerie. It was so very young that its second dentition had not yet commenced. It died during the development of its second canines, which generally forms a crisis very painful and dangerous to wild animals in a state of captivity. Without evincing malignity, it was not familiar. Its graceful form, the facility of its motions, and particularly the soft and brilliant colours of its fur, would have constituted it a very agreeable animal, had it not been for the unpleasant odour which it emitted, and which, unquestionably, would have become much stronger with advancing years. The head, across the lower part of the *os frontis*, round the eyes, and thence to the internal edge of the ears, was of a reddish-gray. The rest of the muzzle was white and black. A little white on the upper lip, then a large black spot, and then white under the lower jaw. The sides and under part of the neck were of a brilliant fawn. The upper part of the neck, back, shoulders, crupper, thigh, and part of the leg, were of a beautiful silvery-gray.

The fur was composed of woolly hairs, in great quantity, generally of a pale-gray, but with a red tint on the extremities, and of silken hairs, short on the muzzle and paws, long elsewhere, but scanty in number. Such was the distribution of its colours. Its organization was, of course, that of Dogs in general, with the exception of the elongated pupil which it possessed in common with other Foxes. Its length of body, from the muzzle to the root of the tail, was about a foot and a half. The tail itself one foot, and the mean height about eleven inches.

Our knowledge of the tri-coloured Fox is, as we have observed, due to Schreber. Under his figure, he gives it the name of *Canis Cinereo argenteus*, and in the text he calls it *Canis-Griseus*. D'Azzara, also, it would appear, speaks of this species of the Fox, in his account of the animals of Paraguay, under the appellation of *Agouarachy*. We should be cautious, however, in the admission of this identity, both as his description differs, in some points, from the animal we are treating of, and as the country in which this specimen was found, is so remote from the habitat of the Fox, which we have been describing.

M. F. Cuvier, in an article in the "Dictionnaire des Sciences Naturelles," admitted this identity, but afterwards, in his great work on the Mammalia, he hesitates to do so for the reasons just assigned. As for ourselves, we cannot venture to dissent from so high an authority, but must fully agree with him, that every error of synonymy has a direct tendency to retard the progress of science. The Agouarachy is referred, by M. Geoffroy St. Hilaire, to a gray Fox of Paraguay, recently brought to the French Museum.

The Tri-coloured Fox is easily tamed, if taken young. It is disposed to be playful with those whom it knows, and can distinguish the family of its keeper from strangers. Should a dog enter its master's dwelling, it will instantly expel the intruder, but with the dogs of the house it lives on terms of great intimacy. It has a great propensity to sleep during the day, a sufficient indication of its nocturnal habits in a state of nature. It is not easily compelled, notwithstanding its tameness, to enter or quit any place, but rather than do so, will submit to blows, which it answers by growling.

The particulars, in this last paragraph, are taken from D'Azzara, whose animal, as we have seen, is not completely ascertained to be the same with the *Canis Cinereo-Argenteus*. But all the manners and cunning of the European Fox are, in general, attributed to this its congener on the other side of the Atlantic. This is, probably, the Virginian Fox of Catesby.

The *Silvery Fox* (*Canis Argentatus*) is a species which has been known for a long time, and in high estimation, on account of the beauty and richness of its fur, which becomes very valuable when manufactured. Notwithstanding this, we have had no figure of the animal until lately, when one was published by the Baron's brother, in his Lithographic work, and another in the "Dictionnaire des Sciences Naturelles." Naturalists did not even possess very clear notions on the animal, until latterly. Brisson

Linnæus, Erxleben, and Gmelin, have not admitted it as a species, and the two last have confounded it with the *Canis Lycaon.* All that Pennant has said upon the subject is founded on the relations of Charlevoix and Du-Pratz. It is to M. Geoffroy St. Hilaire that we are indebted for an exact description of the Canis Argentatus, which was given from a stuffed specimen in the French Museum. The animal itself was brought from North America.

This animal is of the size of the ordinary Fox, and it entire organization is precisely similar to that of the same animal. The organs of sense, of motion, of dentition, and of generation, are the same, and its gait and movements exactly alike. It walks like the Canis Vulpes, with its head and tail depressed. Its glances are pregnant with distrust and penetration, and, in a word, it would be completely our European Fox, if it were fawn-coloured instead of black. It is, altogether, of this latter colour, with which is mixed, in certain points, and in greater or less proportions, a small quantity of white. The extremity of the tail is almost entirely of this last colour, the fore part of the head and the sides are whitish, and some white hairs are detached, as it were, from all the other parts of the fur, and have no other effect than to set off to better advantage the lustrous brilliancy of the black, of which it is generally composed. The hair of the body and of the tail is long and tufted. Silken hairs, widely dispersed, extremely fine, and of a gray, approaching to black, form the immediate covering of the skin, and the colour of the animal is owing to silken hairs, which are generally of a brilliant black, though occasionally terminated by a white point, and sometimes, but rarely, altogether white. On the paws, the hair is short, and on the muzzle still more so. The eyes are yellowish.

This animal plays in the manner of Dogs, and expresses,

like them, its displeasure, by growling. When it has satisfied its hunger, it conceals the rest of his aliments, lies down, and goes to sleep. Its odour is extremely disagreeable, but differs a little from that of the Common Fox. The exhaustion which it suffers from heat, sufficiently indicates the countries of which it is a native.

Almost all authors, who have travelled in the northern parts of the old world, speak of Black Foxes, which has led to an opinion, in which there appears probability, that the species which we have been just describing is to be found in both continents. Some doubt, however, must be preserved respecting their identity, until such time as it shall be confirmed by new observations, and a more exact comparison of characters. The relations of travellers, hitherto, are deficient in the degree of precision necessary for such a purpose.

The *Cross Fox* (*Canis Decussatus*) is described by Geoffroy, as a South American animal, of the size of the Common Fox. The fur is variegated with black and white, which gives a gray appearance to the upper part of the body, and it has a black transversal stripe over the shoulders : the muzzle, lower part of the body, and paws, are black : the flanks and parts about the anus have a yellow tinge : the extremity of the tail is grayish-white.

This animal, in all probability, is a mere variety of the Canis Argentatus of Geoffroy.

The *Arctic Fox* (*Canis Lagopus*) exhibits, in a remarkable degree, the mutation of colour which Polar animals generally undergo on the change of seasons.

It is an inhabitant of the mountainous and open countries of the Arctic region, where it burrows under ground, during the short time that the earth is soft enough for this purpose.

There was one brought from Spitzbergen by Captain Ross, when on his voyage of discovery. The animal looks very rough and ragged during the process of changing its colour, and until near its conclusion. When the change is completed the colour is uniform. In winter it is of a pure white. In summer a dorsal line of a darker colour is observable, with transverse stripes upon the shoulders, from which peculiarities it has been occasionally confounded with the Cross-Fox. The paws are entirely covered with long hairs, and those on the other parts of the body are about two inches in length.

It is with some hesitation, however, that we place Captain Ross's Fox under the Lagopus. In general make and appearance it approaches more to the Common than to the Arctic species ; but in colours and their change, it assimilates to the latter.

There is also a Variable Fox, an inhabitant of the Arctic regions, which has the tips of the ears and tail black, like those of the Variable Hare.

The *Cape Fox* (*Canis Mesomelas*) is distinguished from the others by a more pointed muzzle, a long bushy tail, and elliptical pupils. It is usually been called the Cape Chacal, but the characters we have instanced constitute it a Fox, on the authority of our author. M. Desmarest, however, places it amongst the Caninæ with circular pupils. It is about the size of a small Dog. These animals are found at the Cape of Good Hope, and are represented to have the manners and habits of the Chacal.

We shall simply refer to the table for several other Foxes that have been named by writers, without attempting, for want of data, to determine their respective claims to a distinct classification. Much difficulty indeed exists on this subject, from the changes which many of these animals undergo in colour at different seasons.

Major Smith's drawings include a great many individuals, particularly of the transatlantic species, some of which, from their dissimilarity to described species, may probably be entitled to a specific separation.

In the Museum at Paris there is a nondescript animal, which seems to differ from the Fennec principally in dimensions. It is about the size of the Common Fox. The ears are preposterously large and long. The fur is an iron-gray, slightly tinted with yellow ; along the dorsal line the hair is rather longer than elsewhere, and darker in colour. The ears are gray on the outside, with the edge black, bordered with a few white hairs. The tail is very villose, black, with some gray at the upper end. The head is gray, with the forehead to the extremity of the nose blackish. The belly is pale white ; the four paws black.

It was shot by M. Delalande at the Cape, and as to all its generic characters is decidedly a Dog, most probably of the Vulpine section. The Baron has named it *Canis Megalotis*. Major Smith added *Lalandi*, to distinguish it from the Megalotis or Fennec of Bruce.

This species seems to form a natural gradation in the transition from the Common Foxes to the Fennec, which appears likely to terminate the genus ; so far at least as size and relative proportions may be connected with the commencement or termination of any series of animals.

In the Museum at Frankfort is a specimen of the *Fennec, C. Megalotis,* which we are enabled, by Major Smith's kindness, to engrave from his drawing. It was sent from the interior of Nubia by the German naturalists at present in that country. Professor Grætzmer and M. Temminck, after mature examination, believe this not to be the same as Bruce's Fennec, but a congener. The Major inclines to a contrary opinion. The skull has not been taken out, and the teeth remain to be examined.

Authors are too justly stigmatized as a jealous race, an observation often verified among zoological writers in particular, in the extreme eagerness evinced in describing and naming new or pretended new species, and the arts employed in procuring the means to do so, an abundant source of repetition, inaccuracy, and confusion in zoological

C. Hamilton Smith. del.

THE FENNEC. OF DE LALAND

MEGALOTIS LALANDI.

catalogues. Our countryman Bruce, and a Swedish gentleman, Mr. Skioldebrand, each claim the honour of introducing the Fennec to the scientific world. The latter, as Bruce
asserts by the exercise of a petty and unworthy artifice,
certainly got the start of the former. Neither of their
descriptions, however, has been sufficient to determine
the generic character of the animal, and it has accordingly
been appropriated in turn to almost every genus of Mammalia.

Buffon gave a figure of it in his work, from a drawing
sent him by Bruce. He seems to place it between the
Squirrel and the Hare.

Bruce, subsequently, in his travels, describes it; and M.
Blumenbach, from his description, refers it to the Civets.

Sparman identified it with a South African animal, called
at the Cape, Zerda, a name adopted in consequence by
Gmelin, Pennant, Boddaert, and others. The real Zerda
seems likely to be a Fox.

Mr. Pennant, without attempting to determine its generic
appropriation, intimates his opinion that it is a Vulpine
animal.

Illiger describes the teeth, but does not state his authority,
or where he inspected his type. He makes a new genus of
the animal under the name Megalotis, and places it by the
side of the Hyænas.

M. Geoffroy St. Hilaire, notwithstanding Bruce's description, which he assumes to be incorrect, and with all
that ingenuity and research of which he is so capable, with-
draws the Fennec from among the carnivorous animals, and
makes it a Galago, to which we have alluded in our observations on the animals proper to that subdivision.

M. Desmarest, on the contrary, gives Bruce credit for
the accuracy of his description, and opposes the deductions
of M. Geoffroy. It does not appear whether the former
naturalist had the opportunity of inspecting a specimen,

before he wrote his observations in the *Encyclopedie Metho-dique*, but as he there renounces his assertions made in the *Nouveau Dictionnaire d'Hist. Nat.*, that the nails were re-tractile, which assertions were merely hypothetical, grounded on the facility evinced by the animal in ascending the Palm-tree, it seems probable that he had.

The authority of M. Geoffroy is of such weight, that it may be proper to allude to the points of difference between this animal and the Galago, in all of which it approaches the Caninæ. The Galagos, then, have the hind extremities much longer than those before. The Fennec and the Dogs have not. The Galagos have four fingers and a thumb like that of Monkeys, perfectly opposable to the fingers. The Fennec and the Dogs are tetradactylous, with the mere rudi-ment of a thumb on the fore feet. The Fennec and the Dogs, as well as the Felinæ, have a slit or fold on the lower part of the external edge of the ear, which the Galogos have not. The extraordinary development of the external ear in the Fennec is analogous in kind, but not in degree to what is met with among the Dogs, particularly some newly dis-covered species or varieties of the Vulpine division ; whereas, in the Galogos, the ears never exceed the length of the head. The strong whiskers found on the upper lip of the Fennec and the Dogs are not observable in the Galagos, and lastly, the tail is shorter and more villose or foxy in the Fennec than in the Galagos.

We have already said that this singular animal seems to terminate the canine series. It is true that some of its most important generic characters are still in doubt : for instance, it has six incisors and two canines in each jaw, and six cheek teeth in the upper jaw, but the number of those in the lower jaw, and the character of them in both, have not been described. The character of its tongue, whether rough or smooth, and the certain presence or absence of anal folliculi remain also to be stated. Subject

THE FENNEC.

CANIS MEGALOTIS.

to the accordance of these characters with those of the Caninæ, there seems no good reason for separating this animal from that genus: diminutiveness certainly affords none, and we have already mentioned a Domestic Dog not exceeding five inches in length.

The Baron confining his most useful exertions to actual observation as an operative naturalist, treats on nothing that has not fallen under his notice in a state of nature, for which reason we are as yet deprived of his observations on this curious little animal.

The stuffed specimen whence the figure was taken is so small that it might be concealed conveniently in a pint mug.

After the Caninæ, or at least as a distinct section of the race, and before the Hyænas, must be placed a newly-discovered or described animal, partaking in several points of both these genera, and consequently intermediate between them; the number and character of its teeth corresponding with those of Dogs, would place it in that subgenus of the "Animal Kingdom," in which, as may be observed, dentition is selected as the most influential distinctive character.

This, and such like intermediate animals, appear to claim the particular attention of the zoologist, as affording curious matters of fact, from which results remain to be deduced—they form the connecting links, which, as it were, chain organization together: they seem to multiply the extent and enlarge the influence of secondary causes in the great work of creation, and stand decidedly opposed to a host of other facts which display the impassable barriers interposed by nature between the several creatures and their respective races.

The colonists at the Cape, as well as the aboriginal inhabitants there, appear to have been long acquainted with this animal, under the name of the Wild Dog, but its peculiarities remained unobserved, until Mr. Burchel pointed them out. He brought with him, from South America, a specimen, from which, as we believe, M. Temminck published a curious and interesting memoir, in the *Annales Générale des Sciences Physiques*, treating it as a Hyæna, under the name of the painted Hyæna, *(Hyène peinte.)* He afterwards presented a scull to the French Museum, and M. Desmarest, adverting to its dentition alone for its generic character, has placed it in his catalogue among the genus Canis.

The Baron afterwards notices it in the second edition of his Ossemens Fossiles, under the synonymes of Painted Hyæna, Wild Dog, and Hyæna Dog, which last appellative seems most descriptive, and is analogous to the Hyæna Civet, a species also holding a corresponding station between those genera.

Since these notices, Mr. Burchel, in the second volume of his Travels, has more particularly described it under the name of Hyæna Venatica, which we submit, at least in the Cuvierian arrangement, should be rather abandoned for that of the Hyæna Dog.

It is smaller, says Mr. Burchel, and of a more slender make than either the Common Striped Hyæna, or the Spotted or *Crocata.* The general, or ground colour, is a sandy bay or an ochreous yellow, shaded with a darker hair. The whol body is blotched and brindled with black, intermingled in various parts with spots of white ; and the legs are generally marked in the same manner. All these spots and markings are exceedingly irregular, and in some degree vary in different individuals.

We refer to the figure more particularly for the external character and description.

The osteology of this animal throws the principal difficulty in the way of its classification. In the teeth it agrees with Canis, except that the little lobe in front of the false molars is rather more developed. In the ribs and lumbar vertebræ it also agrees with Canis, but it differs from that

THE HYÆNA DOG. HYÆNA VENATICA.—BURCHELL.

genus in approaching Hyæna, in having but four toes on each foot, and it is said in other essential particulars.— Mos eorum copulandi mos canum non est, u. d. If this be so, the absolute separation of the species seems absolutely necessary.

Mr. Burchel had a living subject in his possession, for thirteen months, chained up in a stable yard. During this time its ferocious nature deterred every body from all attempts at taming it; but it became at length so much softened in manner, as to play with a common dog, also chained up in the yard, without manifesting any desire of hurting its companion, but the man who fed it dared never to venture his hand upon it.

They hunt in regular packs, whence Mr. Burchel's specific epithet: though in general a nocturnal animal, it frequently pursues its prey by day; and as it is well formed by nature for speed, none but the fleeter animals can escape. Sheep and oxen are therefore more particularly subject to its attack, the first openly, but the latter only by stealth, surprising them in their sleep and suddenly biting off their tail, which the large opening and great powers of its jaws enables it to do with ease. The large cattle, it appears, are assaulted by them in no other way, but the loss of their tail is a great inconvenience to cows and oxen, in a country where the warmth of the climate subjects them to great annoyance from flies.

We now come to that subdivision of the Carnassiers which is called the VIVERRÆ, whose generic character is detailed by our author, in the text. The first species is the *Civet* itself. Authors have so imperfectly marked the distinctive characters between this animal and the *Zibeth*, that Buffon was inclined to suspect that there was no essential difference between them, but that they were at the most mere varieties of the same species: in fact, the figures and descriptions which had been given of them appeared to represent animals altogether similar; and in some systematic catalogues, where they had been distinguished from each other, it was by characters which had no foundation in truth. The differences, however, between these two animals have now ceased to be a subject of doubt or controversy; they have both been possessed, alive, by the French menagerie, and their external appearance alone is quite sufficient to prevent any possible chance of their ever being confounded together in future.

They are animals which, certainly, in physiognomy and form, exhibit many mutual relations. In the principal points of organization they entirely resemble—they have the same teeth, the same organs of sense, motion, and generation. Respecting these, under the article of the Genet, we shall add as much to our author's description as may be necessary. Here we propose to institute a comparison between the Civet and the Zibeth, which will form the subject of our next description. This last animal has the body pretty generally covered with black spots, which are round and small, upon a gray ground, occasionally tinted with brown. The Civet has transversal bands, upon a gray ground, narrow, and parallel with each other on the shoulders, larger on the body and the thighs, and which are sometimes so much approximated and curved as to form eye-like spots, like those of the Panther; eight or ten rings, of a blackish brown, cover the tail of the Zibeth, while on that of the Civet there are but four or five, and its extremity, for about six inches, is entirely black, while the tail of the Zibeth is of the same colour only for about two inches at the tip; this last has on the sides of the neck four black bands on a white ground: the Civet has also a white with black bands, but only three in number, and there are some trivial differences in the position of the bands in each of these animals. The Zibeth has a white spot under the eye, and the muzzle

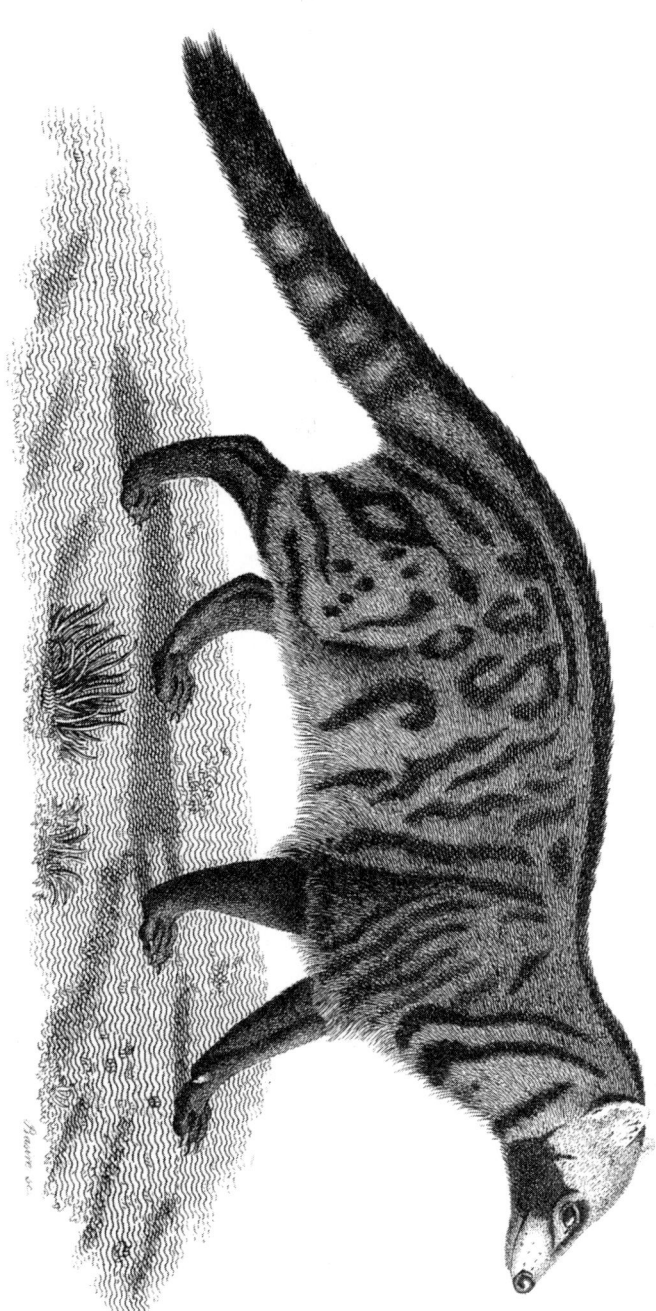

THE CIVET.

VIVERRA CIVETTA. Gm.

London. Published by G.B.Whitaker. Sepr 1825.

is gray; this part of the head in the Civet is entirely black, except the upper lip, which is white, and there is no spot under the eye; the limbs are black in both animals, and, in general, there is a greater quantity of brown in the Zibeth, than in the Civet, whose clear tints are of a pure white. The dorsal mane of the Civet is stronger than that of the Zibeth, and its coat is in general rougher, from the stiffness of the silky hairs; the woolly hairs are of a grayish brown, and considerable in number : the fore part of the ears is of a grayish white, and the hind part black ; the under part of the belly is white, but the hairs are brown at their base, and sometimes black : such are the differences of colour between these two animals. We shall now cite the observations of the Baron upon the Civet, from a work entitled " Menagerie du Museum d'Histoire Naturelle."

" The most remarkable peculiarity in the anatomy of the Civet, is the organization of the bag, containing its peculiar scent. It opens externally by a narrow cleft, situated between the anus and the parts of generation, and is exactly similar in both sexes, which renders their apparent difference but trifling. This cleft conducts into two cavities, which might each of them contain an almond Their internal surface is slightly covered with fine hair, and pierced with many holes, each of which conducts into an oval follicle, of very slight depth ; the concave surface of which is again pierced with innumerable pores. The odoriferous substance comes from these pores. It fills the follicle, and when this is compressed, it proceeds from it something, in form, like vermicelli, and enters the larger bag. All these follicles are enveloped by a membraneous tunic, which receives many of the sanguineous vessels; and this tunic, in its turn, is covered by a muscle, which comes from the pubis, and has the power of compressing all the follicles, and with them the entire bag, to which they are attached. By means of this compression, the animal gets rid of the superfluous part of its perfume. Beside this odoriferous matter, there is another secreted, which assumes the form of stiff silken threads, and is mingled with the first. The Civet has, besides, a small hole on each side of the anus, from which a blackish and very fœtid liquid issues.

The odoriferous substance produced by the Civet, and to which this animal owes its common name, forms, especially in the East, an object of considerable commerce. " Its virtues," says the Baron, " are greatly vaunted among ourselves, and it was once the fashion among those who piqued themselves on their elegance, to use it as a perfume, as it has since been to use musk and amber for the same purpose. It still enters into the composition of some medicaments and perfumes, but its consumption is prodigiously diminished. It used to be brought from the Indies, and from Africa, into Europe, by the way of Alexandria and Venice."

Africa, and a part of Asia, appear to be the native habitat of this animal. In the East the Civet is brought up in a state of domestication, for the purpose of gathering its perfume. Father Poncet says, that Enfras, a town of Abyssinia, is celebrated for the Civet-trade, and that an immense number of these animals are there domesticated. He has seen upwards of three hundred with some merchants. Buffon reports, that a similar practice was prevalent in Holland. Certain it is, that this animal has been repeatedly brought into Europe, and seen by many naturalists ; but as they did not distinguish it from the Zibeth, it is impossible to refer what they have said on the subject to one animal rather than the other.

The Civet sleeps continually, and is roused with much difficulty. They are animals of the greatest possible indolence, and, in this respect, not even the Sarigues can be compared with them. They differ very much in this point

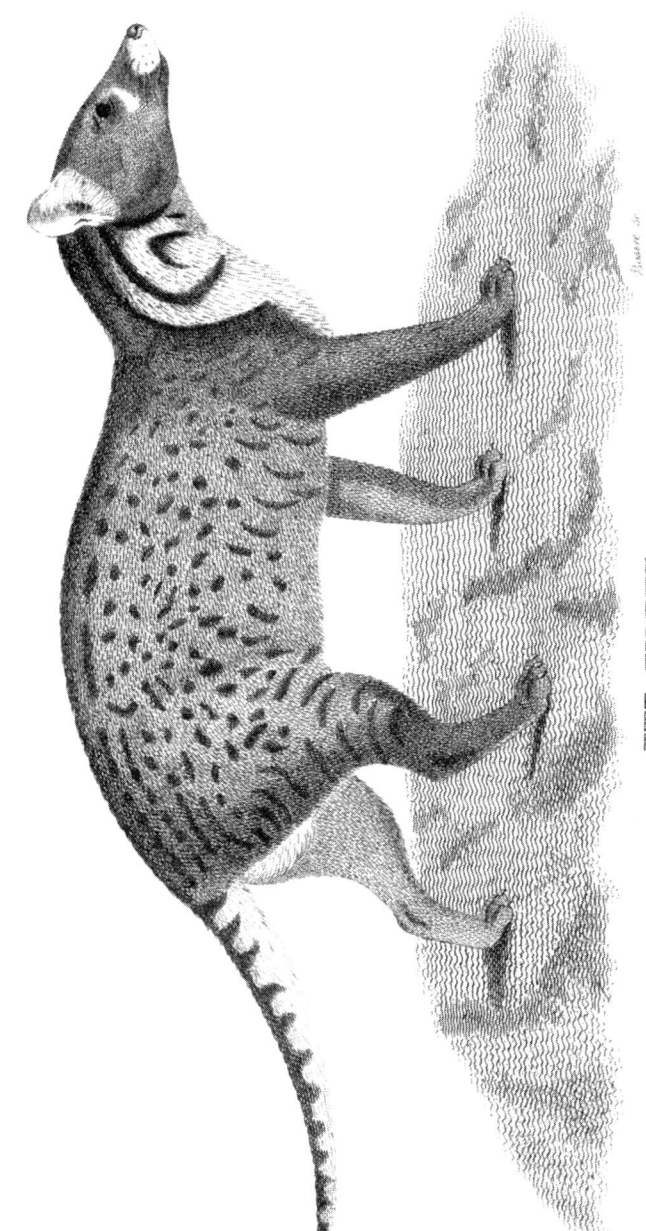

THE ZIBETT

VIVERRA ZIBETTA. 6m.

London.Published by G.B.Whitaker. Sep.r 1825.

from the Mangoustes, with which, however, they have a close analogy in the structure of their organs. This lethargic state does not permit us to discover any thing like intelligence in the Civets: it is probable, however, that they have less of it in their wild state, but in captivity they do nothing but eat and sleep. When they are irritated, the musky odour which they spread becomes stronger; and from time to time it falls from the pouch, in small pieces, about the size of a nut.

The Civet is nearly two feet and a half in length; the tail is more than one foot, and the mean height of the animal is about one foot three inches.

The ancients appear to have been acquainted with no species of this animal. As for the names of Civet and Zibeth, they are but one and the same name, spelt differently—*Viverra Civetta* is the scientific name.

Though the *Zibeth* appears to have been many times described, and known for a long period before Buffon, it was, nevertheless, only from that illustrious writer that naturalists learned to distinguish it from the Civet, with which it had hitherto been confounded. Buffon himself found so much resemblance between those animals, that he doubted whether they should be considered as distinct species or merely as varieties. This doubt might still subsist, although Gmelin cut the difficulty short by separating them, but for the observations of M. F. Cuvier; for the Zibeth was not the subject of any other since the time of Buffon. Buffon's uncertainty may be explained by the bad state of the animals which he compared. The Civet had been kept for a long time in a spirituous liquor, and the Zibeth he did not see until after its death. It is well known how much antiseptic liquids and disease will alter and deteriorate all spotted furs. This alteration is quite discoverable in the vague and indeterminate descriptions which Daubenton gives of these animals, descriptions in which we find nothing of the clearness and precision which generally characterizes all the labours of this worthy coadjutor of Buffon.

We shall not follow M. F. Cuvier (from whom the substance of these observations is taken) in his very minute account of the colours of the Zibeth, as we have already sufficiently marked its distinction from the Civet, in treating of this latter animal. Its musk-pouch, in most respects, is similar to that of the Civet, and constitutes a genuine sac, the bottom of which, divided into two parts, is terminated by two collections of glands which secrete the odoriferous matter.

Few or no observations have been made on the natura. character of the Zibeth. It is a sleepy animal, which sees badly during daylight ; and, like the Fox, forbears to provide for its necessities until the approach of twilight, or of night. It preys upon the smaller mammalia, upon birds and reptiles, and will occasionally eat the sweeter kind of fruits. It is in general silent, but when irritated, it manifests its anger by scolding and hissing, something like the domestic cat, and bristling up the hairs along its back. The animal on which these observations were made, was brought from the Philippine Islands. Notwithstanding that these countries have been so long discovered, they are as yet but imperfectly known ; and it appears from the accounts of travellers, especially of such as have recently visited them, that few regions of the globe are better adapted to enrich Natural History. Like all large and isolated lands, such as Madagascar, New Holland, and New Guinea, we might pronounce them to be the result of a new creation. Most of their productions exhibit new characters, and discover principles of existence before unknown. Accordingly, naturalists who may explore them will be certain to make numerous and important discoveries in every branch of zoological knowledge.

We have already ventured to express an opinion against the needless multiplication of specific distinctions, founded upon trivial, and merely external differences. A better place for a reiterated expression of the same opinion can hardly be, than after the description just given of the Civet and the Zibeth. Here we have seen two animals, in size, in form, and in organic structure, precisely similar. Their habits, too, as far as we can observe, are exactly the same,—both use the same food, both are great sleepers, and both are nocturnal. What then is the difference between them? A very slight variation in the shades of their colours, and in the mode of their arrangement. Is this a sufficient ground of specific distinction? If it be, then assuredly the Negro and the European are different species. The particulars in which they differ from each other are far more numerous and important than the variations of the Zibeth and the Civet. The fact is, that no certain and universally applicable criterion, by which species is to be distinguished from variety, has been yet discovered. The want of such a criterion is more especially felt in the classification of the minor tribes. As specific distinctions are the arbitrary creation of man, perhaps such a criterion cannot be found. At all events, in the case of multitudes of animals, we are working in the dark. Ages of observation will probably be insufficient to establish any thing completely determinate on the subject.

Dr. Horsfield has figured and described the *Viverra Rasse*, the Rasse of the Javanese, as another species of the first sub-division of the Viverra, *i. e.*, the Civets, properly speaking. If the Civet and the Zibet be with propriety treated as mere varieties, the former of the African, and the latter of the Asiatic world, the Rasse, according to Doctor Horsfield's description, would appear to us to stand in the same degree of relationship as a Javanese variety. It is, indeed, almost painful so frequently to have occasion to recur to similar observations, particularly when in some degree opposed to such powerful authority as that of the eminent zoologist just mentioned ; but so long as the true distinctive test between mere varieties of the same species and absolute diversity, remains so indeterminate as it is at present, uncertainties on this subject must prevail. An undue multiplication of species is so much the more earnestly to be deprecated, as it is injurious to zoological science, by swelling the catalogues of proper names, if not to the actual number of individuals, at least of the several countries or degrees of latitude and longitude in which the same species may be enabled to exist under various influences.

A comparative slight diversity of size and colour seems to constitute the principal differences between the Rasse and the Zibet ; but as Doctor Horsfield himself, with his usual learning and ability, compares the two, we shall conclude our observations on these animals in his words.

" The entire length of the Rasse, from the end of the muzzle to the root of the tail, is one foot eleven inches ; the head measures five inches and one-fourth, and the tail twelve inches ; the distance between the ears, at the base, is ten lines. A very perfect specimen of the Viverra Zibetha, the Tanggalung of the Malays, forwarded from Sumatra by Sir Stamford Raffles, affords the means of shewing more distinctly the peculiarities of the Rasse by a careful comparison. The Tanggalung is two feet six inches long ; the head measures six inches and three-fourths, and the tail eleven inches. The space between the ears is two inches. The proportion of the parts of the body of the two species are very different. The Viverra Zibetha is comparatively a stout animal ; the neck is short and thick, and the breast full and distended. The head, which in the Rasse is regularly attenuated, in form of a wedge, in the Tanggalung, is swelled, rounded, and bulging before the ears, and then very abruptly contracted to a short muzzle.

The ears are ten lines distant in the Rasse, and two inches in the Zibetha; this character gives a very different physiognomy to the two animals. The tail is nearly cylindrical in the Tanggalung; in the Rasse, it is regularly and uniformly attenuated to a point. In the hairy covering, or fur, these two animals are essentially different; while it is rigid, coarse, and rather scantily disposed in the Rasse, it is close, soft to the touch, and provided with much down at the base in the Tanggalung, and its thickness affords a peculiarity to the tail of the latter.

" I shall now concisely enumerate the distinctions afforded by the external marks. The Viverra Zibetha has a single black line of considerable breadth, in the highest part of the back, bounded on each side by a white line; exterior to this, is an interrupted line of a dark colour, while the rest of the back and sides is covered with smaller spots, disposed in such a manner, as to give the appearance of these parts of being transversely undulated. In the Rasse, eight regular parallel lines are clearly distinguishable. The upper parts of the head and neck present no difference in these two animals; but the marks on the lateral and interior parts of the neck, are very dark in the Zibetha, while they are faint and indistinct in the Rasse. The rings are strongly marked, and pass uniformly around the tail in the Rasse; in the Viverra Zibetha, they are irregularly defined, and scarcely perceptible on the under side of the tail.

" The name Rasse, like many other Javanese names, is derived from the Sanskrit language; and it is therefore entitled to be employed as a specific name, with the same propriety as Civetta and Zibetha, which are derived from the Arabic. Rasse, as employed by the Javanese, is a modification of Rasa, and is applied to our animal as producing an odoriferous substance. In the original, Rasa has various significations, of which flavour or taste appears to be the primary meaning; the others also relate chiefly to the senses, or to emotions that arise from them ; fluids or juices are comprised among its meanings, and many applications of the word Rasa and its compounds, to odoriferous substances, perfumes, &c., might be adduced.

" The Viverra Rasse supplies in Java the place which the Viverra Civetta holds in Africa, and the Viverra Zibetha on the Asiatic continent ; from Arabia to Malabar, and in the large islands of the Indian Archipelago. I have endeavoured to show that, by its form and marks, it is essentially distinct from the Viverra Zibetha, and it differs as much in its natural disposition as in external characters. The Viverra Zibetha is an animal comparatively of a mild disposition : it is often found among the Arabs and Malays which inhabit the maritime parts of Borneo, Macassar, and other islands, in a state of partial domestication, and, by the account of the natives, becomes reconciled to its confinement, and in habits, and degree of tameness, resembles the common domestic cat. The Rasse, on the contrary, preserves in confinement the natural ferocity of its disposition undiminished. As the perfume is greatly valued by the natives, it is frequently kept in cages ; but as far as I have observed, must always be obtained for this purpose from a wild state, never propagating in a state of confinement.

" The Rasse is not unfrequently found in Java, in forests of a moderate elevation above the level of the ocean. Here it preys on small birds and animals of every description. It possesses the sanguinary appetite of animals of this family in a high degree ; and the structure of its teeth correspond strictly with the habits and modes of life. In confinement, it will devour a mixed diet, and is fed on eggs, fish, flesh and rice. Salt is reported by the natives to be a poison to it. The odoriferous substance, the *Dedes* of the Javanese, or *Jibet* of the Malays, is collected periodically ; the animal is placed in a narrow cage, in which the head and anterior extremities are confined ; the posterior parts are then easily secured, while the Civet is removed with a simple spatula.

" The substance obtained from the Rasse agrees with the civet afforded by the Viverra Civetta and Zibetha, in colour, consistence, and odour. It is a very favourite perfume among the Javanese, and applied both to their dresses, and by means of various unguents and mixtures of flowers to their persons. Even the apartments and the furniture of the natives of rank are generally scented with it to such a degree as to be offensive to Europeans, and at their feasts and public processions the air is widely filled with this odour."

The next sub-division of the Viverræ are the GENETS. It will be sufficient for our purposes to notice here the *Genet* of Barbary. Two animals of this species were presented to the French menagerie, by M. Adanson, brother to the celebrated naturalist and traveller of the same name. These animals were very young when first received at Paris, and they lived for more than ten years. When they died, it was discovered that they had lost all their teeth. Whether this was accident, or the effect of age, is doubtful. They were kept in a cage not very spacious, in a corner of which they passed the day fast asleep, and rolled up in a ball. It was during the night that they watched, took their food, and satisfied all their other wants.

From their slender and elongated body, pointed muzzle, short limbs, and entire physiognomy, we might feel inclined to refer them to the family of the Martens. But a more attentive examination, and a more detailed study of their organization, prove their approximation to the Civets, by the side of which they are accordingly ranged in a particular group.

The teeth of the Genet are exactly similar to those of

the Civets. It is, like them, only a semi-carnivorous animal. If it can be fed with meat, it can also be supported
on bread, milk, &c., without any intermixture of animal
substances.

The Genets have two tuberculous molars in the upper
jaw, and one in the lower, together with three carnivorous
teeth, which are very thick, and are themselves tuberculous.
There are also three false molars in the upper jaw and
three in the lower.

The Genet's organs of motion are also similar to those
of the Civet. They have five toes on each foot. That which
we may call the thumb has but two phalanges, the others
have three. The three middle toes are the longest. The
middle one is the longest of all; next comes the little toe,
and the thumb is the shortest. They are armed with slender and semi-retractile claws, which are very sharp, and
well adapted for climbing. The walk of this animal is digitigrade, and the tail is semi-pendulous and susceptible of
voluntary motion, but not adapted for seizing or involving
objects.

On each side of the organs of generation are twoglands
raher thick and projecting, which are joined together at
their upper part, that is at the side of the anus, by a strip
of skin which covers them, and give to these parts the appearance of a pouch, though in reality they do not form
one. These glands produce a thick matter, and of an odour
approaching to that of musk. This forms an additional
relation between the Genet and Civet.

The Genet is a nocturnal animal, and the pupil resembles
that of the Domestic Cat. There is no other peculiarity in
the organ of vision. The nostrils open at the extremity of
the muzzle. The lips are susceptible of very limited movements. The tongue is covered with horny papillæ. The
external ears are large, elliptical, and provided with a small
lobule. Their aperture is very large.

The ground colour of the Genet is a yellowish-gray, and the body is covered with blackish spots. These are long on the neck and shoulders, and generally rounded on the sides and limbs. They form almost a continuous line along the dorsal ridge. The tail has about ten or eleven dark brown or black rings. There is no difference in colour between the males and females.

The Genet may be about a foot or more in length, the tail is about nine inches. Its mean height does not exceed five inches.

It would appear that the habits of these animals are pretty similar to those of the Weasel tribe. They live, we we are told, in low grounds, and in the neighbourhood of small rivers. They are easily tamed, as are all semi-carnivorous animals. Belon tells us that they are found at Constantinople in a domestic state, and like Cats, are employed to take Rats and Mice. This we may easily believe, as both are nocturnal animals. Those which belonged to the French Museum were a male and female : they coupled, and one young one was produced, which was immediately killed by the male on its appearance in the world. Gestation continued about four months. This young one was about five inches long, and the colour of its parents.

So many changes have lately been introduced into the arrangement of the Viverrine animals, that it is extremely difficult to determine, from books, what is the real number of them, or their varieties already described. The introduction, by the Baron Cuvier, of the new sub-genus paradoxurus, to which we shall shortly have occasion to refer more particularly, renders it difficult to ascertain the identity of the several animals noticed by authors. We shall, however, endeavour, to follow the authority of the Baron on this subject, so far, at least, as the materials before us will elucidate his intentions.

The *Fossan* (*Viverra Fossa* of Gm.) is very much assimilated to the Common Genet, both in form and the disposition of the colours of its fur. The ground colour is reddish-gray, the upper part of the head is brown, mixed with red and gray; there is a pale yellowish-white spot above each eye; four brown bands pass from the neck to the middle of the back; a series of brown spots is continued from them to the tail; similar bands and spots are found on the posterior part of the sides of the neck, shoulders, flanks, and outside of the thighs; the tail is semi-annulated.

Others, marked with transverse bands on the upper part of the body, have been designated as distinct species, but with what propriety is a question that must be left to future investigation. We have now before us no less than seven original drawings of different individuals, all differing very materially in external characters; but in the present state of knowledge of these animals, we really fear to increase perplexity, rather than diminish it, by publishing them, particularly as they cannot be accompanied with verbal descriptions of minute examination. Passing over, therefore, several that have been described as distinct, by Gmelin, Buffon, Geoffroy St. Hilaire, Pallas, and Blainville, we shall proceed to notice one very singularly formed animal, which the Baron has placed, conditionally, in this sub-division of the mammalia, and of which we are enabled to give a figure.

The extreme caution of the Baron on the subject of multiplying the genera and species, has, probably, induced him to place this animal with the Genets; but it must be remembered, that this classification is only conditional, and probability strongly indicates that a more intimate knowledge of the species will fully warrant its generic separation.

The animal in question is the *Hyæna Genet* (*Viverra Hyænoides*, Cuvier). The specimens in the Parisian Mu-

THE HYÆNA GENET.

VIVERRA HYÆNOIDES. *Cuvier.*

seum are young, so that we are, as yet, unacquainted with the minute detail of its organization in an adult state.

The Hyæna Genet has, in all, thirty teeth: the six incisors, in both jaws, are flat, trenchant, and divided by a furrow on the external side; the canines are very sharp, straight, and form a much elongated cone, those of the lower jaw being slightly bent. There are four cheek-teeth above, very small, and separated from each other, the three first being false molars, each with a single point, and the fourth a small tuberculous tooth, with two tubercles. In the lower jaw, there are three false molars, the first with a single point, and a single root; the second with two roots, a single point, and a little posterior heel; and the third with two little points, and a small heel. The condyles of the lower jaw are on a line with the teeth, as in the felinæ.

The general form of the skull, without the integuments, is intermediate between those of the Genets and the Dogs.

The general appearance of the animal is perfectly that of the Hyæna, except in size, its maximum of development not appearing to exceed half that of the Spotted Hyæna; the individuals in question are not larger than a Fox.

But to describe its characters rather more minutely, the ears are long and pointed; the nose is assimilated to that of the Dogs. There are five toes on the fore feet, and four on those behind, armed with strong pointed claws. The foot seems constructed nearly for the plantigrade mode of motion. The ground colour is yellowish-gray, varied on the body with six or seven black bands, passing from the dorsal line to the flanks; three small longitudinal bands mark the fore part of the shoulders, and one the upper part of the crupper. The thighs and legs, both before and behind, have some imperfect annuli. A black mane, like that of the Hyæna, passes down the neck and dorsal line. The tarsi and fingers are of a deep gray-black before. The tail is nearly as villose as that of the Fox, and stronger toward the end than at its base; its colour is grayish near the body, and a brown-black as it approaches the termination. The muzzle is blackish; the upper part of the head and outside of the ears gray.

The boundless diversity of nature must ever be a subject of admiration and astonishment to the limited faculties of the human mind. Every creature of organization is different from all others. Whatever analogies may exist in the various genera and species, absolute similarity is no where to be found, and the observation, however trite, and however familiar, that no two faces are alike, is not the less true or the less astonishing.

The highly singular animal now before us, appears, to a certain extent, however limited, to militate against the universal application of this observation: one might be almost tempted, at first sight, to suppose that nature tired of novelties, and at a loss for further diversity of form, had been, as it were, compelled in multiplying her works, to have recourse to her own created models, and no longer to draw upon the exhausted resources of original powers; but a deliberate conclusion to that effect would be equally at variance with our better notions of Omnipotence, and with the wonderful phenomena of creation we see around us.

We proceed to the third subdivision of the Viverra which includes the MANGOUSTES, the first of which is the celebrated *Ichneumon*. If, in the mythological system of the ancient Egyptians, the various living beings which people the surface of the earth were each entitled to particular reverence in consequence of the influence which they exercise over the economy of nature, and the part which they contribute to the general harmony of the universe, the Ichneumon unquestionably possessed more claims than any other animal to the homage of that singular people. It presented a lively image of a beneficent power perpetually engaged in the destruction of those noisome and dangerous reptiles which propagate with such terrible rapidity in hot and humid climates. The Ichneumon is led by its instinct, and obviously destined by its peculiar powers, to the destruction of animals of this kind. Not that it dares to attack Crocodiles, Serpents, and the larger of the Lizard tribe by open force, or when these creatures have arrived at their complete development. It is by feeding on their eggs that the Ichneumon reduces the number of these intolerable pests. The Ichneumon, from its diminutive size and timid disposition, has neither the power to overcome nor the courage to attack such formidable adversaries. Nor is it an animal of the most decidedly carnivorous appetite. Urged by its instinct of destruction, and guided at the same time by the utmost prudence, it may be seen at the close of day gliding through the ridges and inequalities of the soil, fixing its attention on every thing that strikes its senses, with the view of evading danger or discovering prey. If chance favours its researches, it never limits itself to the momentary gratification of its appetite: it destroys every living thing within its reach, which is too feeble to offer it any effectual resistance. It particularly seeks after eggs, of which it is extremely fond, and through this taste it proves the means of destruction to so many Crocodiles. That it enters the mouth of this animal when asleep is as much true as that it attacks it when awake. This is either a fable which never had any foundation, or, like other miracles and marvels, it has ceased in our unbelieving and less favoured era. The time when animals abounded with the strangest generic mixtures, and the most extraordinary propensities, when every grove and every stream were haunted by natures in which the divinity and the brute were incongruously commingled, has long gone by. The well-spring of faith, the grand source of the miraculous. is nearly dried up ; and men have ceased to witness these sorts of wonders precisely at the period when they ceased to believe them.

The Ichneumon exhibits the utmost perseverance in the pursuit of its prey. It will remain for hours in the same place watching the animal which it has marked out as its victim. This quality renders it an exceedingly proper substitute for a Cat, in the office of ridding a house of such parasitical animals as may have chosen it for their retreat. For this purpose the Ichneumons are constantly domesticated. They acquire an attachment to the house which they inhabit, and to the persons with which they are brought up ; they never wander, or make the slightest attempt to regain their original state of wildness and liberty. They know the persons, and recognise the voices of their masters, and are pleased with the caresses which are bestowed upon them. These animals, however, lose a considerable part of the mildness of their character in the act of eating. They seek the most hidden retreats, and manifest the utmost anger if any one approach them when they are satisfying their appetite.

When an Ichneumon penetrates into a place unknown to it, it immediately explores every hole and corner. Its instrument of research appears to be chiefly its sense of smelling, which is uncommonly powerful and acute. To this it seems principally to trust, for its other senses, particularly those of sight, taste, and touch, are comparatively feeble, and present no peculiar characteristics. The external ear, indeed, is remarkable, by its considerable breadth, and the extension of the orifice.

The organs of the Ichneumon are the same with those of all the Mangoustes. There are six incisives in each jaw and two canines. The upper jaw has three false molars, the carnivorous tooth, and two tuberculous ; the lowest jaw has one tuberculous tooth less, but its carnivorous tooth is remarkable for two tubercles on its internal face. There are five toes on each foot, the thumb is very short, and apparently useless. All these toes are armed with strong and crooked claws. The sole is naked, and covered with a very fine and delicate skin. The eye has a long transversal pupil, but no other particular character. The nose passes the lower jaw, but is not mobile. The tongue is rough like that of Cats. The walk of this animal is completely digitigrade, though occasionally, in standing, it rests on the entire tarsus.

Some differences from the Civets, &c., which form the ground of our author's subdivision of the viverræ, are observable in the anal pouch of the Ichneumon. In them this appendage is found below the anus. In the Ichneumon, on the contrary, the common integuments, elongated and folded over, form, beyond the sphincter muscle, a sack, which the animal can open or shut at pleasure.

The colour of the Ichneumon is a deep brown picked out with dirty white. The tail is terminated by a tuft of hairs entirely brown. The Ichneumon is about one foot three inches in length, and the tail an inch longer. The mean stature of the animal is about eight inches.

The Ichneumon was well known to the ancients; but they have mingled so many fables with their recitals of it, and their authorities on the subject are so very contradictory, that little can be collected from their writings with any certainty concerning the natural history of this animal. The moderns have also been long acquainted with the Ichneumon. From Belon we have had its first description and figure. Afterwards, it was confounded with the other Mangoustes, and little reliance can be placed on the figures which have been given of it, and which are also so generally incorrect, as to convey no idea whatever of the animal. Buffon's is tolerably good, but the best is that of Marechal.

Brisson, Linnæus, Buffon, and all naturalists before Schreber, admitted but a single species of the Mangouste or Ichneumon, though Edwards had expressed a doubt concerning the identity of his Indian Ichneumon with that of Egypt. Schreber was the first who established three species: *viz.*, the Egyptian, the Mangouste of Buffon, and a species which Gmelin has called the *Viverra Cafra*, whose habitat he refers to Southern Africa, though Geoffroy makes it Asiatic. Buffon, indeed, has given, in his supplements, the figure of a large Mangouste, which, however, he does not describe, and that of a smaller species, which he calls *Nems*, now identified with the V. Cafra. Vosmaër, on the other hand, has represented a Mangouste of the Indies, not at all resembling that of Buffon. Such was the state of the history of those curious animals, when M. Geoffroy, in the " Menagerie du Museum," described the Ichneumon, and separated it more forcibly from the other two species than Schreber or Gmelin had done. The French Menagerie has, since that time, received a great number of Mangoustes from Africa and the East. Among these were found distinctly characterized : 1st, the Egyptian Ichneumon, *Ichneumon Pharaonis ;* 2d, the great Mangouste of Buffon, *Ichneumon major ;* 3d, the least, *Ichneumon griseus ;* and 4th, the Mangouste de l'Inde, *Ichneumon mungo* of Buffon, distinct from all the others by transversal bands across the back. Having distinguished these, five others remained from Pondicherry, the Cape, the Isle of France, and Java, differing from each other by almost insensible shades of gray and brown, so that those nearest each other seemed only varieties, while the opposite extremities of the series were so unlike as to look like different species. M. Geoffroy, in his description of Egypt, has designated them by different names, and characterized them with his usual accuracy. How far they may properly be considered as distinct species, we cannot, but express, though with all deference and respect, an humble doubt.

We shall give a description of one of this indeterminate series from M. F. Cuvier. It was brought from the penınsula of Malacca. This Mangouste, which is the generic name of these animals in the East Indies, was rather more than a foot in length, the tail about a foot, and his height at the most elevated point of the back, five inches and a half. These animals have a peculiar faculty of elongating or shortening their bodies some inches, which renders it a difficult matter to measure them correctly.

The organs we have already described. This animal drank by lapping, and held its prey to the earth, like Dogs, for the purpose of devouring it. Its voice was generally hoarse and croaking, but became sharp and sustained when eager for food.

The general colour of this animal is a dirty gray, resulting from the black and whitish yellow rings, which cover the hairs—the circumference of the eye, the ear, and the extremity of the muzzle are naked and violaceous ; the tail is the same colour as the body, very thick at the root, and terminating in a point with yellowish hairs.

This Mangouste, though extremely tame, permitting itself to be handled, and taking pleasure in caresses, grew extremely ferocious at sight of those little animals which constitute its prey. Birds it was particularly fond of, and when they were put into its cage, which was very large, it would spring forward with a rapidity which the eye could not follow, seize them, break their heads, and then devour them with the utmost voracity ; as soon as its appetite was satisfied, it would lie down in the most obscure corner of its retreat. When irritated, the hairs of its tail used to bristle up. Its cleanliness was extreme. We are informed that in the Indies these animals inhabit holes in the walls, or small burrows in the neighbourhood of habitations, and cause as much devastation there as the Weazels or Pole-Cats among ourselves.

We shall now speak of the *Mangouste of Java.* We have seen, in our last description, that after the separation of all the species of this genus, which are distinguished by precise characters, several remained, and formed a kind of series, the graduations of which were marked by almost insensible shades of colour.

The Mangouste of Malacca, which we have just described, may stand at the head of this indeterminate series, if we commence it with the grayer tints, and the Javan Mangouste may close it, as being of the brownest shade.

In fact, the Mangouste of Java differs from the other only by a fur picked out with black and brown, instead of black and white, and by its somewhat larger size; both have the muzzle blackish, the back more deeply shaded than the sides, the extremities, and the head.

This species (if a species) is found not only in Java, but also on the continent, and is probably dispersed through a large portion of the East Indies.

We have a description of it in Dr. Horsfield's Zoological Researches in Java, rendered much more valuable by the observations which accompany it in relation to the viverrine animals in general, and the minute comparison he makes between it and his Felis Gracilis, which he makes a distinct genus of the Feline family. After what has been already said on these animals in general, and on the Javanese species or variety in particular, we shall not extract his descriptions of physical peculiarities, but confine ourselves shortly to those of mental impulse and local interest

It is known in Java by the name of Garangan, and is found there most abundantly in the large teak forests; its agility is greatly admired by the natives: it attacks and kills serpents with excessive boldness, and in this operation it is said that when the snake involves the Garangan in its

Busier sc

THE SURIKATE.

HYÆNA SURICATA. ILIGER.

London. Published by G.B.Whittaker. Feb.1.1825.

folds, the latter inflates its body to a considerable degree, and when the reptile is about to bite, again contracts, slips from between the folds, and seizes the snake by the neck.

It is very expert in burrowing the ground, which process it employs ingeniously in the pursuit of rats : it possesses great natural sagacity, and from the peculiarities of its character willingly seeks the protection of man it is readily tamed, and, in a domestic state, is docile, and attached to its master, whom it follows like a dog ; it is fond of caresses, and frequently places itself erect on its hind legs, regarding every thing that passes with great attention : it is of a very restless disposition, and always carries its food to the most retired place in which it is kept, to consume it : it is very cleanly in its habits : it is exclusively carnivorous (at least, as we conjecture, in a state of nature,) and is very destructive to poultry, employing great artifice in the surprising of chickens—for this reason it is rarely found in a domestic state among the natives, as one of their principal articles of food is the common fowl, and great numbers are reared in all their villages. The Javanese also, like Mahomedans in general, have a great partiality for cats, and they are unwilling, in most cases, to be deprived of their society for the purpose of introducing the Garangan. It has also been observed that its sanguinary character shews itself occasionally in a manner that renders it dangerous in a family as a domestic animal, and it indulges,at intervals, in fits of excessive violence.

The Suricate has been known hitherto only by the figure and description of Buffon and Daubenton. Its organization was inferred to be similar to that of the Civets, Genets, and Mangoustes ; and in methodical catalogues, it was accordingly united with those animals. Even M. Fred. Cuvier, in his valuable observations on teeth in the animals of the

Museum, was led, from the general similarity between its dentition and that of the Civets, to attach it to the same group.

Its general figure and habits bear little resemblance to those animals: Buffon compared it with the Civet, and if his idea was not followed up, it was because more importance was attached to the colours of the fur and glandular pouches of the anus, than to the physiognomy and habits of body. Yet these last characters are often of greater consequence than the former, and the general analogies of nature may be established on them with greater certainty.

Erxleben, Gmelin, &c., made the Suricate one of the species of their *Viverræ*, and placed it next the Mangoustes and Coatis. The Suricate is now a sub-genus of the reformed subdivisions of the Viverræ.

Among our own animals the Pole-Cat and Ferret are those which, in external appearances, are closest to the Suricate. Among foreign species, it most resembles the smaller Mangoustes; but it differs considerably from both, by its slender and elevated limbs, compact body, extreme length of nose, plantigrade walk, &c. Its physiognomy is indeed altogether peculiar, and has no type among the known Mammalia.

Like all the Carnassiers the Suricate has five incisors and two canines in each jaw : its upper molars on each side are five, two false, one carnivorous, of the form of an isosceles triangle. The internal tubercle of this tooth is so thick, that we can scarcely consider the tooth as trenchant ; then come two false molars of the same form, but smaller than the carnivorous, and having a tubercule on each angle of their triangles.

The lower jaw has three false molars ; the third of these has anteriorily an elevated point, and posteriorily a sort of heel composed of two soft tubercles. After these comes the carnivorous tooth, with a thick joint, divided into two small tubercles, and terminating behind like the preceding tooth. The series of teeth in this jaw is terminated by a tuberculous one, which greatly resembles the carnivorous tooth just described.

From this description it is easy to see that the Suricates are less carnivorous than the Mangoustes, and approach more to the omnivorous character, not from the number of the tuberculous, but from the form of the carnivorous, teeth. The mode of action in these teeth confirms this notion. Among animals of prey the carnivorous teeth act one upon the other, like the blades of a pair of scissors, and the lower carnivorous passes completely behind that of the opposite jaw. In proportion as the appetite is less sanguinary, the upper carnivorous tooth advances, and the lower recedes, so that they act but partially on each other. In the Mangoustes the whole anterior part of the lower carnivorous acts against the entire extent of the internal face of the upper. But this same part in the Suricate corresponds to the vacuum left between the opposite teeth, while its posterior part is in opposition with the first tuberculous. Finally, the anterior part of the upper carnivorous is opposed to the posterior part of the last false molar. Thus we see that both the action and the form of these teeth have many relations with those of the insectivora.

The Suricate has a very fine sense of smelling, which neither detracts from the extent of the brain, nor causes that preponderance in the sense of taste, which is usual where the former organ is much developed. Among the Cats, we find that where the brain is extended, the sense of smell is feeble. The anterior extremity of the cerebal cavity advances so as to correspond with the middle of the orbit, and all the parts of the olfactory organ are much limited, and likewise those of that of taste. In Dogs, the brain also advances to the middle of the orbit, but the bones of the nose are elongated, and the parts of the mouth extended in a similar proportion. With the Suricate, the brain, as in Cats and Dogs, corresponds, anteriorly, with the middle of the orbit, but as the bones of the nose remain very short, the animal's sense of smell would be very feeble, but for the extension of the cartilaginous parts in the same organ. This incontestably favours the exercise of smell, while the power of taste is still limited by the shortness of the aforesaid bones. The nose is terminated by a glandulous organ, in which open the nostrils, formed somewhat like those of Dogs. There is nothing particular in the other organs of sense.

The cerebral cavity is remarkable by its extent, breadth, and rounded form. This, again, distinguishes the Suricate from the Mangoustes, which are characterized by the narrow and cylindrical form of this cavity.

The mammæ are three in number on each side, and the anus is surrounded by a naked skin, which covers a glandulous apparatus, leading by two orifices to the internal edge of the rectum. The limbs are terminated by four toes, armed with long and digging nails. The feet are characterized by certain tubercles, and are covered with a fine skin, like that of the human hand

With these differences of organization, certain differences of habit are found to correspond. The Suricate does not move like the Polecats or Mangoustes, with the head low, the body elongated, and the rapidity of an arrow. Its body is arched, and though it proceeds quickly, it has not that uniformity of motion which makes the others appear to glide rather than run. It places the entire sole on the ground, and can easily stand upright on its hind legs. Sometimes it will carry its provisions to its mouth with its fore-paws. Its sense of smell is its principal guide. It ferrets about, thrusting its mobile nose into every hollow place, and when it finds an object which strikes its sense f smell, it seizes it instantly, and devours it. Sweet fruits are not disagreeable to it, but it prefers animal matters, milk, eggs, and the flesh of birds. It laps in drinking. It cannot bear light, and sees but in obscurity.

Its sense of hearing must be but feeble, from the small extent and mobility of the auditory conch, and also from the extreme predominance of the sense of smelling over the rest, a result not only of the great development of those parts in which it is situated, but also of the frequent use which the animal makes of it. That of touch resides, like as in other Mammalia, chiefly in the silken hairs in the mustaches, and, probably, in the soft and naked skin, which covers the soles of the feet.

The Suricate, as might be prejudged from the great development of the brain, is easily tamed. It soon acquires a clear notion of the circumstances in which it is placed, and learns to estimate the degree of confidence it should repose in all that surrounds it. Like a Cat, it traverses the house which it inhabits, and will never leave it. It is, in a high degree, susceptible of affection, and also of hatred ; though we cannot agree with M. Fred. Cuvier, that one is always a necessary consequence of the other. Cats are less susceptible of affection than Dogs, and more so of hatred. The Suricate recognises those who tend it, is pleased with their caresses, and becomes permanently attached to them ; but it preserves rancour against those who have offended it, and will seize the first favourable occasion for vengeance. It will even conceive prepossessions so powerful as not to be removed by the kindest treatment. This species, erroneously attributed by Buffon to America, is found in the south of Africa.

The fur of this animal is a dull brown, inclining to fawn underneath, and crossed by slight transversal bands, principally on the back. On the limbs there is a silvery tint. The skin itself is of a tan colour on the naked parts. The tail is brown. The length of the body from tail to muzzle is about eleven inches; the tail itself about seven, and the mean height of the animal may be about six and a half.

From the details now given, we may conclude that this animal fills the void between the genuine carnivora and the plantigrades. The teeth are more tuberculous than those of the Mangoustes, and less so than those of the Coatis. The organization of the hind-foot, the number of toes excepted, is the same as that of the Mangoustes, but the sole with the latter is only half uncovered, but in the Suricate it is entirely so, as in the Coati. Like the last-mentioned animal, the muzzle of the Suricate is prolonged considerably beyond the jaws, but its tongue, furnished with horny papillæ in the middle and soft at the sides, approaches it by the first of these characters to the Mangouste, and by the second to the Coati. If the Suricate does not completely fill the void we have mentioned, it requires only some very slight modifications to do so. The discovery of a new genus might so completely unite the two groups, as to leave nothing abrupt between them. In consequence of what we have now detailed, M. F. Cuvier seems to think that the subdivisions of the Plantigrades is not a natural one, and that these animals ought to terminate or commence the series of one of the branches of the genuine Carnivora.

This animal is the *Viverra Suricata* of Erxleben, the *Viverra Tetradactyla* of Schreber and Gmelin, the *Suricate Viverrin* of M. Desmarest, and the *Ryzena* of Illiger. This last, as a generic name, seems more eligible for adoption than names which time and usage have consecrated to specific designation.

We now come to the last subdivision of the Digitigrades, the first snb-genus of which is the formidable HYÆNA. Rounded spots, scattered in small number over a fur of a yellowish-dun colour, and the Southern part of Africa as its habitat, are the only characters which distinguish the *Spotted* from the *Striped* Hyæna.

We begin to fear that, from frequent recurrence to the same position, it may be thought we are rather broaching a particular hypothesis than concentrating by compilation to one focus the various labours of others, which, with occasional comments and reflections, and original graphic illustrations, forms in reality the more humble object of our endeavours. Renouncing, therefore, more lofty pretensions, and deprecating the anger of those who may know better and think differently, we again advert to the too great readiness with which some naturalists have established diversity of colour as a ground of diversity of species. There are cases, however, we must allow, in which there is no other obvious criterion of distinction, in the present limited state of our zoological knowledge. Two species may have always remained distinct, though the only point of dissimilarity between them may be a very slight variation in the arrangement of their colours. Yet even here we should be cautious, and hesitate to pronounce any more than a provisional judgment on the subject. Many causes may prevent the intercourse of animals, which, from their not intermixing, we refer to different species. When we see two races of animals inhabiting the same country never intermix, and always preserve the same external differences, we may with confidence pronounce them distinct. But we must always be liable to some error in our judgments concerning animals which inhabit different countries, and whose characteristic differences are slight and external. Even if when we bring them together, they refuse to intermix, it would be no sufficient proof of specific difference, for we know how much the instincts of wild animals are weakened or perverted in the unnatural state of captivity. This is true even of those animals of whose specific identity no doubt can be entertained. Transpose the habitats of the two species of Hyæna, and let them breed each in the country proper to the other, and if their progeny did not depart from the specific character of their ancestors in assuming those of the other species, we should have strong evidence of diversity of species. In short, if no external influences are found to approximate their specific characters, and the joint offspring of the two turn out to be hybridous and sterile, that doubt upon their distinctness would then be removed, which appears to us to exist in the absence of such evidence.

We must be contented, however, on this subject, with simple observation. The *modus operandi* of those causes which act so differently on different species of animals, is to us buried in the profoundest obscurity. Many of the consequences, therefore, which we are fond to deduce from the phenomena of nature, and to elevate into general laws, may, for aught we know, be forced, foundationless, proofless. To be assiduous in observation, and cautious in deduction, is the golden rule of philosophy.

Notwithstanding, then, the slight differences between the Spotted and the Striped Hyæna, we must consider them as separate species, until we are in possession of certain proofs of their intermixture and identity; until we see the characters of one confounded with those of the other, until we see the spots of the former lengthen into stripes, and the stripes of the latter shorten into spots.

The *Spotted Hyæna*, in stature and corpulence, resembles a large Mastiff. The head, however, is more thick and less elongated, and its motions have less freedom and elasticity. The hinder part of the body it carries very low, owing to its constantly keeping the articulations of the hinder legs considerably bent. Its glance is unsteady, for it is dazzled by a strong light, and this gives an additional indecision to its movements. Not that the animal is by any means deficient in force and vivacity. It is susceptible of very violent feelings, and on such occasions is capable of acting with equal promptness and erergy. The sentiments, indeed, which it manifests, however opposite in their nature, are all of a violent character: its hatred and its affection are both equally strong. An individual of this species, described by M. F. Cuvier, showed the utmost confidence in all its keepers; and for one in particular, evinced an affection very unusual in wild animals, and parallel to nothing but what we witness daily in the common domestic Dog. On the other hand, his hatred was extremely violent, and he often would exhibit excessive rage against persons who had done him no kind of injury. On such occasions he would tremble with rage, the foam would issue in abundance from his mouth, the hairs of his back would bristle up, and blows had no other effect than to exasperate his anger. He was taken very young at the Cape of Good Hope, and had been tamed without difficulty. On his arrival in France, his cage having been left partly open, he walked out, and went away before he was observed. As soon as his flight was known, his keepers went to take him, and saw him enter the cottage of a peasant very quietly, where he suffered himself to be retaken without the least opposition. This docility is not peculiar to some individuals of this species, but common to all. Barrow informs us, in his journey to the Cape, that the Spotted Hyæna has been tamed in the district of Schneuburg, where it is considered more serviceable for the chase than the Dog, and fully equal to that animal in intelligence and fidelity. The relations between the Hyæna and Dog, led Linnæus to class them together in the same genus, but a more attentive examination has shewn, that the Hyænas form a genus in themselves, as distinct and natural as that of the Dogs. Their molars are five in number, in the upper jaw: three false ones, one carnivorous, and one tuberculous. Four only in the lower:

three false molars, and one carnivorous. All these teeth
are remarkable for their size and strength; and are particularly adapted thereby for breaking large bones, a physical
fact, in perfect harmony with their singular liking and
appetite for bones as food. In each jaw are also two
very strong canines, and six incisors. The fore-feet, as
well as the hinder, have four toes armed with claws,
adapted for digging. The pupil, when half closed, has
an elongated and pyramidal form. The nostrils resemble
those of Dogs. The tongue is covered with rough papillæ.
The ears are large, very open, possessing great mobility,
and habitually directed forwards. There are mustachios
on the upper lip and cheek. The hairs of the coat are
generally long, copious, and rough. Under the anus is a
longitudinal cleft, which produces an unctuous and fetid
matter. The Hyæna is digitigrade. Its sense of smell
seems most acute, yet it prefers flesh which has began to
turn to that which is fresh. Four or five pounds a day
seem to satisfy it. It drinks lapping. Its voice resembles groaning, or loud wailing. The reproduction and
habits of these animals are equally unknown. The usual
colour is a dirty fawn, bordering on a blackish-brown
towards the lower parts. The extremity of the muzzle is
black. The Spotted Hyæna is about four feet and upwards
in length, and îts height to the shoulder about two feet
three or four inches; to the loins not two feet.

The *Striped Hyæna* presents a remarkable example of
the facility with which errors are admitted and spread
abroad, even when the truth might be easily ascertained.
Aristotle, who knew the animal, has left us a very succinct
description of it, and combated fables on the subject which
were even current in his own time. Pliny, however, relates
these fables in preference to following the text of Aristotle, and he has been copied by most of those who have
written subsequently on natural history, even by those

authors to whom the restoration of the science in Europe has been owing. To Busbec and Kœmpfer we are indebted for a new and more correct acquaintance with the Hyæna. Since them it has been often described, and it is at present one of the best known of the carnivorous tribe. The length of the body, measured on the skeleton, is, from tail to occiput, about four feet. The head ten inches, and the tail seven. The posterior height is above two feet, the anterior about a foot and a half. In the upper parts of the body the colour is a yellowish-gray, varied by transversal bands of a black-brown. The mane is gray, with some black spots. The muzzle and the external face of the ears are violet-brown. The woolly hairs are small in number, the others rough, long, and somewhat thick, except on the limbs, where they are short and close. The mane is composed of them, and is considerably larger than that of the Spotted Hyæna. There are long mustaches on the upper lip, above the eyes, and on the cheeks. It is unnecessary to repeat the generic characters. The ears, however, it may be observed, have two folds at their base, one on the internal, the other on the external, edge.

The voracity of this animal, its preference of the flesh of carcasses to living prey, and its consequent propensity to disinter the dead, have bestowed upon it a character for ferocity not founded in truth. Ill treatment will render it extremely furious, but under opposite circumstances, it will exhibit the most remarkable degree of mildness and docility. Its cage may be entered with impunity, it will approach to fawn upon those it knows ; and were it not for the prejudices of the public on this subject, a Hyæna thus tamed might be intrusted with as much liberty as a common Dog. There is in these respects a remarkable coincidence of character between the two Hyænas, and both in a domestic state would doubtless render to Man services of the same kind and degree as the canine species.

Among the Striped Hyænas are found considerable varieties, in the shades of the fur and of the transversal bands.
In some the ground-work of the coat is clear fawn, with
spots of the same colour, but deeper.　In others the ground
colour is a deep brown, varied with slight bands of gray
and black.　Bruce thought he could distinguish the Hyæna
of Syria from that of Barbary : but additional observations
are necessary to ascertain the nature and importance of
those differences.

Notwithstanding the facility with which the Hyæna is
tamed, there can be no doubt of its excessive ferocity in
its savage state.　Its partiality to corrupted flesh leads it
to the habitations of Man, and the enemies which it finds
there to encounter must unquestionably contribute to maintain the fierceness of its character.　In the East it is commonly observed by night to traverse the suburbs of towns,
and even penetrate into the interior, to feed on the remains
of animals which may be found there.　It is by no means
uncommon to see Chacals, Hyænas, Dogs, and Vultures
fasten on the same carcass, and agree together tolerably
well, until the portion becomes too small to satisfy the voracity of each.

As we have observed, the ancients were acquainted with
this animal.　It is mentioned by Aristotle, Pliny, Ælian,
and Oppian.　The earlier modern naturalists did not recognise the identity of the Hyæna with that described by the
ancients.　Belon gives the name Hyæna to the Civet, and
calls the Hyæna itself by the strange appellation of Marine
Wolf.　Busbec and Kœmpfer were the first to recognise it.

" There is in the French Museum an Hyæna, whose country is unknown, on which I am in doubt," says the Baron,
" whether to call it a variety of the Striped Hyæna, or to
consider it a distinct species."

The hairs, not only down the spine but on the whole of
its back and flanks, are long and rough, longer even than

those on the spine of the Striped Hyæna. They hang down on each side, whitish-gray at their base, and blackish-brown thence to the tip, so that the whole fur appears of an uniform brown colour, only on the fore-legs and hind-feet there are some transverse bands, whitish-brown ; the insides of the legs under the belly and tail are whitish-gray, and there is gray and brown on the head.

This individual is a small degree less than the common striped species.

There appear also to be two varieties of the Spotted Hyæna.

One whitish-gray, a little inclining to yellow, with circular pure brown spots on the flanks and thighs ; those of the shoulder form a band, which is continued as a longitudinal brown line on each side of the neck ; the feet whitish, a little red towards the bottom ; the tail annulated, whitish, and brown at the base, and blackish two-thirds down ; the head, of the same general colour as the back, has a little brown on the cheeks, and some red on the tip.

The other has more fur, of a red gray colour ; the under part of the neck and body only is whitish ; indistinct black spots are to be seen on the flanks, crupper, and thighs, and there is also a black band on each side the throat ; the legs and feet are blackish, but the inner side of the fore-legs is reddish-white ; the tail, red the first half, is black for the remainder ; the head is red, with blackish on the forehead and between the eyes ; the under part of the forehead is reddish-brown. This appears to be the most common variety in the vicinity of the Cape.

The Baron Cuvier, who describes these varieties, is decided in his opinion upon the diversity of the striped and spotted species. He observes it would be important to determine the limits of the country which each species occupies ; but we know too little of the natural history of inter-tropical Africa for this purpose.

It seems certain that in the Levant, in Persia, and in

Egypt, none but the striped species are known, (Bruce's large Hyæna being, in all probability, merely a great individual,) and, it is said, they are as big in Barbary; but however erroneously modern Zoologists may, in some cases, have located the spotted species in northern Africa and western Asia, contrary to better observation and authority, it is still very unaccountable, says Cuvier, how it came to be represented in an ancient manuscript of Oppian. Chance could hardly have induced so singular a coincidence.

We now proceed to notice two new sub-genera of mammiferous animals; first the PARADOXURUS of the Baron Cuvier and his brother; and, secondly, the PRIONODON of Dr. Horsfield.

The type under which M. F. Cuvier describes this new genus is the *Pougouné,* a name modified from its Malabar appellation of Pounougar-Pouné, that is, Civet Cat, (*Paradoxurus typus,* Cuv.) This animal has been long known to naturalists. It was described by Buffon and Daubenton, under the erroneous name of Genette de France; but of the true nature of it, we have been, till lately, in great ignorance. It is the Palm Marten of M. Lechenault.

To form a correct estimate of the real peculiarities of this animal, it seems quite necessary to observe it during life, as its skin and skeleton are very similar to those of the preceding sub-genera.

But however allied to the Civets, Genets, and Mangoustes, this animal, and its congeners, may be in certain of its organs, it differs from them nevertheless, not only in parts of its external form, which may or may not have an influence on the character of the animal, but also by modifications, which must necessarily, arguing from analogy, have a strong and decided effect on its character. Indeed, when seen alive, it is said to be easily recognised as distinct from those groups to which it is more or less assimilated.

THE POUGOUNÉ.

PARADOXURUS TIPUS — CUV.

London, Published by G.B. Whittaker, Feb.y 1825.

In the influential characters of dentition, and of the toes, the sub-genus Paradoxurus approximates to the Civets. In its heavy form, general physiognomy, and especially the plantigrade mode of walking, it approaches the Badgers, and in the sub-prehensile power, or spiral twist of its tail, it is assimilated, in some degree only, to the transatlantic Monkeys, the Poto, and other genera.

The Pougouné is entirely plantigrade. Its toes, five in number, on each foot, are furnished, at their extremity, with a thick tubercle, which hinders the point of the nails from touching, and wearing on the ground ; these nails are also nearly as retractile as those of Cats, and are slender and sharp, so that they are enabled to use them like the felinæ, not merely as offensive weapons, but also as an effective mean of ascending and descending trees, &c. The feet are nearly palmate, united by a membrane, even to the last phalanx. They are furnished with four naked fleshy tubercles. The relative length of the toes decreases in the following order : the middle toe, annular, index, little toe, and thumb.

The tail is one of the greatest peculiarities in the animal, and seems, indeed, to have nothing corresponding exactly to it in any other. When the tail is as near to a straight line as the animal can make it, it is, nevertheless, twisted from right to left towards its extremity, so that by a particular disposition of the vertebræ, no doubt, the upper part of the tail is turned downwards, whence results the following phenomenon : when the tail is curled by the action of the upper muscles, this movement is made, at first, from the upper to the under side ; and if the muscles cease to act when the curling is only half effected, the tail, in that case, seems organized like the ordinary prehensile tails of other animals ; but if the muscles continue their action, the tail returns to its natural state, and the curling goes on, but from below to the upper part, as far as the insertion, and in which state it is represented on the plate.

The mammæ are three on each side, one pectoral, and two ventral. The parts of generation present some peculiarities, and there is no appearance of a pouch, or even folliculi of skin near the anus, as in the Viverræ.

The eye has a third lid at the internal angle, which can be drawn entirely over the ball. The nostrils are similar to those of the Dog; the snout has a swelling, divided by a vertical furrow, stretching from the edge of the upper lip, but not entirely dividing the part. The tongue is long, narrow, thin, and aculeated with horny papillæ, globular at their base, but terminated by a spiny thread; between these, are other tubercles, which are round and soft, and at the base of the tongue is a particular gland.

The ear is rounded, with a deep slope on its posterior edge, covered by a strong lobule; the whole internal surface of the ear is furnished with tubercles of various sizes; the auditory canal is covered with a sort of little valve, which appears to be destined to close it, and all the projections appear disposed to shut into one another, so as the more effectually to close the passages when the animal shuts the ear, by drawing together the anterior and posterior parts of the conch, an operation which takes place, and continues so long as the animal sleeps.

The fur is of two sorts, silky and curling, but the former is the most abundant; there are long mustachios on the sides of the upper lip and over the eyes.

The colour of this species is remarkable, inasmuch as it varies in different positions and under different angles of light; hence, the animal has been alluded to under the name of the Palm Marten, and others as of different colours. The prevailing tint is yellowish-black, that is, when viewed sideways, and so as that the extremities of the hairs only are visible; but if looked upon so as to show any length of fur, it is yellowish; on this ground, may be perceived three ranges of spots on each side of the spine, and others sprinkled on the thighs and shoulders; but if contemplated in its black position, the latter spots are no longer visible, and the former assume the appearance of stripes or lines. The silky hairs are entirely black, so that when they only are to be seen, the animal is black; but as these, though long, are few in number compared with the others, which are yellow, hence results the difference in the general tint, according as the eye may view the surface only, or penetrate further into the fur.

The limbs are black, but the tubercles on the feet are flesh-coloured; the tail is black for one-half its length, as well as the head, which becomes paler; however, towards the muzzle, there is a white patch above each eye; the inner side of the ear is flesh-coloured, except round the edge; the external surface is black, except at the edge, which is white.

The individual described by M. F. Cuvier, kept in the Paris Museum, slept during the whole day, rolled into a ball. It was with much difficulty it could be raised from this lethargy; towards the decline of day, he roused himself for a short period, but as soon as he had eaten and drank, returned to his bed again, which he kept very neat and clean. His motions were slow, nor did he appear capable of lively sensations. On being troubled, he would utter a sort of grunt, by way of menace. Contrary to the Civets and Genets, he exhaled no scent. His tail, though curled, had no prehensile power. It must be observed, however, that this individual died of accumulation of fat, resulting, no doubt, from his sedentary life, whence we may infer that, in a state of nature, his habits would have been very different.

The nails of Cats, the teeth of Civets, and the very singular manner of twisting its tail, have induced MM. Cuvier to apply to this animal and its congeners, the generic epithet of Paradoxurus.

The *Musang*, *Viverra Musanga* of Dr. Horsfield, is identified by the Baron with the species last described; the Doctor, however, informs us that it is subject to several varieties, and that in the most common variety of Java, the Viverra Musanga is of a much lighter grayish colour.

Of its manners and habits, Doctor Horsfield says, that they are very similar to those of the Genet. If taken while young, it becomes patient and gentle during confinement, and receives readily animal and vegetable food. It requires little attention, and even contents itself with the scanty remains of the meals of the natives, with fish, eggs, rice, potatoes, *&c.* It prefers, however, delicate and pulpy fruits; but when pressed by hunger, also attacks fowls and birds.

It is most abundant near the villages situated at the confines of large forests. It constructs a simple nest in the manner of Squirrels, of dry leaves, grass, or small twigs, in the forks of large branches, or in the hollow of trees. From these it sallies forth, at night, to visit the sheds and hen-roosts of the natives, in search of eggs, chickens, *&c.* Its rambles are also particularly directed to the gardens and plantations, where fruits of every description, within its reach, and particularly pine-apples, suffer extensively from its depredations.

The coffee plantations, in Java, are greatly infested by the Viverra Musanga; in some parts of the island it has, on this account, obtained the name of the Coffee-Rat. It devours the berries in large quantities, and its visits are soon discovered by parcels of seeds, which it discharges unchanged. It selects only the ripest and most perfect fruits, and the

THE BINTURONG. *RAFFLES?*

PARADOXURUS BINTURONG?

Baire sc

London. Published by G.B Whittaker Feb? 1825

seeds are eagerly and easily collected by the natives, as the coffee is thus obtained without the tedious process of removing its membranaceous arillus.

The injurious effects occasioned by the ravages of this animal in the coffee plantations are, however, fully counter-balanced by its propagating the plant in various parts of the forests, and particularly on the declivities of the fertile hills ; these spontaneous groves of a valuable fruit, in various parts of the western districts of Java, afford to the natives no inconsiderable harvest, while the accidental discovery of them, surprises and delights the traveller, in the most sequestered parts of the island.

Sir Stamford Raffles has inserted, in his valuable catalogue of animals, contained in the thirteenth volume of the Linnæan Transactions, an account of the Binturong of Malacca, and it must be observed, that some time previous to the separate classification of the Paradoxurus typus, as a distinct genus, by MM. Cuvier, Sir Stamford must have seen the impropriety of arranging the animal with the Viverræ, as although he does so in point of fact, he adds a mark of doubt on its propriety.

Immediately between *Viverra* and *Ursus* is an animal, says Sir Stamford, called Binturong, found at Malacca, by Major Farquhar.

The body of this animal is about two feet and a half in length, tail nearly the same, bushy and prehensile, height from twelve to fifteen inches. It is entirely covered, with the exception of the legs and face, with a thick fur of strong black hair. Its general appearance and habit is slow and crouching ; the body long and heavy, and low on the legs ; the tail is thick at the root, gradually diminishing in size to the extremity, where it curls inwards. The muzzle is short and pointed, somewhat turned up at the nose, and is covered with bristly hairs, brown at the points, which lengthen as they diverge, and form a peculiar radiated circle round the face, giving the countenance a striking and remarkable aspect. The eyes are large, black, and prominent ; and the ears are short, rounded, edged with white, and terminated by tufts of black hair. There are six incisors, &c.

It climbs trees, assisted by its prehensile tail, in which it has uncommon strength. Major Farquhar kept one alive many years ; it lived both on animal and vegetable food ; was particularly fond of plantains, but would also eat fowls' heads, eggs, &c. Its movements are slow, and it is rather of a timid disposition ; it sleeps much during day, but is more active at night.

M. Desmarest, perhaps too hastily, makes a distinct species of this animal. We may, perhaps, gather from Doctor Horsfield, that he considers it only as a variety. " In Sumatra," says he, " the Musang assumes, agreeably to the description of Sir T. S. Raffles, a dusky fulvous-colour, and the point of the tail is uniformly white. The stripes on the back and sides are more distinct than in the dark Javanese variety."

In addition to the figure of the Paradoxurus typus, from M. F. Cuvier, we present a figure of a specimen in the Museum at Paris, which is referred to the Binturong of Sir Stamford Raffles. It is a female, and certainly does not accord in colour with the above description, being almost uniformly, except about the face, forehead, and ears, of a slate colour. It is also much larger than the Paradoxurus typus of Cuvier.

There is also another specimen, in the Paris Museum, of a uniform, fine golden-yellow colour, which was sent to that establishment, preserved in spirits ; we have a drawing of it before us, but as it differs exteriorly, at least from the others, in colour alone, we have not engraved it. It is treated, also, by M. Desmarest, as a new species under the name of the Golden Paradoxurus, (*Paradoxurus aureus*). It is a very young specimen.

Doctor Horsfield, also, gives us a description and figure of a rare Javanese animal, the *Delundung*, which appears to hold, analogous to some others, an intermediate station between two genera or sub-genera of our author, *Felis* and *Viverra*. He, at first, placed it among the Felinæ, under the name of Felis Gracilis, but he was afterwards confirmed in his original intention of making it a distinct genus, under the name of Prionodon. The French naturalists, knowing no more of the animal, apparently, than is to be obtained from the description of Doctor Horsfield, and the short, previous, and doubtful allusion of General Hardwick, have placed it, unhesitatingly, among the Viverræ, and, accordingly, we find it in M. Desmarest's catalogue, forming a species of Civet.

Intermediate genera we have had frequent opportunities of noticing, and shall have frequent occasion to mention others, and they may be the more deserving of particular attention by the Zoologist, as affording matters of fact, from which important results may be elicited. The inviolability of genera, on the one hand, and their gradual gradations on the other, are facts from which no very determinate conclusions have hitherto been drawn. Certain it is, that some one or more species of most genera exhibit a declination, as it were, to other very different animals, and when such facts are placed in conjunction with what we know of the tendency to varieties, from location, or whatever other cause arising, the greater becomes the natural curiosity to ascertain the probable number of primary organic creations, and the consequent number of those springing from secondary causes : while, therefore, the utmost caution should be exercised, on the one hand, not to destroy the utility of zoological science by an undue multiplication of names, it is equally necessary, on the other, to avoid a commixture of species essentially different, and separated, however slightly, yet effectually, by the hand of nature herself.

In the number of toes on the hind feet, and of the teeth, as well as in the form of the head and body, the Delundung resembles the Viverræ, but the character of the claws, and peculiar structure of the teeth, indicate an affinity to the Felinæ. Altogether, therefore, the animal is neither, and is properly separated from both.

The Felinæ and the Viverræ have the incisive teeth of the same number, and alike. In this the Delundung agrees with both, nor does there appear any essential difference in respect to the canine teeth of the three. The Cats have three or four cheek-teeth above, and three below; the Viverræ have six; the Delundung differs from both, in having five above, and six below. The eye-pupil in the first genus is, in some species, circular, in others, oblong; in the Viverræ it is elongated transversely, and in the Delundung it is circular.

In the ears and form of the body, the former short and round, the atter long and low, the Delundung diverges from the Cats in approaching the Viverræ, and in the number of toes the same thing occurs, the Felinæ being pentadactylous before, and tetradactylous behind; the Delundung and the Viverræ having five toes on all the feet. The character of the tongue, whether soft or aculeated, and the number and situation of the mammæ, in the Delundung, are not known; and the non-existence of an anal pouch, or anal folliculi, in the Delundung, is asserted, by its describer, with a mark of doubt.

The ground colour of the animal may be said to be a pale yellowish-white, but there are various large spots or patches of a rich deep-brown over the back, with smaller spots of the same size on the sides and thighs; and the tail has nine annuli of the like colour, the tip being of the paler ground-colour.

Dr. Horsfield found this animal in the district of Blambangan, in Java, nor could he ascertain that it was known in any other part of the island. It inhabits, though rarely, the extensive forests which cover this district. Of its particular habits and manners, the natives could give but little information ; the Doctor obtained a second individual, which soon escaped, and he was never afterwards enabled to get another.

If the Paradoxurus be placed with propriety in a distinct genus, there seems equal reason to do so with the Delundung, to which genus, as we before stated, the Doctor has given the name *Prionodon*, and he has added the specific epithet *Gracilis*, from its elegant appearance.

We are inclined to suspect that there are other species of Viverrine Cats ; that figured in Daniel's Sketches of the native tribes, animals, &c., of Southern Africa, pl. 36, if we may judge from the plate, has the appearance of it. *Vosmaer's Chat Bizeam*, and others, also, may eventually turn out to belong to this sub-division.

We are now arrived at the genus FELIS, the most prominent of this terrible order of animals, a genus more distinct and isolated, more obviously characterized to the eye of common observation, and more easily defined by its systematic characters, than most others. A similarity in physical and moral character, nearly approaching to identity, prevails throughout almost all the species, from the dauntless Lion and ferocious Tiger, to their common domestic congener the Cat : size and colour form their leading specific distinctions. It is true, indeed, that one species at least, and probably another or two, exhibit a slight approximation to the Dogs, whence they have been called Canine Cats ; and a similar aberration from the common type has also been observed in one or two species to the Weasel family, instances of partial exception to the above general observation.

To avoid repetition, we shall not dilate generally on the physical characters of this genus, but merely in recapitulation of the text, remind the reader, that their bent trenchant retractile claws, drawn into a sheath, when inactive, and thus constantly preserved sharp for use, the small number and carnivorous character of their cheek-teeth, the number of their toes, five before, and four behind, their short muzzle, powerful jaws, and aculeated tongue, added to their moral character of natural ferocity, and appetite for a living prey, prevail in all the species.

A more particular description, however, of the fourth or flat cheek-tooth, found in the upper jaw of some of the Felinæ, may not be unacceptable, to which we shall, with all deference, add a few observations on the eye-pupil of the genus.

In the upper jaw of most of the species is found a flat cheek-tooth, altogether differing from the rest, and which, from its singular shape, position, and apparent office, we should be inclined to call an auxiliary tooth. It is so situated as not to be seen, except by opening the mouth wide, and looking upwards. It does not protrude from the edge of the jaw, like the other teeth, but a little way up the inner inclined surface of it, and takes a direction across the lower part of the last carnivorous tooth. It is flat at the top, and seems to be intended as an anvil to receive the cutting edge of the large lobe of the last lower carnivorous tooth, so as to render it more available in acting on the food. From its situation in the mouth, it may easily escape observation; whence it is not unfrequently said, that the cats have only three cheek-teeth in each jaw. The second figure on the opposite plate is intended to show this auxiliary tooth.

The pupil of the eye is in some species oval, and in others

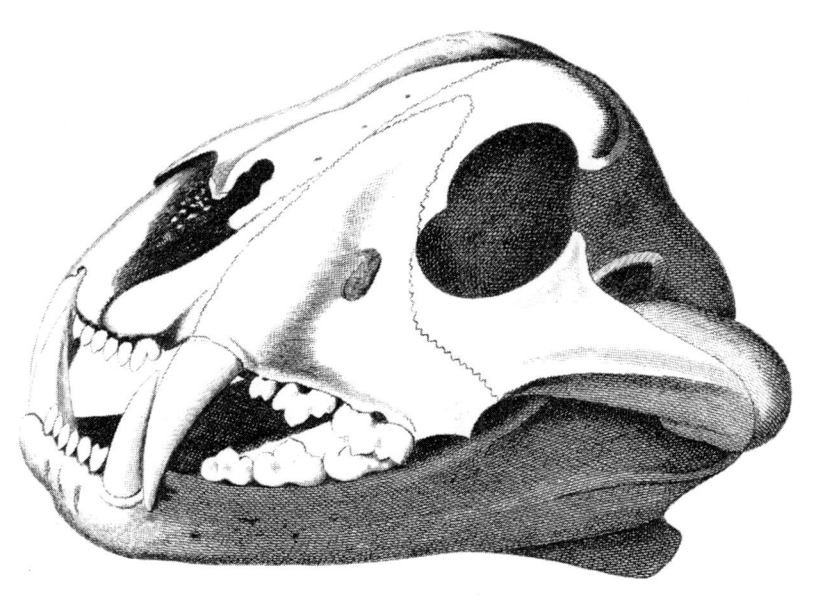

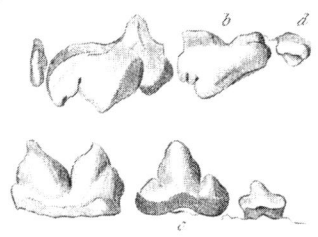

TEETH OF THE FELINÆ.

SPECIES F. LEO.

b c

The cheek teeth in each jaw

separately viewed.

d

The situation of the fourth cheek

tooth in the upper jaw.

London Published by G.&W.B.Whittaker. Sept.ʳ 1824.

circular. It is also capable of much alteration, not only in size, but also in figure, resulting from the degree of light acting upon it, and occasionally from some sudden mental impulse, so as to be sometimes round, sometimes oval, and sometimes a mere vertical line in the same animal.

There are some positions so universally considered as true, that no one ever thinks of doubting them ; and it is, indeed, on such, that all reasoning must be grounded : but we cannot be over scrupulous in admitting, or too nice in investigating any proposition, before it is classed with those fundamental axioms as self-evident, and therefore not requiring to be demonstrated.

That the pupils of Cats are oval, and that therefore they are enabled to see in the dark, is an assertion very generally made, and seldom questioned ; and some naturalists, observing that the felinæ vary in this particular among themselves, have separated them into diurnal and nocturnal species ; distinguishing the former by the circular pupil, and the latter by that of an oval figure. It may, nevertheless, be doubted, whether the shape of the eye-pupil be at all connected with the extent of the power of vision: the size of it must, in all probability, be materially so ; but it does not appear certain, that those animals which dilate the iris, so as to elongate the pupil, have also the greatest power of contracting the former, and consequently of enlarging the latter, more than others which have the pupil at all times circular.

Major Smith has observed, on this subject, that the diskous or circular eye-pupil is believed to be diurnal, and the Lion and Tiger are both, in general, associated together on this account; but the Lion, although he sees by day, may be said, probably, never to hunt his prey while the sun is above the horizon, unless pressed in an extraordinary degree. The pupil of his eye is also at all times circular, and always of a yellowish colour. The Tiger, on the contrary will seek his prey by day and by night ; and his eye-pupil is capable of either shape, and in the twilight or dark, its colour is like a blue-green flame. This remark he made, while drawing a specimen of a large Bengal Tiger at New York. The room of the menagerie in which it was placed was generally rather dark, and at the time was rendered more so by the gloominess of the weather. The animal was exceedingly vicious, endeavouring, occasionally, to strike his keeper; yet he lay in a stately, and, seemingly, unconcerned attitude, with the cleft pupils of his eyes fixed upon the Major while drawing: but if a person passed near him, they were changed instantly into a disk, and their colour altered from yellow-green to blue-green. To this facility of expansion and contraction of the pupils, which, in this instance, resulted from a mental excitement, and not from any alteration in the degree of light, may be attributed the diurnal habits of the Tiger, as also his disregard of night-fires; while the Lion, whose eyes are not calculated for the glare of day, cannot bear the effect of firelight in the dark.

The Puma has the pupil constantly circular; yet this animal is as dangerous by day as by night; or, to speak more correctly, he will hunt his prey while the sun is above the horizon.

Of the Lynxes that are found in the United States, that called by the furriers the Chat Cervier, has complete Cats' eyes; while the Felis Canadensis, which is so nearly allied to it, as at most to be a mere variety, has round pupils, yet the habits of both are similar.

The Angora Cat, when in little light, has the eye-pupils nearly, if not quite circular; they form an ellipsis more and more narrow as the light increases, till when exposed to the sun they are almost linear.

If we refer to other genera, we find considerable variety in this particular. To select a few: the genus canis has some species with circular, and others with oval pupils;

the Hyænas have them extremely narrow ; the Zibet, Civet, and Genet, have the pupils elongated transversely ; the Ichneumons and Caffrarian Weasel have them like the Cat ; and yet, perhaps, none of these are more particularly nocturnal than the rest of the carnivora, all of which appear to prefer the twilight or night for their predatory excursions. The cloven-footed animals, the Horse, and the Whale, have transverse elliptical pupils ; and the frugiverous sort, as the Lemurs, Squirrels, and Loris, have them much larger than any other animals, but always circular ; in the genus Delphinus it assumes the figure of a heart ; in the Todes it is triangular ; and in the Alligators and Sharks it is lozenge-shaped.

If, then, it be considered, that the Lion has round eye-pupils, though it is generally inactive by day, and hunts principally after sunset ; that the pupils of the Tiger assume either shape, and that it is equally active day and night ; that the Puma, like the Tiger, is equally disposed for action at all times, though its eye-pupils, unlike those of that animal, are always circular ; that one Lynx has the pupils changeable as to shape, while the other has them only varying in size, and that both their habits accord ; and lastly, that the common Cat, which has the pupils varying greatly in shape, though we know it sees with little light, seems to possess in the day a vision as perfect as those animals which merely increase or decrease the size of the pupil, though it continues always round ;—we seem led to a probable conclusion, that there is a fallacy in adopting the form of the pupil as a physical characteristic of the disposition and habits of the animal.

The eyes of Cats nd of some other animals are frequently much illuminated with a prismatic sort of light in obscurity. On this subject, says Sir Everard Home, there are two opinions ; one, that the external light only is reflected, and the other, that light is generated in the eye.

Professor Bohn, at Leipsic, made experiments, which

proved, that when the external light is wholly excluded, none can be seen in the Cat's eye. The truth of the results was readily ascertained; it therefore only remained to determine, whether the external light is of itself capable of producing so great a degree of illumination : this was attended with difficulty; for, when the apartment is darkened, and nothing but the light from the Cat's eye seen, the animal, by change of posture, may immediately deprive the observer of all light from that source.

This was found to be the case, whether the common Cat, the Tiger, or the Hyæna, was the subject of the experiment. On the other hand, when the light in the room is sufficient for the animal to be seen, the light in the eye is obscured, and appears to arise from the iris.

As these difficulties occurred in making experiments on the living eye, others were made immediately after death, and it was found, that a strong light thrown upon the cornea illuminated the eye, as in the living animal, but when the cornea was removed, the illumination disappeared. The iris was then dissected off, and the tapetum lucidum completely exposed to view, the reflection from which was extremely bright; the retina proving no obstruction to the rays of light, but appearing equally transparent with the lens and vitreous humour.

These experiments prove, that no light is generated in the eye, the illumination being wholly produced by the external rays of light collected in the concave bright-coloured surface of the tapetum, after having been concentrated by the cornea and crystalline lens, and then reflected through the pupil. When the iris is completely open, the light is the greatest, but when the iris is contracted, the illumination is more obscure.

The influence, if any, the animal has over this luminous appearance, depends on the action of the iris: when it is shut, no light is seen; when the animal is alarmed, the eye glares, the iris being opened.

A very prevailing, if not a generic character, distinguishes a large proportion of the Felinæ,—which is, a white spot on the back of the ears. Those that are uniform in colour, as the Lion, the Puma, and the black species, as well as the common varieties of the domestic Cat, seem to be without it ; but it is certainly to be found in the Tiger, Panther, Jaguar, Ounce, Hunting Leopards, Ocelots, Lynxes, and several others. In the spotted species, a black bar across the chest seems equally prevalent.

It might well be presumed, that the natural history of a genus of animals playing so conspicuous a part on the theatre of life as the Felinæ, would be by this time clearly known, and the species accurately defined ; but such a conclusion would be very wide of truth. The Baron Cuvier commences his learned observations on the species of the Cats, to be found in the fourth volume of his Fossil Osteology, from which we shall extract largely in the present essay, by observing " that the large Carnivora with retractile claws, and spotted fur, have been for a long time the torment of naturalists, by the difficulty in distinguishing with precision their several species." Some of the species are certainly sufficiently notorious, as the Lion, the Tiger, &c., others are known to zoologists, principally by specific descriptions, which sometimes seems to agree, and sometimes to vary, as different types successively come to light, as the Panther, Leopard, the little group of Ocelots, Servals, &c.; and many more, hitherto hidden from the eye of critical examination, exist to the scientific world, at least, only in the pages of voyages and travels : these, though named and described, are properly mere candidates for insertion in systematic catalogues, rather than recognised species. Such, however, according to our plan, will be noticed either here or in the table, with the exception only of those that are clearly negatived ; and as the authorities for their insertion will be given with them, the reader will sit in judgment upon their admissibility. Our own collection of drawings would also furnish us with figures of many apparently new species, a small selection from which will be engraved and inserted.

They may be divided into small groups by certain peculiarities of colour ; the first of these groups includes the single-coloured uniform large species, and includes only the Lion and the Puma.

The superior bodily powers possessed by the *Lion*, joined to his carnivorous regimen and consequent predacious habits, while they place him at the head of the beasts of prey, and make him also the undisputed tyrant and universal dread of the plains and forests, point obviously to the situation in which he must be regarded as the first and most important species of the carnivora.

Verbal description, or even the best of figures, will convey but a very inadequate notion of this tremendous animal. The rhetorician and the painter alike fail in describing and depicting the terrific work of nature exhibited in this king of beasts.

The specific characters are thus given by our author, "a large Yellow Cat, with tufted tail, the neck of the adult male furnished with a thick mane." These vary in slight degrees, as we shall notice presently. If the ancients had any authority in nature for the male Lion without a mane, sometimes figured in their statuary, the species is lost, or at least unknown to modern research *

* Major Smith was lately informed by Professor Kretschmen of Frankfort, that he was in expectation of receiving from Nubia the skin and jaws of a new species of Cat, larger than the Lion, of a brownish colour, and without mane. The invoice of the articles from Cairo was already received, but the objects themselves were not arrived. This will probably, says the Major, prove to be the maneless Lion of the ancients, known to them by their acquaintance with Upper Egypt; and not unfrequently observed in the hieroglyphic sculptures of that country.

THE LION

The period of gestation of the Lioness is about one hundred and eight days, and the young, when first born, are very small in proportion to their adult size. They arrive at maturity in about five years, and are then nearly eight feet in the length of the body, with a tail of about four feet. If we judge from the length of their nonage, and from their size and general constitution, as observed by Buffon, it should seem probable, that the average life of this animal does not exceed twenty-five years ; though it has been said, that some have been kept in a state of confinement for nearly three times this period. The mane appears to increase as the Lion advances in age, and not to depend for its growth on that of the animal. The female is without it altogether. The Lion laps in drinking, but turns the tongue downwards, contrarywise to the Dog.

It seems needless to enter into any general description of an animal so well known. We shall only therefore, in reference to its organs of sense, observe, that the pupil of the eye is circular ; the external ear small, rounded with a lobule on the outer edge like that of the common Cat. Their other organs of sense in common with those of loco-motion and generation, the peculiar retractibility of the claws, and system of dentition, correspond also with the Domestic Cat.

When young, the Lion has no trace of the mane or of the tuft at the end of the tail. These appear at about three years old. The hair of their body is then partially curled and tufted, and not smooth as in the adult state of the animal. It is remarkable also, that when young, they have a dark dorsal line, together with several transverse parallel dark stripes and faint spots, which give them the appearance, to an inexperienced eye, of being young Tigers. They are born with the eyes open, but the external ear is semipendant and does not become erect for two months.

The talons also do not attain their retractile power till the animal is nearly a year and a half old. At about a year old the canine teeth appear, a period very frequently fatal to the young, at least to those born in confinement.

The characters of the Lion and the Tiger have been of late considered as perfectly similar. This assertion, contradicted by the ancients and early moderns, has wholly arisen from some remarks made by travellers to the Cape. No doubt, where similar appetites, similar propensities, similar means, and similar circumstances occur, a great similarity of character must be found. Although individuals are observed to be more undaunted and ferocious, in proportion to the increased distance at which they may be found from the habitations of mankind, more especially the civilized races, yet the Lion, we should submit, when compared with the Tiger, is a noble animal ; he possesses more confidence, and more real courage ; he likewise differs in his permanent attachment to his mate, and protection of his young ; while the Tiger shows no partiality beyond the period of heat in the female, and is himself frequently the first and greatest enemy to his own offspring. The former of these *traits* of character is substantiated by a great variety cf authors and testimonies, and denied only by the assertion of the colonists of the Cape, who report that the Lion, when he fancies himself unperceived, will flee from the hunters ; but it must be remembered, that the Lion is generally pursued by day, and it is probable that he bears the glare of an African sun, reflected from a sandy soil, with great inconvenience. It is, therefore, as unjust to tax this animal with cowardice, because he wishes to avoid a contest, at a period when his sight is much deteriorated, as it would be to rate the hunter for his timidity, because he will not chase the Lion in the dark.

Major Smith has met with eleven instances of different

Lions, which have protected and fostered Dogs, and but a single one of the Tiger exhibiting a similar kindness of disposition.

In a state of confinement, they have frequently shown unequivocal marks of gratitude and affection toward their feeder and keeper, as in the case mentioned by Seneca, of which he was personally witness, of a Lion, to whom a man, who had formerly been his keeper, was exposed for destruction in the amphitheatre at Rome, and who was not only instantly recognised, but defended and protected by the grateful beast. Indeed, those animals which are exhibited as public shows, when they have been for some time accustomed to restraint, will, in general, not only become obedient to their feeder and keeper, but even show a considerable degree of liking toward him, though, in such cases, it is necessary for the man to exercise caution and discretion, and not to expose himself to the animal when feeding, or when its irritability is at all excited.

The keeper of a Lion, which was exhibited about the country, at fairs, a few years ago, was in the habit of putting his head into the mouth of the beast, having previously put on a worsted cap, to defend himself from being lacerated by the animal's tongue; and Major Smith has seen a young man stand upon a lioness, drag her round the cage by the tail, open her jaws, and thrust his head between her teeth.

" A keeper of wild beasts, at New York," says the Major, " had provided himself, on the approach of winter, with a fur cap. The novelty of this costume attracted the notice of the Lion, which, making a sudden grapple, tore the cap off his head, as he passed the cage; but perceiving that the keeper was the person whose head he had thus uncovered, he immediately laid down. The same animal once hearing some noise under its cage, passed its paw through the bar, and actually hauled up the keeper, who was clean ing beneath ; but as soon as he perceived he had thus ill-used his master, he instantly laid down upon his back, in an attitude of complete submission."

The Lion, while feeding, will exhibit a more disinterested courage than most of the Carnivora. When the prey is thrown to him at one corner of the cage, and the keeper holds up a stick at the bars of the opposite side, the animal will instantly quit his food to attack the disturber of his meal ; but if the same thing be done to the Tiger, he will lie close upon his food, snort, give shrill barkings, and, at most, just rise to fly at the stick, and then drop upon his meat again."

Unlike some of the carnivorous animals, which appear to derive a gratification from the destruction of animal life beyond the mere administering to the cravings of appetite, the Lion, when once satiated, ceases to be an enemy. Hence very different accounts are given by travellers of the generosity or cruelty of its nature, which result, in all probability, from the difference in time and circumstances, or degree of hunger, which the individual experienced when the observations were made upon it. There are, certainly, many instances of a traveller having met with a Lion in the forest during day,

> Who glared upon him, and went surly by,
> Without annoying him ;

but when urged by want, this tremendous animal is as fearless as he is powerful ; though in a state of confinement, or when not exposed to the extremity of hunger, he generally exhibits tokens of a more tender feeling than is to be met with in the Tiger, and most of the Felinæ.

The effect of the voice of the Lion, to be properly felt, must be heard. During sexual excitement, its noise is perfectly appalling, and produces on the mind of the by-stander, however secure he may feel himself, that awful admiration commonly experienced by us on witnessing any of the grand and tremendous operations of nature. When in the act of seizing his prey in a natural state, the deep thundering tone of the roar is heightened into a horrid scream, which accompanies the fatal leap on the unhappy victim. This power of voice is said to be useful to the animal in hunting, as the weaker sort, appalled by it, flee from their hiding-places, in which alone they might find security, as the Lion does not hunt by scent, and seek for it in ineffectual flight, which generally exposes them to the sight of their enemy, and, consequently, to certain death.

The Lion is capable of carrying off, with ease, a horse, a heifer, or a buffalo. The mode of its attack is generally by surprise, approaching slowly and silently, till within a leap of the predestined animal, on which it then springs, or throws itself with a force, which is thought, in general, to deprive its victim of life before the teeth are employed. It is said, this blow will divide the spine of a horse, and that the power of its teeth and jaws will break the largest bones.

The Asiatic variety of the Lion is of a uniform yellow-colour. The mane, which is more scanty than in the African variety, is also entirely yellow. In physiognomy, as well as character, they seem to agree, but the Asiatic is rather the smaller of the two.

The existence of the Lion in south-eastern Asia has not been long ascertained. Two young officers of the 8th light dragoons, during one of the campaigns in India, when out one morning on a hunting excursion, and having quitted their Elephants, were walking near a jungle ; one who was more experienced in the country than his companion, suddenly observed a recent track, of what he took to be a Tiger ; instantly looking back towards the jungle, they hastened forward in the direction of it, and in the middle of a field, found the mangled remains of a Nyl-ghau, (*Antilope picta et Trago Camelus*) They were surprised to observe no less than three distinct tracks, all leading to the prey, but from different parts of the jungle, and justly concluded that there were more than one of these fierce animals near them. While returning to their Elephant and their party, among whom was one of the gentlemen who have charge of the Elephants belonging to the East India Company, they were astonished to see a Lion come out to the edge of the jungle, open his jaws, and stretch himself, and then coolly return into the cover. Having mounted their Elephant, a large female, they proceeded into the jungle, with more courage than wisdom ; but they had scarcely advanced a few yards, when the Lion sprang at them, and the Elephant, wheeling round, fled to the plain, but with a severe wound in the hough. The next day, a regular hunting party, with a considerable force of Ele- phants, was mustered, and when the line was formed, the half-hamstrung Elephant, trembling with anxiety, and giving proofs of her extreme uneasiness, was yet so keen as to be always her whole length before the others in the clearing of the jungle *. Before night, three Lions were killed, and thus, for the first time, probably, the presence of the Lion in India was satisfactorily established.

The distinctness of the two varieties may be inferred from this circumstance, that a Lioness of the Asiatic breed, which was in Exeter 'Change, was frequently offered to the African Lion, which is also kept there, and was constantly refused, while his attachment still remains unaltered for the Lioness of his own country, in the same menagerie, which has produced several litters, the fruits of their inter- course. Major Smith has known two other instances of the same kind.

There is also a South-African variety of the species, with the mane nearly black ; and if the specimen, lately in

* This was, probably, from a desire of vengeance in the saga- cious animal, which continued lame, and was afterwards sold at a considerable loss.

Exeter 'Change, be considered as an ordinary type of this variety, the head and muzzle are broader and more like those of the Bull-dog ; the under jaw more projecting ; the ears larger, more acuminated, and blacker in this than in the more ordinary breed.

The noxious animals yield, if not to the physical, at least to the intellectual powers of man, and, accordingly, their decrease, either generally or locally, may be observed to accord with the progress of refinement in human life. The Baron has, with much learning and research, accumulated instances of the existence of Lions in parts where they are no longer indigenous, and of their former great abundance in countries where they are now but partially known.

" It is true," says he, " that the species has disappeared from a great number of places where it was formerly found, and that it has diminished in an extraordinary degree everywhere."

Herodotus relates that the Camels which carried the bag-gage of the army of Xerxes were attacked by Lions, in the country of the Pæonians, in Macedonia ; and also, that there were many Lions in the mountains between the river Nestus, in Thrace, and the Archeloüs, which waters Arca-nania. Aristotle repeats the same, as a fact, in his time. Pausanias, who also relates the accident which befel the Camels of Xerxes, says further, that these Lions often descended into the plains at the foot of Olympus, between Macedonia and Thessaly.

If we except some countries between India and Persia, and some parts of Arabia, Lions are now very rare in Asia. Anciently they were common. Besides those of Syria, often mentioned in Scripture, Armenia was pestered with them, according to Oppian. Apollonius of Tyan, saw near Babylon, a Lioness with eight young, and, in his time, they were common between the Hyphasis and the Ganges. Ælian mentions the Indian Lions, which were trained for the chase, remarkable for their magnitude, and the blackish tints of their fur.

That the species has become rare, in comparison with former times, even where it is now most abundant, may be sufficiently inferred from the Roman historian Pliny, who tells us, that Quintus Curtius was the first who exhibited many of them, at one time, in the Circus. Sylla caused one hundred, all males, to engage together, for the amusement of the people; Pompey six hundred, of which one hundred and fifteen were males; and Cæsar, four hundred.

The same abundance continued also under the first Emperors; Adrian often destroyed one hundred in the Circus; Antonine, on one occasion, one hundred; and Marcus Aurelius the like number on another. The latter exhibition Eutropius considers as particularly magnificent, whence the Baron infers, that the number of the species was then diminishing, though Gordian the Third had seventy, which were trained; and Probus, who possessed a most extensive menagerie, had one hundred of either sex.

A great deal of interesting matter might be added on the subject of this species, which our need of brevity obliges us to forego inserting.

The *Puma* (*F. Concolor*) is large, and uniformly yellow, without mane or tuft to the tail. It is placed next to the Lion, on account of its corresponding uniformity of colour. It is called, by the Mexicans, *Mitzli;* in Peru, *Puma;* in Brasil, *Cuguaçuarana,* (the word Cougoua, is contracted by Buffon from this latter barbarous appellation;) and in Paraguay, *Guazuara.* The name by which it is most generally known is that of the American Lion; so called from a distant similarity it bears to the Lion of the old world, in the uniformity of colour before alluded to. It seems the more necessary to advert to these synonymes, because the name Cougoua, by which it is most commonly known in Europe, particularly in France, appears, probably, to have been borrowed from that proper to another animal *. The Linnæan epithets, Concolor and Discolor, have, likewise, no appropriate meaning; but Puma is its native name. Its length, from the nose to the root of the tail, is about five feet; and its height, from the bottom of the foot to the shoulder, twenty-six inches and a half; hence it is longer in the body and lower on the legs than the Lion, but it differs from that species more particularly in the shape of the head, which is small and round, and not square, as in F. Leo.

D'Azara says this animal is less ferocious, and more easy to be killed than the Jaguar; it lies concealed in the under-wood, and does not have recourse to caverns for shelter like the Jaguar. Unlike this animal also, the Puma ascends and descends the highest trees with celerity and ease, though it may be considered in general rather as an inha-bitant of the plains than of the forests. He states also that it is not known to attack a man†, or even a dog, but avoids both with great timidity. Its depredations are ge-nerally confined to quadrupeds of a middling size, as calves, sheep, &c.; but against these its ferocity is more insatiable than its appetite, destroying many at an attack, but carry-ing away perhaps only one. If it have more than sufficient for a meal, it will cover and conceal the residue for a se-cond repast; in which it differs also from the Jaguar, which is not so provident.

D'Azara possessed a tame Puma, which was as gentle as any Dog, but very inactive. It would play with any one; and if an orange were presented to it, would strike it with the paw, push it away, and seize it again, in the manner of a Cat playing with a Mouse. It had all the manners of

* See the Eira.

† Buffon states that it will seize a Man if it find him sleeping, which Azara denies.

a Cat, when engaged in surprising a bird, not excepting the agitation of the tail ; and purred when caressed like that animal.

An incident occurred a few years back, not far from New York, which seems to disprove the assertion of Molina and D'Azara, that the Puma will not attack a Man ; and while it shows the ferocity of the animal, evinces that its power is not much inferior to that of the Jaguar. Two hunters went out in quest of game on the Katskill mountains, in the province of New York, on the road from New York to Albany, each armed with a gun, and accompanied by his Dog. It was agreed between them, that they should go in contrary directions round the base of a hill, which formed one of the points in these mountains; and that, if either discharged his piece, the other should cross the hill as expeditiously as possible, to join his companion in pursuit of the game shot at. Shortly after separating, one heard the other fire, and, agreeably to their compact, hastened to his comrade. After searching for him for some time without effect, he found his Dog dead and dreadfully torn. Apprised by this discovery that the animal shot at was large and ferocious, he became anxious for the fate of his friend, and assiduously continued the search for him ; when his eyes were suddenly directed, by the deep growl of a Puma, to the large branch of a tree, where he saw the animal couching on the body of the Man, and directing his eyes toward him, apparently hesitating whether to descend and make a fresh attack on the survivor, or to relinquish its prey and take to flight. Conscious that much depended on celerity, the hunter discharged his piece, and wounded the animal mortally, when it and the body of the Man fell together from the tree. The surviving Dog then flew at the prostrate beast, but a single blow from its paw laid the Dog dead by its side. In this state of things, finding that his comrade was dead, and that there was still danger in approaching the wounded animal, the Man prudently retired, and with all haste brought several persons to the spot, where the unfortunate hunter, the Couguar, and both the Dogs, were all lying dead together *.

D'Azara asserts, that the Jaguar cannot climb trees, but that the Puma can. The last anecdote sufficiently evinces that the latter can mount a tree; but it seems probable, that it is accomplished rather by a vigorous bound in the first instance, than by absolute climbing

Major Smith witnessed an extraordinary instance of the abstracted ferocity of this animal, when engaged with its food. A Puma, which had been taken and was confined, was ordered to be shot, which was done immediately after the animal had received its food: the first ball went through his body, and the only notice he took of it was by a shrill growl, doubling his efforts to devour his food, which he actually continued to swallow with quantities of his own blood till he fell.

Notwithstanding such instances of the violence of disposition of this animal, it is very easy to be tamed. The same gentleman saw another individual that was led about with a chain, carried in a waggon, lying under the seat upon which his keeper sat, and fed by flinging a piece of meat into a tree, when his chain was coiled round his neck, and he was desired to fetch it down; an act which he performed in two or three bounds, with surprising ease and docility.

A tame Puma, which died recently, was some time in the possession of Mr. Kean, the actor. It was quite docile and gentle. After the death of this animal, it was discovered that a musket-ball, in all probability, had injured its skull, which was not known in its lifetime.

* This incident was related to Major Smith by Mr. Skudder, the proprietor of the Museum at New York, where the animal was preserved after death as a memorial of the story.

Many of the actions and manners of the Jaguar and Puma have been confounded by different describers. It may perhaps be observed generally, that the Puma is of the most cruel and sanguinary disposition in a state of nature, though easy to be tamed; but is inferior to the Jaguar in bodily powers, and still more in energy and courage.

Though this species is found from Patagonia to California, the Baron has not been able to ascertain any hereditary varieties. Some individuals indeed seem of a deeper colour, and others exhibit indications of spots, the colour in places being more opaque in certain angles of light: but such may be attributable to age, sex, or circumstances, and not probably to actual varieties.

The Black Tiger of Laborde is erroneously applied, according to Cuvier, to a Black Couguar, by Buffon, the existence of which he doubts.

In arranging the various species of Felinæ into groups, the *Bengal* or *Royal Tiger* will be found nearly isolated. He is easily distinguished from al. other species by his transverse dark stripes. Compared with the Lion, he is thinner an l lighter, and has the head rounder. The upper part of the body is yellow, the under part white. The whole internal face of the ears, and a spot on the external surface round and over the eyes, the end of the muzzle, cheeks, throat, neck, chest, belly, and internal sides of the limbs are white ; and the tail is annulated with black on a whitish-yellow ground. The eye-pupils are generally said to be round, and indeed we have never observed it otherwise ; but in the instance already mentioned, witnessed by Major Smith, they assumed an elliptical figure. The black bands are extremely irregular, and vary in different individuals. We shall not describe its person further, but merely refer to our figure of a specimen lately at Exeter'Change.

Beneficence, however capriciously exercised, may be said occasionally to exhibit itself in the lion ; but the ferocious

Griffith del.

THE TIGER

F. TIGRIS

character of the Tiger, in its natural state, presents no such palliation. When its appetite is satisfied, the former seems no longer delighted with blood; but butchery appears to afford gratification to the latter, even after its hunger has been satiated.

This animal is the scourge of Asia and the Indian Islands. Equal to the Lion in stature, though generally inferior in strength, it wants not courage and ferocity to attack that animal; but although the combat is sometimes furious, it generally falls a victim to its temerity in so doing, unless some disparity of age or other circumstance should bring the strength and power of the two animals more to a level. Its swiftness and strength enable it to seize a man while on horseback, and to drag, or rather to carry him in its mouth by bounds and leaps into a jungle or forest, in spite of all efforts to prevent it, short of musket-balls: indeed, the weight of a man, or even of a more ponderous animal, in its mouth, does not appear to incommode or delay the ordinary swiftness of the beast.

Mr. Marsden informs us, that the Tigers in Sumatra prove to the inhabitants there, both in their journeys and even their domestic occupations, most fatal and destructive enemies. The number of people usually slain by these rapacious tyrants of the woods is almost incredible. Whole villages are sometimes depopulated by them. Yet from a a superstitious prejudice, it is with difficulty they are prevailed upon, by a large reward which the India Company offers, to use methods of destroying them, till they have sustained some particular injury in their own family or kindred, and their ideas of fatalism contribute to render them insensible to the risk. Their traps, of which they can make a variety, are very ingeniously contrived. Sometimes they are in the nature of strong cages, with falling doors, into which the beast is enticed by a goat or dog enclosed as a bait. Sometimes they manage so that a large beam is made to fall in a groove across the Tiger's back; at other times it is noosed about the loins with strong ratans, or led to ascend a plank nearly balanced, which, turning when it has passed the centre, lets the animal fall upon sharp stakes prepared below. Instances have occured of a Tiger being caught by one of the former modes, which had many marks in its body of the partial success of this last expedient The Tigers of Sumatra are very large and strong. They are said to break the leg of a Horse or Buffalo with a stroke of the fore-paw, and the largest prey they kill is, without difficulty, dragged by them into the woods. This they usually perform on the second night, being supposed on the first to gratify themselves with sucking the blood only. Time is by this delay afforded to prepare for their destruction ; and to the methods already enumerated, be-sides shooting them, may be added that of placing a vessel of water strongly impregnated with arsenic near the car-cass, which is fastened to a tree, to prevent its being car-ried off. The Tiger having satiated itself with the flesh, is prompted to assuage its thirst with the tempting liquor at hand, and perishes in the indulgence

Buffon's assertion, however, that the nature of the Tiger is perfectly incapable of improvement, is rather too strong, as many instances have evinced since the time that Buffon wrote. A full-grown Tiger was lately in the possession of some of the natives at Madras, who exhibited it held merely by a chain; it was indeed kept muzzled, except when it was allowed (which was occasionally done) to make an attack on some animal, in order to exhibit the mode of its manœuvring in quest of prey. For the purpose of this exhibition, a sheep in general was fastened by a cord to a stake, and the Tiger being brought in sight of it, imme-diately crouched, and moving almost on its belly, but slowly and cautiously, till within the distance of a spring from the animal, leapt upon and struck it down almost instantly dead, seizing it at the same moment by the throat with its teeth; the Tiger would then roll round on its back, holding the sheep on its breast, and fixing the hind claws near the throat of the animal, would kick or push them suddenly backwar s, and tear it open in an instant. Notwithstanding, however, the natural ferocity of these animals in general, the individual in question was so far in subjection, that while one keeper held its chain during this bloody exhibition, another was enabled to get the carcass of the sheep away by throwing down a piece of meat previously ready for the purpose.

Three specimens in the Museum at Paris became extremely tame, and Mr. Cross has had instances of Tigers, taken quite young, and bred up in a state of confinement, exhibiting nearly as much gentleness as the Lion under similar circumstances; by showing attachment to their keeper, and, in one instance, to a Dog, which was exposed to one of them; so that their nature appears, in some degree, capable of training and education; and the furious character attributed to the Tiger must be considered as applicable to it only in a wild and unfettered condition.

But however the disposition of the Tiger when in confinement may yield to judicious treatment, it is very different in a state of nature, and may be deemed a sanguivorous animal, as it sucks the blood of its victim previously to tearing or eating it; and, after having so done with one animal, will leave the carcass and seize on any other that may come in sight, suck the blood of this also with the most horrid avidity, which will induce it almost to bury its face and head in the body of its prey. · We are, therefore, not disposed to palliate its natural ferocity, but only to give it credit for its capability of improvement.

Tigers' skins vary as to the number of stripes and brightness of the colours, which latter abates in some degree when the animal is living under restraint, and much more when a skin is dried and prepared for commercial purposes.

The Tigress is pregnant about fourteen weeks, produces four or five at a time, and has been known to breed when confined. When first born, the young do not exceed the size of a Kitten about three months old. They are of a pale-gray colour with obscure dusky transverse bars, like those proper to the Lion of the same age.

A white variety of the Tiger is sometimes seen, with the stripes very opaque, and not to be observed except in certain angles of light. We have engraved from a specimen of this variety, formerly in Exeter 'Change.

We have already had occasion to offer a few observations on the production of mules, considered quite independently of the standing mystery of fecundation. The instance quoted was that of a breed between the Common Chacal and the species or variety of Senegal. We have at this place to notice a similar reproduction between two species of a more prominent character on the great theatre of life, between, in fact, the two tyrants of the world—the African Lion and the Asiatic Tiger.

The recent discoveries of intermediate genera, discoveries resulting principally from the superior degree of attention lately paid to comparative anatomy, have puzzled and confused the strict systematic zoologists. The small portion of the animal kingdom contemplated in the previous pages, has furnished several instances of these intermediate animals partaking, in many points of character and peculiarities, of distant and strongly-marked genera. It has also furnished ample proofs of the effects of secondary causes on animal organization, and demonstrated that all the varieties of animated nature before us are not the result of distinct acts of original creation, but are, in fact, from time to time, springing up before our eyes.

Few branches of natural knowledge are more obscure than the excitements to these modifications of animals.

C.H. Smith Esq.r del.t

WHITE TIGER.

The subject is one of legitimate inquiry, though it has hitherto almost entirely eluded demonstration ; we may still therefore venture to hope, that as we improve in detecting many of the modes of action of nature in her wonderful operations, we may eventually arrive at some partial insight into the principles to which she seems subjected, or, in other words, the secondary causes which induce the phenomena in question.

The point more particularly intended in relation to the subject before us is, whether the interfecundity of animals differing generically, in the artificial language of zoologists, has ever produced a third permanent distinct genus, and whether the like interfecundity of animals differing specifically, in the same artificial language, has ever produced a third permanent distinct species, so as in fact to account for the origin of any or many of the creatures around us.

It seems very difficult to conclude that the race of every genus and species originated in a distinct act of creation, especially as we know their varieties do not ; and however we may amuse ourselves and puzzle others with attempted definitions of genus, species, and variety, we must remember that these distinctions are of man's making, not of nature's, whose works are so interlinked together, as by no means, in all points, to conform to artificial separation and arrangement. These reflections are not intended to prejudice the utility of zoological systems in the assistance they afford to our limited faculties ; they merely point to the impossibility of separating, by strong distinct lines of demarkation, the varied creatures of the earth, which appear to us to be interlaced and linked together like the meshes of a piece of network, though not like it in regard to the regularity of the points of contact.

If this position be correct, if, in fact, one species of a given group of animals be found to approximate to the physical characters of another group, on the one hand, and a second species be found assimilated to a third group, on the other, so that in fact all organization, including both the animal and vegetable kingdom, be irregularly reticu lated together, we seem naturally led to the inquiry whe ther these assimilated individuals have ever reproduced a third race differing from both, and the zoological cata- logue has been thus enlarged, and to what extent.

Although we by no means intend to enter into the dis- cussion of a subject of this nature, but merely to allude to it *en passant*, it may be proper not to quit it precipitately without adverting to the ordinary-admitted conditions of hybrid animals.

Sterility seems one of the most determined of these con- ditions. It appears to be the mean employed by nature to prevent the very effects to which we have just ventured to allude, and if universal in its application, must at once put an end to the possibility of factitious races springing from these unnatural intercourses.

It is from the domesticated or tamed races that we prin- cipally draw our facts upon this subject. Those specifi- cally different, as the Horse and Ass, being the most fami- liar to us, produce the ordinary Mule, which is steril in effect, though it may occasionally produce offspring; for even then such offspring is found to degenerate from its parent, and to be perfectly incapable of reproduction. But the events of the sequestered forest and trackless wilder- ness are less easy of detection. In relation to these, we must be content with few facts and many analogies.

The course of reproduction of the larger Felinæ has, of late years, been frequently open to observation in the me- nageries, and one cannot but feel some surprise that the breeding these animals in a state of confinement should be comparatively matter of modern experiment only. We have now not only instances of these animals breeding in the ordinary manner, but we have instances of varieties so

CUBS BRED BETWEEN A LION AND TIGRESS *THREE MONTHS OLD.*

distant as to be inserted in zoological catalogues as specifically different, which have reproduced, and in the instance we are more particularly about to notice, of a breed between two decidedly different species, the Lion and the Tiger.

Mr. Cross, at Exeter 'Change during the few years he has possessed that establishment, has bred twenty-one Lions, six Tigers, four Jaguars, and four Leopards. In the French menagerie they do not appear to have been so successful, though M. F. Cuvier gives us the particulars of some young Lions bred there with his usual interesting detail.

In addition to the instances already enumerated at Exeter Change, Mr. Cross has succeeded twice in producing a cross breed among these animals. These were between the Black Leopard, generally treated as a distinct species, and the common African Leopard. The male was the black variety. Three or four individuals of this breed are still exhibiting about the country.

The offspring of varieties are generally observed to partake of a middle character, as (if we may be allowed to instance them here) in the human species. The Negro and the European produce an offspring as certainly partaking of the physical character of both as does the issue of parents both of the same colour. If the offspring of these two varieties breed with either the black or the white, the product will still be certain, and will partake in its physical appearances and colour the same proportions of which it actually partakes in blood ; and again, if the offspring of the black and white continue to be united in succession to either pure black or pure white, it is not until after the fourth generation that the traces of the intermixture will be lost, as is evinced constantly in our West India settlements.

And so it is with the lower Mammalia. If a thorough

bred Dog be crossed with a different variety, the consequences will be manifested in the offspring for several generations.

But this did not appear to be the case with the cubs bred between the black variety and the Common Leopard; they were in all respects like the ordinary cubs of parents both of the common breed.

But to return to our Lion Tiger-cubs. Mr. Atkins, an itinerant exhibiter, and dealer himself, bred the Lion, the father of these cubs. He was a very fine and very valuable beast, for his beauty and the docility of his disposition, the ferocity of which had never been entirely developed by natural habits. At the period in question, he was about four years old. The Tigress, the mother of the cubs, is supposed to be about four or five years old. She has been in Mr. Atkins's possession about two years, and was, probably, taken very young, as the gentleness of her disposition seems to evince.

These two animals, ever since the arrival of the Tigress, have been confined in one den, and have always agreed well together. From the beginning of their being so placed, there had been frequent possibility of issue, though the first, consisting of a litter of three cubs, was not born till the 17th of October, 1824, the result of a more particular intercourse, which lasted ten or twelve days, in the beginning of the previous July. They were born at Windsor, and were shortly afterwards honoured by a visit from his Majesty. The Lion, unfortunately, died about six weeks afterwards.

The cubs were taken from the Tigress immediately after birth, and were fostered by several bitches and a Goat; they are all alive, and promise, at present, to attain maurity.

In regard to their personal appearance, we feel constrained, after what has been already said, to be very brief in our descriptions, an omission, however, we hope to compensate by the figures. These show them at three months old, and we have added figures of the young, both of Lions and Tigers, that they may be compared with this singular breed.

The young of the Lion are calculated to deceive an inexperienced observer, from the fact of their being striped transversely, so as to induce the opinion, at first sight, that they rather belong to the Tiger, and, in this respect, the cubs in question agreed with those of the Lion. In the young Lions, however, these stripes soon become obliterated, but in those before us, they appear to be getting more decided and permanent, and, in fact, to be assuming the permanent Tigrine character. Our Mules, in common with ordinary Lions, were born without any traces of a mane, or of a tuft at the end of the tail. Their fur, in general, was rather woolly ; the external ear was pendant toward the extremity ; the nails were constantly out, and not cased in the sheath ; and, in these particulars, they agreed with the common cubs of Lions. Their colour was dirty-yellow or blanket-colour ; but from the nose over the head, along the back, and upper side of the tail, the colour was much darker, and, on these parts, the transverse stripes were stronger, and the forehead was covered with obscure spots, slighter indications of which appeared also on other parts of the body. The shape of the head, as appears by the figures, is assimilated to that of the father (the Lion) ; the superficies, of the body, on the other hand, is like that of the Tigress.

In our collection of drawings, are three figures from a curious and unique specimen, which was for some months in the possession of Mr. Polito, at Exeter 'Change. Major Smith also took a hasty sketch from the same ; and Mr. Landseer made another. We have engraved this animal from the last mentioned drawing, under the name of the Nebulose or Clouded Tiger, in reference to the peculiarity of the spots or rather patches, which covered its skin. It acquired the name among the keepers, at the menagerie, of the Tortoiseshell Tiger.

The head of this animal was small compared with its general bulk. The throat was thick; the body long, heavy, and cylindrical; and the legs very thick, short, and muscular. The tail also was remarkably thick and long. The extreme irregularity of the markings of this animal render a compressed description difficult. The forehead and the limbs, both outside and within, were covered with numerous small, close spots; on the sides of the face, were a few diagonal stripes; the sides of the throat also, and the dorsal line were marked with long, irregular, black stripes, with but little parallelism or regularity of angle. The whole sides of the animal were covered also with black stripes, forming a few large irregular enclosures, some nearly round; others approaching a long square or oblong, thus assuming something like the irregular uncertain figures of a passing cloud, or the bright yellow and rich brown of tortoiseshell, when viewed by refracted light. The tail was marked with many annuli, almost from beginning to end; and the whole appearance of the animal was excessively beautiful.

The irregular, circular, and oval open patches on the sides of this animal, might approximate it, in some measure, to the group of Ocelots; but its Asiatic habitat and large size, separate it entirely from these American Cats, and bespeak its relative situation among the Felinæ to be after the Striped Tiger, and before the group designated by circular open spots.

We are unable to give the particulars of his dimensions, but in the bulk of his body, and the size of his head, he was said to be nearly equal to the Bengal Tiger, though his legs were shorter, and appeared still stronger; his tail was also much thicker, and appeared much browner and duller. He was fierce in disposition, but was less active and lively than the Bengal species; nor did his eye convey that

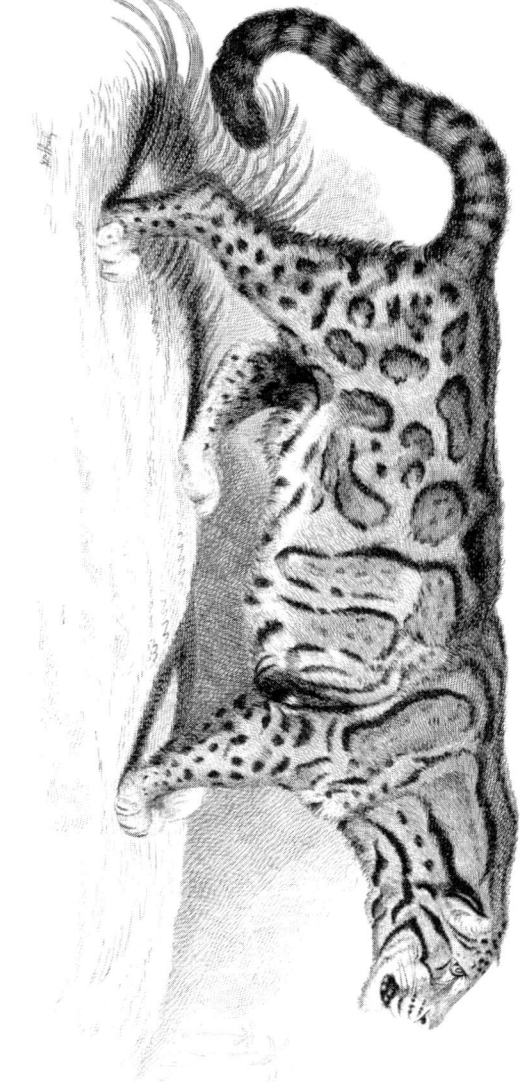

THE CLOUDED TIGER.   F. NEBULOSA.

London, Published by G.B. Whittaker. 1824.

treacherous watchfulness of glance observed in the latter animal. He was said to have been brought from Canton.

The specimen in question, was taken into the country with an itinerant exhibition, and died there, and so little attention did Zoology, at that time, receive here, that, as far as appears, its skin was cut up to make caps for the keepers, and no vestige of the animal is now known to remain.

It seems, however, there is no doubt of the distinctness of this species, as we are informed Sir Stamford Raffles is acquainted with the animal as indigenous in Sumatra. We may, therefore, hope for some more detailed particulars of it from that distinguished officer, and able writer *.

* After the above observations on this animal were printed, No. 4 of the *Zoological Journal* came to the hands of the Editor, in which is amply fulfilled his anticipations of further and satisfactory particulars of the species, at least, presuming the identity of that, there described, with the one noticed in the text.

These particulars are furnished by Doctor Horsfield, in his usual detailed and masterly manner, with the addition of various interesting remarks by Sir Stamford Raffles. Under present circumstances, we have only the opportunity of inserting Sir Stamford's notice of the animal, with a few additions, by way of explanation, which seem to be required.

A specimen of this species, that described by the Doctor, arrived in England in August last, and is lately dead. Sir Stamford refers to this and to another individual under the native name. In regard to the dimensions, he says, " A small RIMAU-DAHAN, lost in the Fame, which had been living in my possession about ten months, and might have been four months old, when he first came into my possession, attained a size of about one-third larger than the specimen which was brought to England last August, (length from nose to tail, three feet; length of the tail, two feet eight inches; height one foot four inches.) The colours and marks were nearly the same, but more defined, and nothing yellow or red about it, the black having a striking velvety appearance. The tail was longer and more bushy than in the latter specimen.
This

The *Jaguar* is so named in Brazil. The Portuguese have called it *Onça*, which Linnæus adopted as its specific name.

This was obtained a few days before I last left Bencoolen, in April. It was then smaller than the Common Tiger Cat, and only distinguishable from that animal, by the length of the tail, breadth of the paw, and colours. The natives assert that they do not attain a much larger size than the first specimen, and, perhaps, the full size of the wild and full-grown animal may be fairly taken as half as large again as the present specimen."

To the preceding remarks on the dimensions of the Rimau-Dahan, Sir T. S. Raffles has added the following particulars regarding its manners : " Both specimens, above-mentioned, while in a state of confinement, were remarkable for good-temper and playfulness ; no domestic kitten could be more so ; they were always courting intercourse with persons passing by ; and in the expression of their countenance, which was always open and smiling, shewed the greatest delight when noticed, throwing themselves on their backs, and delighting in being tickled and rubbed. On board the ship, there was a small Musi Dog, who used to play round the cage and with the animal, and it was amusing to observe the playfulness and tenderness with which the latter came in contact with his inferior-sized companion. When fed with a fowl that had died, he seized the prey, and after sucking the blood and tearing it a little, he amused himself, for hours, in throwing it about and jumping after it, in the manner that a cat plays with a mouse before it is quite dead.

" He never seemed to look on man or children as prey, but as companions ; and the natives assert that, when wild, they live principally on poultry, birds, and the smaller kinds of deer. They are not found in numbers, and may be considered rather a rare animal, even in the southern part of Sumatra. Both specimens were procured from the interior of Bencoolen, on the banks of the Bencoolen River. They are generally found in the vicinity of villages, and are not dreaded by the natives, except as far as they may destroy their poultry. The natives assert that they sleep and often lay wait for their prey on trees ; and from this circumstance, they derive the name of *Dahan*, which signifies the fork formed by the branch of a tree, across which they are said to rest, and occasionally stretch themselves.

" Both specimens constantly amused themselves in frequently

It is peculiar to America, and is sometimes called the Tiger
of that continent. In size and powers, indeed, it is but
little inferior to that formidable beast.

jumping and clinging to the top of their cage, and throwing a
somerset, or twisting themselves round in the manner of a Squirrel
when confined, the tail being extended, and shewing to great advantage when so expanded."

Dr. Horsfield refers to a figure by Howitt, published by the editor of the present undertaking, some time ago, in an incomplete
work, (the remainder of which is cancelled,) and also to the figure
already given in a previous number of this work, under the name
of Felis Nebulosa; and having compared these with his specimen,
he doubts the identity of the species of both individuals intended,
and, therefore, drops the name of F. nebulosa, and in anticipation
of M. Temminck, appropriates that of F. microcelis, which that
gentleman had given to an inedited species in his possession, said
to be the same as that of Dr. Horsfield. A comparison of our
figure, here, merely (for of Howitt's accuracy, in general, little
can be said,) with that by Mr. Daniel, which illustrates the Doctor's description, has led us, we confess, respectfully, to a different
conclusion from that of Dr. Horsfield. Major Smith, it is true, as
we shall see, suspects they are varieties. It will be seen, by
the text, that our figure was taken from the specimen to which,
also, the Doctor alludes, as identified with his species, under the
name of the Fox-tailed and Tortoiseshell Tiger.

Howitt's drawing was purchased by the editor a few years back,
of that artist, and was, it seems, copied, though not at all faithfully, from Major Smith's sketch. To the species intended, the
editor, long since, applied the epithet nebulosus, which Major
Smith adopted. Knowing, therefore, no more of the type, he sent
the *Zoological Journal* to Major Smith, at Plymouth, who has returned, in effect, the following particulars.

He gave, it seems, a copy of his drawing of the animal, together
with his manuscript notes upon it, in 1817, to the Baron Cuvier, in
whose collection, and in that of his brother, M. F. Cuvier, he saw it
during the last summer. M. Temminck he believes, also, was first
made acquainted with the species from his (the Major's) drawing,
in 1820, at Amsterdam, at least, M. Temminck professed himself
to have been previously unacquainted with it. In the absence, therefore, of further particulars from that gentleman, Major Smith is inclined to suspect that M. T.'s inedited species may be, in fact. the

The open circles of black, with a central dot, form a strong specific character. The marks, however, differ much

F. macraurus of Prince Maximilian, mentioned and figured in this work, a specimen of which the Major saw at M. Temminck's house. This conjecture, it is true, seems strongly negatived by Dr. Horsfield, who says, expressly, " no doubt remains as to the identity of the subjects from which the description was made," that is of M. Temminck's inedited species, and that of Doctor Horsfield.

Major Smith inclines, also, to think either that the specimens of Sir Stamford Raffles and of Doctor Horsfield were small, or that they belong to a small variety, if not a separate species from nebulosus. The latter, he says differs from the former in bulk, in colour, and in the marks on the head, no account being given of the zigzag between the eyes, which distinguished his specimen of F. nebulosa, a peculiarity, we must observe, which is noticed in Howitt's drawing, before-mentioned, but not in that made by Mr. Landseer. In bulk, he was, it seems, full as large as the great Jaguar, consequently, not quite equal to the Bengal Tiger. With respect to the habitat, the F. nebulosa was said to have been brought from Canton ; but it is true that an animal, said to have come from China, may very well have, in fact, been brought from Sumatra or Borneo, both being in the line of route of ships from China homeward.

The editor, presuming the identity of the species, and in deference both to Doctor Horsfield and M. Temminck, would most willingly have cancelled the name of F. nebulosa, and have substituted for it that of F. macrocelis. Some slight uncertainty, however, still remaining, as to the identity of the species described in the text, with that of Doctor Horsfield, particularly in reference to colour, and of both with that of M. Temminck, there would, therefore, be an impropriety in doing it, were there no other objection.

But should the identity of the three be clearly proved, it is obvious, that though the first detailed description of it is due to Dr. Horsfield and Sir Stamford Raffles, the first notice and liberal communication of its figure to zoologists long before, both here and on the continent, is attributable to Major Smith. It would, therefore, be a slight, and an injustice done to him, to cancel the name he had adopted, and with it the memorial of his first knowledge and drawing of the animal.

The editor takes the present opportunity of observing, that no small inconvenience presents itself in the progress of this work, in

THE JAGUAR. SMALL OR COMMON VAR.

F. ONCA. L.

in different individuals, nor do the two sides of the same animal always agree.

Some inhabitants of South America describe two varieties, corresponding in colour and general appearance, but one of them stands higher than the other, has the forelegs smaller, a fur not quite so bright, and a more gentle disposition. Azara says, it is called Popé, but he thinks they are but one; but Major C. H. Smith, whose long residence in America afforded him ample opportunity of inquiry, satisfied himself there were two distinct varieties of the Jaguar, differing principally in dimensions.

The opposite figure is from his accurate pencil, from a specimen recently killed in America. The type was, as he believes, of the great Jaguar, which was shot in the act of devouring a Peccary, in the woods of Surinam; it measured two feet ten inches in height at the shoulder, but, from its compact and heavy make, it appeared larger than it was in reality. The spots do not strictly agree with what either the Baron or M. Lichtenstein have fixed as criteria; and Major Smith doubts whether any skin of this variety (presuming it to be the Popé or large Jaguar) has ever come under the observation of those indefatigable and accurate observers. The line of lengthened spots on the back was not quite full, and it seems probable, when they are so, that it arises from nonage. The marks on the sides are very irregular, and indefinable; the eyes were small and sunken; the whiskers very long; and the whole character that of an aged animal. It was a male. The portrait is extremely like that given by Azara, in his Travels, particularly as to the make of the animal.

relation to the several new or uncertain species which may be noticed, and for which, however unwilling, he is in some degree obliged to coin names, while inedited figures of the same may already exist in the portfolios of zoologists, to which some other name may have been appropriated.

Our figure of the small Jaguar is also from a drawing of Major Smith taken in America. It was a male, two feet two inches in height. Its general colour was paler and more ashy than the large variety, with five large distinct rows of annulated spots on the sides. It was excessively fierce and untameable.

The Jaguar is very like the Panther or Leopard of the Old World, but the spots or rings of the former are larger and more oblong, particularly down the back, and those near the dorsal line have a central black dot, which is never seen in the Panther or Leopard; the head is rounder; the animal altogether stouter and stronger; and the tail never reaches farther than to the ground, which last is, perhaps, the most obvious difference between them.

Their young are born blind; those of the Panther or Leopard have their eyes open from the first.

On the whole, we are inclined to conclude that no accurate description has hitherto been given of the large variety of the Jaguar; or otherwise, that the individuals of this species are so subject to vary, as to render any specific character inconclusive.

There is also a black variety* found in the forests on the frontiers of Brazil, which has the same spots and marks as the others, on a ground of a somewhat browner black; so that they are visible only on close examination, and by viewing the skin when inclining at a certain angle from the direction of the light. This appears to be the Felis Discolor of Gmelin, the Couguar Noir † of Buffon, and the Black Tiger of Shaw; although the figure given by Buffon does not correspond with it, inasmuch as the under part is

---

* It is extremely difficult to say what is a variety, and what a distinct species. The Black Jaguar is, probably, only a variety; but as it is not found in the parts where the Common Jaguar abounds, it may be thence presumed, that they are distinct.

† Major Smith thinks this is distinct. See p. 473.

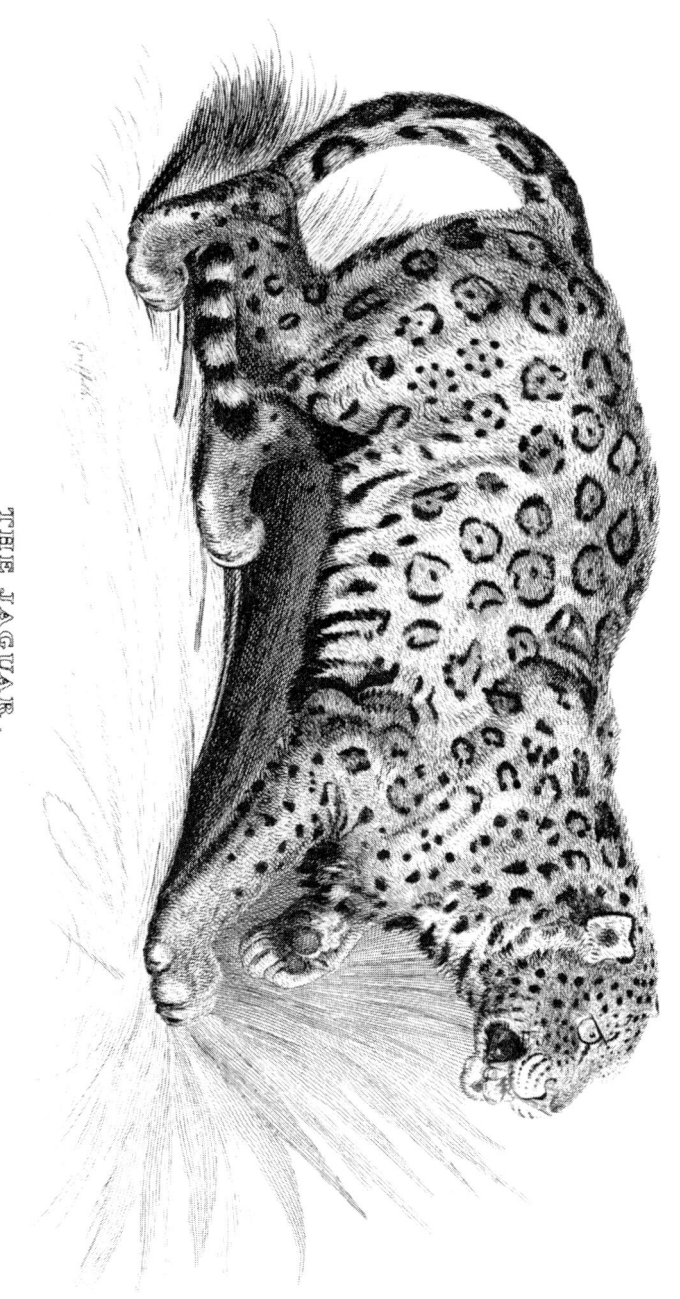

THE JAGUAR.

GREAT VAR? F. ONCA. L.

white. The black variety, however, is extremely rare. One is also mentioned by Azara, perfectly white, with the spots indicated by a more opaque appearance; but this peculiarity was possibly the effect of albinism.

The Jaguars are solitary animals, or are met with only in pairs ; they inhabit thick forests, especially in the neighbourhood of great rivers; and if they be driven by their wants to seek for sustenance in the cultivated country, they generally do so by night. It is said they will stand in the water, out of the stream, and drop their saliva, which, floating on the surface, draws the fish after it within their reach, when they seize them with the paw, and throw them on shore for food.

They will attack Cows, and even Bulls of four years old, but Horses seem to be their favourite prey. They destroy the larger animals by leaping on their back ; and placing one paw on the head, and another on the muzzle, they contrive to break the neck of their victim in a moment. Having thus deprived it of life, they will drag the carcass, by means of their teeth, a very considerable distance, to their retreat, from which their great strength may, in some measure, be estimated.

The Jaguar is hunted with a number of Dogs, which, although they have no chance of destroying it themselves, drive the animal into a tree, provided it can find one a little inclining, or else into some hole. In the first case, the hunters kill it with fire-arms or lances ; and in the second, some of the natives are occasionally found hardy enough to approach it with the left arm covered with a sheepskin, and to spear it with the other ; a temerity which is frequently followed with fatal consequences to the hunter.

The traveller, who is unfortunate enough to meet this formidable beast, especially if it be after sunset, has but little time for consideration. Should it be urged to attack by the cravings of appetite, it is not any noise, or a firebrand, that will save him. Scarcely any thing but the celerity of a musket-ball will anticipate its murderous purpose. The aim must be quick and steady; and life or death depends on the result.

Many parts of South America which were once grievously pestered with Jaguars, are now almost freed from them, or are only occasionally troubled with their destructive incursions.

D'Azara was once informed, that a Jaguar had attacked a Horse near the place where he was. He ran to the spot, and found that the Horse was killed, and part of his breast devoured; and that the Jaguar, having probably been disturbed, had fled. He then caused the body of the Horse to be drawn within musket-shot of a tree, in which he intended to pass the night, anticipating that the Jaguar would return in the course of it to its victim: but while he was gone to prepare for his adventure, the animal returned from the opposite side of a large and deep river, and having seized the Horse with its teeth, drew it for about sixty paces to the water, swam across with its prey, and then drew it into a neighbouring wood, in sight the whole time of the person who was left by D'Azara concealed, to observe what might happen before his return.

The husbandmen frequently fasten two Horses together while grazing; and it is confidently stated that the Jaguar will sometimes kill one, and in spite of the exertions of the survivor, draw them both into the wood. This is a performance Molina also attributes to the Puma. It may be reconciled by supposing, that the extreme terror of the surviving Horse paralyzes its efforts.

Generally speaking, particularly during day, the Jaguar will not attack a man; but if it be pressed by hunger, or have previously tasted human flesh, its appetite will overcome its fears; and during the residence of d'Azara in Paraguay, no less than six men were destroyed by this

THE LEOPARD.   F. LEOPARDUS.

formidable beast, two of whom were at the time before a large fire.

The central spot and short tail of the Jaguar will, with but little observation, soon enable any one to distinguish that transatlantic species from those of the old world, however confused, to which it is nearly allied, and to which we shall now proceed.

We shall treat of the *Panther* and the *Leopard* conjointly, necessarily so indeed, as the distinctness of the two on the one hand, or the identity of both subject only to variety on the other, seems still in some degree problematical.

The history, says our author, in his *Ossemens Fossiles*, of the great Cats with round spots of the Old World, is more difficult to elucidate than that of the Jaguar, on account of their mutual resemblance, and of the vague manner in which authors have spoken of them.

The Greeks knew one of these from the time of Homer, which they named Pardalis, as Menelaus is said in the Iliad, to have covered himself with the spotted skin of this animal. This they compared, on account of its strength and its cruelty to the Lion, and represented as having its skin varied with spots. Its name even was synonymous with spotted. The Greek translators of the Scriptures used the name Pardalis, as synonymous with Namer, which word, with a slight modification, signifies the Panther, at present, among the Arabians.

The name Pardalis gave place among the Romans to those of Panthera and Varia. These are the words they used during the two first ages, whenever they had occasion to translate the Greek passages which mentioned the Pardalis, or when they themselves mentioned this animal.

They sometimes used the word Pardus, either for Pardalis, or for Namer. Pliny even says, that Pardus signified the male of Panthera, or Varia.

So reciprocally the Greeks translated Panthera by the

word Pardalis. The word Panthera, although of Greek root, did not then preserve the sense of the word Πανθηρ, which is constantly marked as different from Pardalis, and by Oppian is said to be small and of little courage. The Romans, nevertheless, sometimes employed it to translate the word Πανθηρ, and the Greeks of the lower empire, induced by the resemblance of the names, have probably attributed to the Panther some of the characters which they found among the Romans, on the Panthera.

Bocchart, without knowing these animals himself, has collected and compared with much sagacity every thing that the ancients and the orientalists have said about them. He endeavours to clear up these apparent contradictions by a passage in which Oppian characterizes two species of Pardalis, the great with a shorter tail than the less.

It is to this smaller species that Bocchart would apply the word Πανθηρ. But there are found in the country known to the ancients, two animals with spotted skins ; the common Panther of naturalists, and another animal, which, after Daubenton, is named the Guepard, (the Hunting Leopard.)

The Arabian authors have there also known and distinguished two of these animals ; the first under the name of Nemer, the other under that of Fehd, and although Bocchart considers the Fehd to be the Lynx, " I rather incline to think," says the Baron, " it is the Hunting Leopard."

The Guepard, then, would be the Panther, and there is nothing stated by the Greeks repugnant to this idea.

Sometimes they associate it with the great animals, sometimes with the small, which seems to imply that it was of middling stature. Its young were born blind, says Aristotle; it inhabited Africa with the Thos, according to Herodotus ; its skin was spotted, and its natural disposition tameable, as we are informed by Eustathius.

The two last *traits* appear inapplicable to any other

species than that secondly indicated by the Arabians : it is true, they are silent on the subject of its being employed in hunting, but this is very natural ; if, as Eldemiri informs us, the first person who so employed them was Chalib, son of Wail.

As to the word Leopardus, its usage is much more recent, and there is no proof that it indicated a particular species. It is met with only in the authors of the fourth age, and was introduced by the fable of the intercourse between the Lioness and the Pardalis, and by degrees was applied to the Pardalis itself ; for, when Vopiscus says, that Probus, when on occasion of the German triumph, he exhibited one hundred Leopards from Lybia, and one hundred from Syria, he could not, doubtless, have meant to say, that they were the produce of such an unnatural intercourse.

Thus abstracting for a moment the Lynx, the Greeks and Romans appear to have known but two species of these spotted animals, notwithstanding the opportunities, particularly of the latter, of becoming acquainted with them.

We know at present of Africa but the two species of the ancients, the Panther and Leopard, ordinarily understood, and the Hunting Leopard, (*Felis jubata.*) The Leopard of modern naturalists, according to our latest researches, comes only from the parts of India the least known by the ancients.

Thus far, in effect, the Baron, with his usual learning and research : to which we shall subjoin a few observations.

Pliny tells us, that in his time the words Variæ and Pardi were applied to all this family ; the former to distinguish the females, and the latter the males: and in a previous passage he observes, that these and the Tiger are almost the only spotted or striped beasts, the rest being uniform in colour, though it varies in the different species. Our author has noticed Pliny's observations, but it may be as well to refer to the passage more particularly, and by the whole context of the quotation from this writer subjoined, it appears probable, that the moderns have been incorrect in applying the word Pardus specifically, as it was originally used only to denote a sexual distinction in the whole genus. " Panthera et Tigris macularum varietate propè solæ bestiarum spectantur, cæteris unus ac suus cujusque generis color est leonum, tantum in Syriâ niger. Pantherus in candido breves macularum oculi. Ferunt odore earum mire solicitari quadrupedes cunctas, sed capitis torvitate terreri. Quamobrem occultato eo, reliquas dulcedine invitatas corripiunt. Sunt qui tradunt in armo iis similæ lunæ esse maculam, crescentes in orbes, et cavantem pari modo cornua. Nunc varias, et pardos, *qui mares sunt,* appellant in eo omni genere, creberrimo in Africâ Syriâque. Quidam ab iis pantheras solo candore discernunt, nec adhuc aliam differentiam inveni." *Plinii Nat. Hist.* lib. x.

In another passage mention is made of the Pardi, Pantheræ, Leones, et similia. Now, unless Pardi and Pantheræ were applied to the two sexes of the Spotted Cats, they could not have been synonymous, as the moderns have made them.

If we turn to modern zoologists prior to the time of our author, we shall find that they have fallen into so many certain errors in describing these species as distinct, that the probability of their identity is rather strengthened by applying to their authority on this subject. To select a few instances.

Linnæus gives as the specific characters of the Panther, " Felis, caudâ elongatâ, corpore maculis superioribus orbiculatis, inferioribus virgatis." With a long tail, the upper part of the body covered with orbicular spots, the lower part with stripes. This short description, it has been well observed, is inapplicable to any known species of the genus. Perhaps it is nearer to the Servals than to any other. His characters of the Leopard are, "Felis, caudâ mediocri, corpore fulvo, maculis subcoadunatis nigris." "With a moderate tail, a fulvous body covered with subcontiguous black spots." Dr. Shaw observes: "In the twelfth edition of the Systema Naturæ, the Panther and Leopard seem to be confounded by Linnæus himself, who appears to have considered them as the same species, under the name of Pardus." And if we consider the description given to the Panther to be irrelevant and factitious, it follows, that Linnæus has only described one species of the large Spotted Cats found in Asia and Africa, which must include the Variæ, and Pardi, and Leopardi, of the Romans.

Buffon, the brilliancy of whose work has blinded mankind to his imperfections, imbibed an idea which he never seems to have lost sight of, that the American animals were degenerate, and less in size than the species of the old world belonging to the same order : hence, probably, he was led into a misunderstanding, or too willingly confirmed in error on this subject. He has mistaken the Jaguar, which he describes from an Ocelot ; and refers the former animal, because, probably, it was a large species to the Panther of the ancients, transposing his figures accordingly. The furriers and exhibiters of wild beasts have imbibed this error ; and the Jaguar of America has altogether usurped the name of Panther from the species of the Old World, to which it was originally applied.

Pennant's description of the Panther so nearly accords with the Jaguar of America, both in person and disposition, that there scarcely seems a doubt of this animal's being the type whence his description was taken.

Dr. Shaw states, that the Leopard is best distinguished from the Panther by its paler yellow colour, and that a true distinctive mark between them is by no means easy to communicate, either by description, or even by figure; but he adds, the Leopard is considerably the smaller of the two.

He therefore makes the principal difference to consist in size and colour.

Pliny says further: " Quidam ab iis Pantheras *solo candore* discernunt, nec adhuc aliam differentiam inveni." It is possible, however, if the F. Uncia be really distinct, that Pliny refers to that species. Major Smith believes him to be distinct, and describes him as whitish-gray, faintly tinted with buff. " He may," says the Major, " have been a Syrian and Armenian animal, and I believe him now a resident of the mountains of Northern Persia." We refer to our figure of the specimen formerly in the Tower. It seems probable, that all those which come from Asia are much brighter in colour than those from Africa, and that the females in general have more white about them than the other sex. Mr. Cross, who has had opportunities of inspecting probably some hundreds of specimens, insists, that he has never observed any specific difference between those brought from Asia and Africa among themselves, except that the Asiatic are generally larger and brighter ; and except, also, that some individuals constantly carry their long tail curved outwards, and others inwards, the latter of which they call ring-tailed Leopards, It seems probable, therefore, that Dr. Shaw's leading specific distinctions of size and colour, apply rather to the Asiatic and African varieties, than to distinct species found in both those continents. The figures, however, in the General Zoology, neither illustrate the author's position on this subject, nor throw any light on the question; for they are merely copied from Buffon, and that which is called the Panther is properly referable to the Jaguar.

M. Lichtenstein, in a note communicated to Major Smith, draws a specific distinction. He describes the Panther as resembling the Jaguar in having the same number of rows of spots, but differing in having no full spots on the dorsal line. But it does not appear that full spots on the dorsal

THE PANTHER.

FELIS PARDUS.

C. Hamilton Smith. Esq.r

London, Published by G.B.Whittaker. June.1825.

line always make a specific character of the Jaguar ; and the Asiatic Leopard is sometimes distinguished by this peculiarity, though it does not in other respects resemble the American animal. When, therefore, it is said, that the Panther much resembles the Jaguar, it is always to be strongly suspected, that the type, whence the observations are taken, is an American animal.

We have selected two from amongst the several drawings before us, as being most opposed to each other, and, therefore, most illustrative of the differences between the Leopard and Panther of naturalists, whether as species or varieties.

The specimen, named the Leopard, was at Exeter Change. Compared both with the Jaguar, and with the Panther of naturalists, it was uniformly of a paler yellow-colour, rather smaller, and the spots rose-formed, or consisting of several dots, partially united into a circular figure, in some instances, and into a quadrangular, triangular, and other less determined forms in others ; there were also, and especially on the outside of the limbs, several single isolated black spots.

The other, or Panther, is from Major Smith's drawing of one of the several Felinæ, called Panthers, now in the Paris Museum. It is the smallest there, and the most closely marked with spots. These spots make a considerable contrast with those of the other figure, the most so of any of the five or six specimens of Panthers in that celebrated collection. This figure, also, it will be seen, approximates very nearly that of the animal next described, particularly when uncoloured, and also, though less in degree, to that of the large and small Jaguar. The differences which distinguish the former of these, will be observed upon in the description of the animal ; those that mark the latter have been already adverted to.

The animal we have figured under the name, conditionally,

of the *Panther of the Ancients,* may deserve particular attention, in ascertaining the diversity of species of its consimilars, especially as it seems to possess traits of a real specific character.

Major Hamilton Smith met with this species, stuffed, at Hesse Cassel. The animal measured five feet three inches from the nose to the insertion of the tail, and stood about two feet nine inches high at the shoulder.

The first and great difference which distinguishes this from all the large-spotted Cats, hitherto described, whose specific characters have been before stated from our author, is that the entire colour of the whole animal is a buff-yellow which assumes a darker tint, approaching to red, on the nose, and more ochery on the back and sides. The belly and insides of the limbs partake of this general colour, but paler, there being no white part about the animal.

There may be said to be seven vertical rows of interrupted or imperfect annuli on the sides of the animal. These, as as well as the like open spots which mark all the Panthers, have, as Major Smith observes, the inner surface of the annuli more fulvous than the general colour of the sides. rn the Leopards no such distinction appears, nor is there oom, as the small congregated dots are too close to admit it. The dorsal line is marked in the same manner, not with close, but open spots. These annuli differ from those of the Jaguar, to which they bear a considerable general similitude, n being all nearly circular, whereas those of the American animal become oblong as they approach the dorsal line; they are also smaller when compared with the size of the animal, and much more numerous, covering not only the back, ribs, and haunches of the animal, but descending on the outside the legs, at least, to the knees.

They differ again from the open annuli of the Jaguar, in being altogether without the spot in the centre, which renders that species so obvious; and the tail is spotted from

C. Hamilton Smith, Esqr del.
Museum. Hase Cassel.

London, Published by G. B. Whittaker Decr 1824.

T. Landseer sc.

THE PANTHER OF THE ANCIENTS ?

beginning to end, unlike that of the Jaguar, which has the open oblong marks some way down, and is terminated by annuli of black, yellow, and white, running round it. The forehead, cheeks, sides of the neck, shoulders, throat, and inside of the limbs, are covered with numerous, close, small spots, and there is a narrow black bar crosses the lower part of the throat.

The animal stands higher than the Great Jaguar, thougl. it is lighter and slenderer, in which respect it approaches the Felis Jubata, though it is much larger, in proportion, than that species. The head is smaller than that of the Jaguar, and, in that respect, agrees with the known species of the Old World.

Its native country was unknown, but it had lived in the menagerie of the Elector.

The characters of this animal, which seem intermediate between the American Jaguar and its large spotted congeners of the Old World, though diverging from both in the uniformity of the ground colour, seem to accord considerably with the prevailing notion of the Panther of antiquity, when considered as distinct from the Leopard. The present apparent rarity of the animal, however, militates against the idea of its identity with the Panther, hundreds of which were frequently collected together at a time in Rome. It may be observed, however, that none of these animals are now imported from Syria, whence the Romans drew a great number, and where they still are, according to Dr. Clark.

We have felt constrained, with Major Smith's permission, who drew the animal, to apply it to the Panther of antiquity, but with a mark of doubt. After all, the ancients, who were no great zoologists, may have applied the words Panther, and Pard, or Leopard, to all the larger Spotted Cats indifferently, to the Common Panther and Leopard of our menagerie, the present animal, the Felis Jubata, the

Felis Uncia, and even the Lynx, in which case, the animal in question, would not be allowed to appropriate to itself, exclusively, the name of Panther. Conjecture must, for the present, supply the place of certainty; we have endeavoured to compress together the sum and substance of what has been said upon the subject, but by no means pretend to determine the question, or even to offer an opinion on a mere question of fact, hitherto not satisfactorily ascertained even by Cuvier himself.

The large Spotted Cats of the Old World, though occasionally found in some parts of Asia, are much more common in Africa, and are, to the latter continent, almost as destructive as the Tiger is to the former. They seem, however, to have more respect, dictated by fear, for the human species, and will seldom attack a Man, unless provoked, or much pressed by hunger; but they are cruelly destructive to the inferior animal creation.

For the purpose of taking them, it is usual for the hunter to construct a hiding-place within musket-shot of a tree, on which is suspended some flesh as a bait for the unconscious beast, which receives the ball while in the act of taking it. The hunter, for greater caution, then waits till the following day, when a Dog, properly trained, is sent forward to track the animal to its retreat. If it be still alive, the Dog generally falls a victim, and saves the hunter from exposing himself, until he is satisfied that the beast is no longer capable of mischief.

The female of the Panther or Leopard is gravid nine weeks, and the young, when born, are blind, and remain so about nine days afterward; but the American Jaguar, which appears to have been confounded so much with this animal, is produced with the eyes open, and the mother is pregnant nearly four months.

In Dr Gmelin's edition of the " Systema Naturæ," the

THE ONCE.

F. O. UNCIA. L

Ounce *, *F. Uncia*, is described with the following specific characters: " Felis caudâ elongatâ, corpore albido; maculis irregularibus nigris:" which accord pretty well with the figure here given. Buffon also describes the Ounce at some length, and gives a figure of it; but Cuvier seems to doubt the existence of this animal as a distinct species. After taking much pains to ascertain the truth, he states his opinion to be, that the *Once* of Buffon is no other than a variety of the Panther, because he has never been able to meet with an animal or a skin corresponding with Buffon's description; he, therefore, omits it in his catalogue of Cats, both in the *Ossemens Fossils* and in the *Règne Animal.*

The figure of the animal here represented, is from a specimen brought from the shores of the Gulf of Persia, which was in the Tower of London. It is very distinct from all the other species in make, mark, and general appearance; and corresponds with Buffon's figure, which has been copied by Schreber, Shaw, and others. It was about the size of a Panther or Leopard.

Major Smith has also met with a skin of this species brought from the Gulf of Persia, from which he has made a drawing in his collection. He conjectures it to be a mountain species; and, from the length of the fur, which is shaggy, one that resides in the higher snowy regions of northern Persia.

It is with the utmost respect for the opinion of our author, that we venture to present this figure, which may, indeed, be only that of a variety of the Leopard. Indeed we are well aware of the extreme risk incurred the moment we depart from his dicta or opinions, more especially in multiplying factitious species, an evil under which Zoology is at present most grievously suffering.

---

* Buffon says the word *Once* is a corruption from *Lynx* or *Lunx*, and that he retained the name because the animal in question has some affinity to the Lynx.

The existing catalogue of species, even as we have stated in our Table appended, in which several that have been treated as distinct are omitted, is still, it is to be suspected, capable of many subtractions ; and many of them, when submitted to the observation of judicious, operative naturalists, and tried by the test of anatomical character, will be found to be mere varieties of others.

Figures, however, if at all accurate, claim a more decided consideration than imperfect verbal descriptions; and it must not be forgotten that Zoology has been neglected to an extraordinary degree in this country, notwithstanding the opportunities possessed here, above all others, of prosecuting the science. The drawings in our possession have occupied many years in collecting, during which time a number of new species have occurred in our different exhibitions, each of which would have been the subject of observation and comment, in the different learned societies of the continent, though they have been treated here purely as matters of pecuniary speculation of the exhibitors, and almost altogether neglected by men of science.

Under these circumstances, therefore, we earnestly deprecate the imputation of that foolish vanity which has induced many men to incumber Zoology by editing, as novelties, species, or even genera, upon the authority of a single type, which, eventually, turn out scarcely to deserve the name of varieties. Such figures as we possess, which seem to claim attention from their novelty, we shall present, as we find them, *Valeant quantum valere possint.*

At the end of the Large Spotted Cats, the Baron places the *Hunting Leopard, Felis Jubata,* the *Chetah* of India, and the *Guepard* of Buffon. The peculiarities of this species might suffice to qualify it for a separate station, at the end of the genus approaching the Dogs, so far as to be called, without impropriety, the Canine Cat.

The Feline family is, in general, very strongly marked,

but inclinations are to be found in certain of its species, both to the Dogs, and the Viverræ; the Chetah or Maned Hunting Leopard, is the type of the former. In the system of dentition, and all the organs of sense, it corresponds with the Felinæ, but in the non-retractibility of the claws, it differs from the genus in general.

In this species, we have again, in a remarkable manner, the opportunity of observing the mutual harmony existing between the mental impulses and the physical powers of animals; their disposition or inclination to destruction is precisely in unison and proportion with their bodily powers. If very weak, they are excessively timid; if extremely strong, they are equally undaunted; while those which hold a medium station, in this respect, seem generally to appreciate, as it were, with more sobriety, the conditions of their existence, and to submit themselves to the dominion and artificial education of Man more easily than the rest. The Hunting Leopard is in this intermediate situation. About as big as a large Dog, its leading weapons of offence, the claws, are in the same situation as those of that animal; incapable of being withdrawn into a sheath for protection, they are constantly exposed to the friction of the ground, by which they become worn and blunt, and so much the less effectual for active warfare; but otherwise the animal has all the suppleness and elasticity, the trenchant teeth, and the powerful jaws of the Cats. Partially deficient, therefore, in the physical powers of its congeners, it is equally wanting in the extreme ferocity of its disposition.

The Hunting Leopard is of a pale yellow colour on the upper part, white underneath, and covered all over with very small spots without regularity; it has a slight erect mane down the neck, whence it is named. The eye-pupil is round at all times. The slim make of the body and limbs of this animal, calculated apparently rather for speed than strength, assimilate it in a remarkable degree to the canine race, with which we have already compared it. In a certain aptness or capability it possesses of being trained for field sports, it is also more like the Dogs than the Cats. It is, therefore, strictly speaking, intermediate, and we apt pear to pass naturally from the latter race of animals through this species to the former. It also exhibits the first step or remove from the perfect fitness for carnivorous and predatory habits in the loss of the retractile power of the talons.

M. F. Cuvier says of the individual he describes, that except in regard to that mistrust natural to the Cats, he had all the habits of those animals, playing in the same graceful manner and with the same address ; and although his nails were not trenchant, he exercised his forepaws in play in the same manner as the Cats, striking any small moving body with his paw, and seizing his food in both. Under all these circumstances, he was altogether a Cat, and differed only from his congeners in his much greater degree of confidence, and in all the consequences resulting therefrom. Familiar with every one, he was always ready to make the slight noise we call purring whenever he was caressed by any person.

He was brought up with perfect liberty, and was accustomed to live even with children and the domestic animals. In his passage from Senegal he was equally as free, and while in France, was kept during summer in a park, where he had the opportunity of amusing and exercising himself. When it was cold, or his inclination induced him to come in doors, he made a uniform frequent mewing, which he also used to express his hunger, or his thirst, or his affection.

This individual was taken in Africa, and the species is said to be indigenous also in India ; but from the inspec tion of several individuals, particularly two that were lately in Wombwell's itinerant collection, and from a drawing in particular in our possession, from a sketch made in

FELIS CHALYBEATA. *Hamilton Smith*

India by the late Mr. Devis, we are strongly inclined to think that there are two distinct species of the Canine Cats, agreeing in general description, particularly in the want of the retractile power of the claws, the one with a mane and the other without ; the former proper to Africa, and the latter to India. The Indian or maneless species appears also to us to be taller, and to have a longer neck, smaller head, and shorter muzzle than the other.

The Hunting Leopard, it is said, is conveyed in a carriage, or on a pad behind the saddle of a horseman, with a hood over the eyes, to the field, and when the game, Antelopes, &c., is started, the hood is taken off, and it is sent out in pursuit. It follows by leaps or bounds, and if unsuccessful in taking its prey after a few efforts, declines the pursuit, and returns to its keeper.

Specimens of large black Felinæ have been frequently quoted as distinct species, as the Felis Chalybeata of Herman, which M. F. Cuvier pronounces decidedly to be the common Panther or Leopard badly coloured ; the Felis Melas of Peron, said by the Baron to be a Black Leopard ; the Jaguaraté of d'Azara, the Black Cougouar of Pennant, &c. Major Smith has a drawing he refers to this last species ; the form of the animal is that of the Cougouar ; the cheeks to the ears, the throat, and belly, are white, the rest of the animal black. " There is," says the Major, " a Black Tiger in the mountains of Chitagong pretty common, probably the same species as that brought to this country by Mr. Hastings ; and there is also an un described Panther, of considerable size and of a dark co_ lour, with very numerous black rose and conglomerate spots, at Erlangen."

We rather apprehend that most of the known Felinæ are apt to vary to a uniform black, and we have drawings which may be attributed to the Black Jaguar and the Black Leopard.

But it may be proper not to pass over a specimen which

was in Mr. Bullock's late Museum, and we believe is now in Germany. It was wholly of a grayish, liver-colour, or chocolate and white mixed upon each hair, marked with more opaque round full dark brown spots and blotches The tail was darker than the body, and annulated with about twelve darker rings, the tip being black. The back of the ears was black, with a white spot in the middle; the insides whitish. Three rows of barbs formed the whiskers. It measured two feet nine inches in length, the tail one foot three inches, and was a male. This may be the Jaguaroundi of the Brasilians, but the peculiarity of its colour seems to bespeak it distinct The drawing is by Major Smith, who, without insisting on it as an undescribed species, names the animal conditionally *Felis Chalybeata?*

Of the Ocelots, a group in the Feline family of middling-sized Cats, distinguished by yellow spots more or less oval, bordered with black, several individuals have been described, but whether any or all of these were varieties or distinct species, may be doubted. D'Azara considers them all as a single species. Our author makes three specifically different; and we shall have occasion to submit the figures of some others which appear to us to be distinct.

The Jaguar of Buffon is evidently and very erroneously one of these. It will be seen by the table that this species is identified by modern zoologists with the Tlatco-Ocelots of Hermandez and the Chibigouazou of Azara, and is figured by F. Cuvier under the name of Chati. His figure is copied here, in order that it may be compared with the others from drawings by Major Smith, one of which that gentleman refers, though not without hesitation, to the Tlatco-Ocelots of Hermandez, and the Jaguar of Buffon, and the other to the Chibigouazou.

We had written some observations on this group of the Felinæ when Major Smith favoured us with his sentiments

C. Hamilton Smith Esq.ʳ. del.ᵗ

OCELOT, Nº 1 of Hamilton Smith.

upon them, which, with his permission, we shall insert in his own words, illustrated with engravings from his drawings, referred to in the description :—

" My present view of the Ocelots," says he, " is that they form a subordinate group in the great family of the Felinæ. As a general character, I would describe them as being of middle size, between the larger and the small Cats, of more slender and elegant proportions, without tufts on the ears, the spots diverging more or less in concatenations or streaks from the shoulders backwards and downwards, and, as far as I have hitherto observed, the pupil of the eye round. (Of this last character, however, I am still very doubtful, and my doubt arises from the probability that all the living specimens which I examined were, from the very circumstance of attentive inspection, under a state of alarm, and therefore with the pupils dilated.) They belong all to the New World, but there are two or three species of the Old that approach them in several particulars, and therefore might make the next group. I shall refer in the following descriptions to my drawings numerically. My appropriation of their types to species hitherto described, must, in our present state of knowledge, be conditional only.

" Felis Ocelot, No. 1.—Of this both the male and female are rufous on the nose, face, neck, and shoulders; on the back and upper part of the tail they are white, very slightly tinged with reddish and gray; on the under parts of the head, throat, legs, breast, belly, and hams, there is white running up the rump and sides between the streaks and spots in dichotomous rows. These two colours, the rufous and the white, are separated by black streaks in spotted lines, forming very elongated rays, and on the rufous colour within the rays are a few black specks. On the hams and thighs there are large black blotches. The tail has blotches variously figured above, and smaller spots underneath; the tip is white. On the forehead two streaks running from the inner angle of the eye, proceed to behind the ears, and the space between is filled up with small spots. The outside of the ears is black, with a white spot. From the external angle of the eye two black streaks with a white space between, running to below the ears; and on the throat, two black streaks in broken forms go downwards to the breast, with some spots between.

" No. 2 is about the same size as the last, but the rufous covers a larger space on the back and hams. The spots on the shoulders are more numerous and smaller. There is one large spot on the cheek, and four or five small open chainlike spots on the hams. There are *no specks* within the large streaks.

" These are both from South America. I have examined several of them alive, and about twelve in a stuffed form.

" I incline to think that one (or both) of these (if varieties) must be the Chibi-gouazou of Azara. The word is South American, where these animals are found. I believe Nos. 1 and 2 to be South American, and Nos. 3 and 4 Mexican, though I do not mean to assert that each may not be found in the country of the other ; and if this be so, Azara's appellative seems most likely to belong to the former.

" No. 3 is smaller, with the nose, forehead, neck, back, shoulders, fore part of fore legs, and rump ashy mixed with ochery. The streak from the inner angle of the eyes to the ears has only one row of spots within it. The long open spots on the neck and back are shorter, less diverging, fulvous within, but without any spot on the fulvous. On the fore legs only there are a few large spots ; on the hams there are some round, open, and a few small black wavy spots. The tail is altogether, or nearly, fulvous, ringed with black, with the tip white. There is a black ring round the eyes, and a streak down each side of the nose ; a large spot on the cheek, and two bars, with white between them,

C. Hamilton Smith Esq.ʳ del.ᵗ

THE OCELOT. Nᵒ 2. of Hamilton Smith

I. Landseer sc

C.Hamilton Smith Esq.ᵗ delᵗ Albany

THE OCELOT. Nᵒ 3 of Hamilton Smith.

T Landseer sc

C. Hamilton Smith Esq.ʳ del.ᵗ

OCELOT. Nº 4. of Hamilton Smith.

from the outer angle of the eye to below the ear.   Four black broad bars cross the throat.

" A young female of this is now in Mr. Bullock's Mexican collection.   It came from Mexico.

" I have examined five or six specimens, and believe I have sufficient grounds for considering the differences between this and the preceding not to arise from nonage.

" I believe that at Paris there are stuffed specimens of male and female Ocelots, with specks in the centre of the open rings or spots, but am not certain of this, or whether this character be any indication of sexual difference.

" No. 4 I have seen but twice.   It is fulvous on the nose, forehead, shoulders, fore-arm, back, rump, and paws.   The temples are ochery; the rest of the animal white.   There are no black streaks on the forehead, but instead of them a number of small round spots covering the whole surface. Two broken streaks run from the outer angle of the eye to below the ear.   On the shoulders and flanks there are four or five long open fulvous spots, bordered with a chain of black.   On the rest of the back, rump, and hams, there are small open spots.   The tail is annulated, the tip black. On the fore legs and the lower part of the hind legs are small black spots.   The specimen figured was formerly in Bullock's collection, supposed to belong to southern Mexico, Honduras, &c.

" This appears to me to be the animal from which the figure in Buffon and Shaw was taken, under the name of the Jaguar*.

" Whether these few are specifically different or hereditary varieties, I do not mean to determine; but from the number of specimens of each that have fallen under my observation, there seems little doubt that one of these alter-

---

* Buffon gave two figures for his Jaguar: the one is the Chati of F. Cuvier, the other the species here alluded to.

natives is correct, and the several figures are not mere individual differences.

" These that follow bear a still more decided appearance of specific distinction, and have been or may be named.

" The *F. Catenata*, is an undoubted species, of the size of a Wild Cat. The legs are in proportion shorter than those of the before-described; the head and body heavier. The nose, forehead, under the eyes, arms, shoulders, back, rump, hind legs, and tail, are of a reddish-yellow colour; the temples ochery; the cheeks, throat, belly, and inside of the legs white. Several rows of black spots from the ears converge on the forehead. There is a single streak from the outer angle of the eye to below the ear. On the shoulder, back, side, rump, and hams, there are long chain-like streaks of black and reddish-brown, intermixed; the belly and throat have black streaks, and the tail has imperfect black annuli.

" Of this I have observed two specimens, one in Bullock's former collection, the other in the Museum at Berlin, which I examined with Professor Lichtenstein, and which proving by the teeth to be an old specimen, convinced him of the reality of its being a distinct species, and not a young Ocelot as he had previously conjectured.

" *Felis Macrourus* of Prince Maximilian, of Neuwied. This is about the size of the former, but higher on the legs; the neck is long and thick; the face very short; the tail nearly a fourth longer than the former. The face, neck, back, shoulders, rump, and hams are ochery gray, streaked and marked with $\frac{3}{4}$ rows of large black spots, describing somewhat regular figures. The tail is semi-annulated, with the tip black. Two streaks under each eye run to the angle of the jaw, and one above to the ear. There are some spots on the forehead and cheeks, and others still larger on the paws.

" Of these I have seen two specimens, one in Mr. Tem-

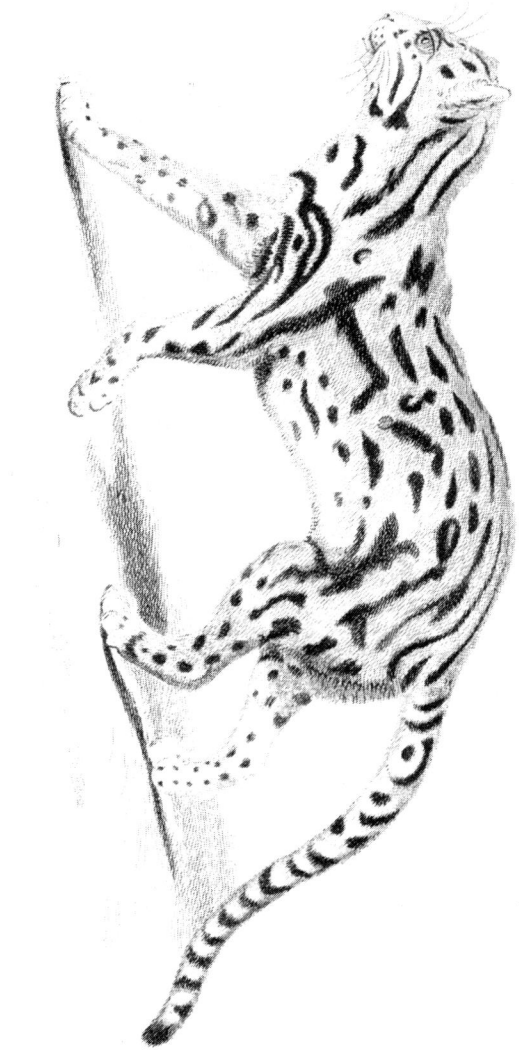

F. MACROURUS. FR. MAXIMILIAN AND TEMMINCK.

London Published by C. B. Whittaker. Feb.y 1825.

C. H. Smith Esq.r Pel.

C. Hamilton Smith Esq.r del.t

THE LINKED OCELOT.

FELIS CATENATA. Hamilton Smith

London, Published by J. B. Whittaker, Sep.r 1825.

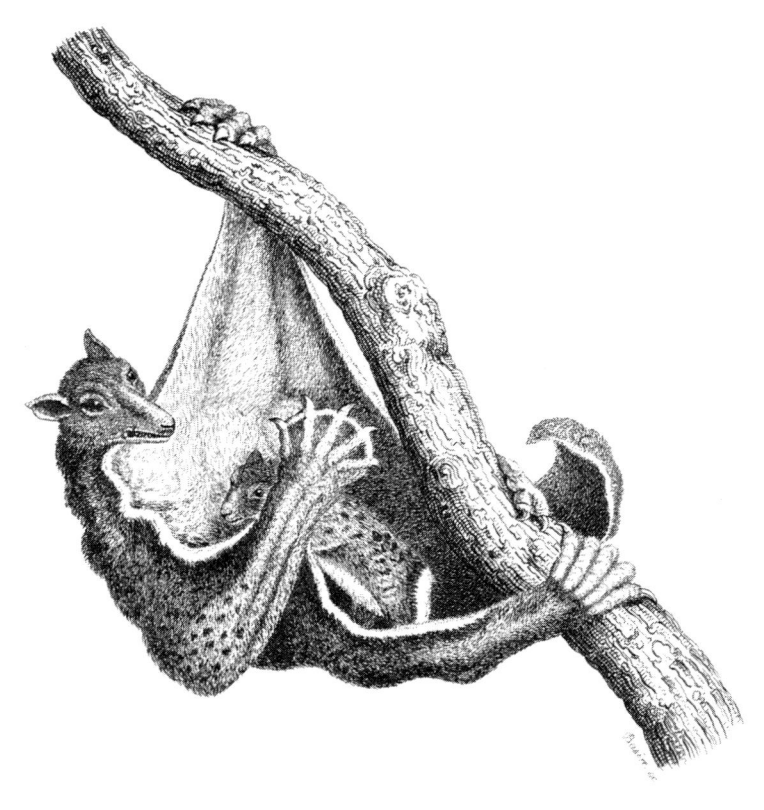

THE COLUGO.

GALEOPITHECUS RUFUS. PALLAS

London. Published by G.B. Whittaker Dec.r 1824.

minck s museum, and the other in that of Prince Maximilian, who, I believe, brought both from Brazil.

" I insert here, as distinct, the Chati of M. F. Cuvier.

" The specimen I have named conditionally Colocolo? from Molina, seems to terminate this little group, and by the character of its markings, to approximate to the Servals and Tiger-Cats of the Old World.

" It does not appear certain, though it may be probable, that this is the animal Molina indicated as the Colocolo, as he calls the marks spots, and not streaks; at least, the word is so translated.

" This fierce animal was shot in the interior of Guiana, by an officer of Lewenstein's riflemen, and by him stuffed and sent to England for his Royal Highness the Duke of York; but probably never reached its destination. A whimsical occurrence took place with it. The gentleman who had shot it, placed it on the awning of the boat to air, as he was descending the river to Paramaribo; the boat often passed under the branches of large trees which overhung the river, and on which were the resting-places of numerous Monkeys, sometimes hanging to the extremest branches above the water. Although the vessel would on other occasions excite but little attention, no sooner was the stuffed specimen in sight, than the whole community would troop off with prodigious screams and howlings. It was of course surmised from the excessive terror of these animals, that this species of Cat must be an active enemy to them.

" This animal was larger than the Wild Cat. The head was remarkably flat and broad. The ears large and round. The body slender. The tail just touched the ground when the animal was standing. The legs were very strong. The colour of the neck and back was whitish gray. The head, throat, shoulders, sides, belly, and inside of the limbs white. The back was marked with lengthened streaks of black, edged with tawny; and towards the shoulders and thighs, with streaks of tawny. There was a black streak from the corner of the eyes to the jaws, and some barry marks on the forehead. The outside of the ears were dark gray; the insides pink and naked, as well as the nose. The tail was semi-annulated with black, having a black tip, and it exhibited a great peculiarity in the legs, which were all of them of a very dark gray colour up to the knees.

" This may, perhaps, be the New Spain Cat of Buffon, before alluded to.

" To these may perhaps be added the Jaguarondi of Azara, which seems to be the black species of this subdivision. In size and proportions, he belongs entirely to the Ocelots, and I could just detect something of darker streaks running lengthways on the flank of the only specimen I have ever examined."

Of the *Chati* of F. Cuvier, the Baron observes, that it is more than one-fourth less than the Chibi-gouazou; less even than the Wild Cat, not having the head longer than four inches and a half, the body eighteen inches, the tail ten, and the height eleven.

The ground colour of its fur is brownish-gray, paler on the flanks, and white on the cheeks and under the body.

The spots both black and white of its head and ears are the same as in the Ocelot. Three series of black spots pass along the back; those of the flanks, shoulders, and crupper are deep yellow, bordered with black all round, except at the anterior edge. There are seven or eight, one above another. Some of those on the shoulders unite and form an oblique band; they are smaller on the feet, and there are none on the toes; those on the belly are full but cloudy; the tail has ten or eleven black rings.

" This species is from Brazil, and it appears to me," continues the Baron, " that it is the same as M. Sching in the German translation of the *Règne Animal*, names Felis

THE CHATI.

OF F. CUVIER.

London Published by G.B Whittaker Dec.' 1824

Wiedii, after an individual which had been communicated to him by the Prince Maximilian."

The Chati and the Felis Macrourus are very different, as the comparison of our two figures sufficiently evinces.

We shall now recross the Atlantic, and proceed to the Felinæ of the Old World, which bear some analogy to the small group of Ocelots of the New.

We shall abbreviate the Baron's observation on the *Serval* or *Tiger-Cat* of the Furriers, (*F. Serval*, Gm.,) which, by the longitudinal bands of the neck, seems to announce the following species.

Perrault described it (*Mem. de l'Acad.* t. iii. pl. 13,) from a very fat specimen under the name of *Chat Pard*, which Hernandez had given to the *Tlatco-ocelot*, and again, in Part III. of the same volume, under the name of *Panther*, much more exactly. Buffon named it Serval, applying very arbitrarily a passage of Vincent Marie on an Indian Cat less than the Civet, and which assuredly cannot be said of the Serval

" The fact is, that the skins of the Tiger-Cats of the Furriers come to us by hundreds from the Cape of Good Hope, and after the information I have received from merchants, I have no longer any doubt," says the Baron, " on the African habitat of this animal ; I am therefore convinced that M. d'Azara was wrong when he thought he recognised his *Mbaracaya* in one of the Tiger-Cats of the museum."

As we give a figure of this species from M. F. Cuvier, we shall not detail its specific characters, except to observe, that the ground colour of the fur is bright yellow, more or less gray. Round the lips, the throat, the under part of the body, and the interior of the thighs, is whitish.

The bands and spots are larger or smaller, and more or less numerous in different individuals. They are in general about thirty inches long, and the tail nine or ten inches Their height from fifteen to twenty inches.

It is manifest, says our author, that the pretended Ca-racal of Barbary, without pencils of hairs to the ears, with stripes and black spots which Buffon describes from Bruce, is no other than this Serval. The Cinereous Cat of Pennant and Shaw he also refers to the gray variety of the Serval.

Buffon refers to the Serval his New Spain Cat. It was of a bluish-ash colour, spotted with black in pencils. If this notice which was sent to Buffon by an anonymous author be genuine, the species must be very distinct. Pennant refers this species to Seba, f. ii. pl. 48, which is no other than a bad figure of a very young Panther or Leopard.

The *Margay* of Buffon, (*Felis Tigrina*, Gm.,) has the lines and spots of the head similar to the Chati of Cuvier. The upper part of the body is yellowish-gray, the under part white ; four black lines pass from the vertex to the shoul-ders, and then change into series of long streaks. The spots on the flanks are long and oblique ; on the shoulders there is one vertical ; on the crupper and limbs they are oval and scattered ; on the inner sides of the limbs are some transverse bands ; the feet are gray and spotless ; the tail has some irregular unequal rings, to the number of twelve or fifteen ; the middle of the lateral spots is paler than the edges. It measures about twenty inches in length, and the tail is about eight or nine.

The individual described by Buffon came from Cayenne. The Brasilian word *Margaia*, is the root of Buffon's appel-lative Margay.

D'Azara describes, says the Baron Cuvier, an animal under the name of *Mbaracaya*, which the Baron for a long time took to be the Serval, but which he now believes to be an adult Margay.

The *Kenouk* or *Javan Cat* of Dr. Horsfield, comes next

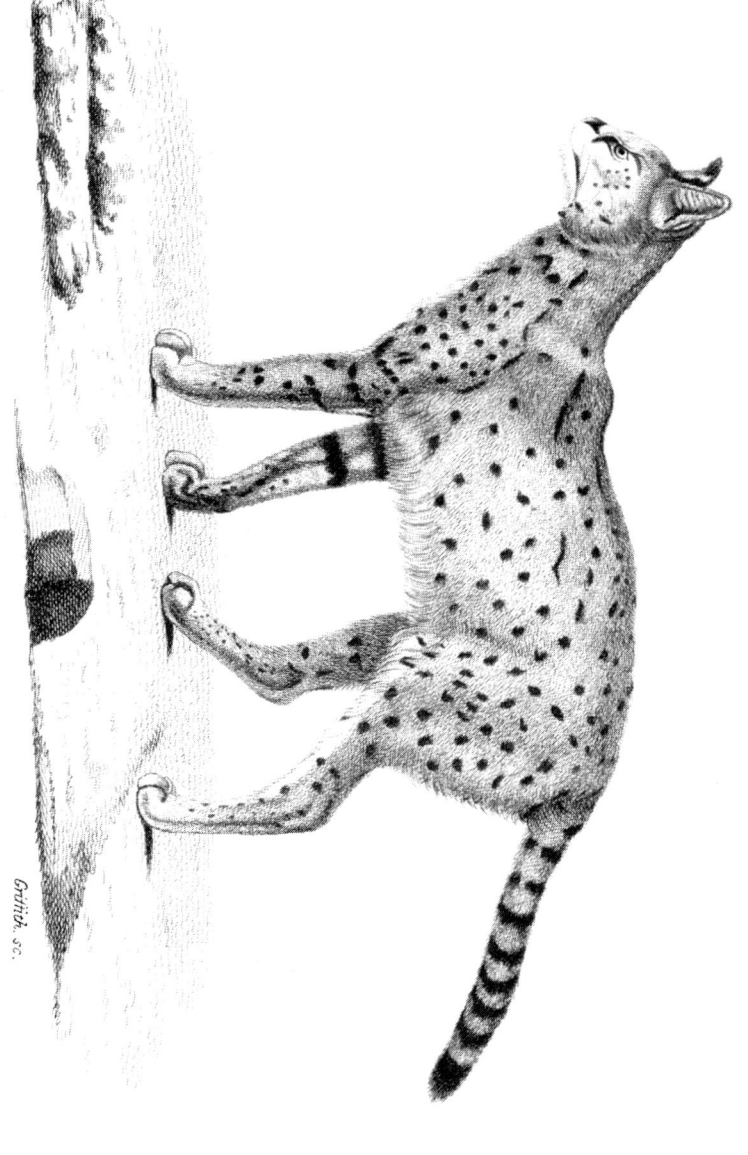

THE SEEVAL.

OF F. CUVIER.

Griffith. sc.

ın the Baron's catalogue of these species. The general colour of this animal is grayish-brown, exhibiting on the body, neck, and limbs, a delicate mixture of gray of different shades. The upper parts are more intensely coloured and inclined to tawny; the throat, cheeks, fore-part of the neck, the breast, belly, and tail underneath are whitish. Although it resembles the common Domestic Cat in many points, the smallness of the ears, and their distance from the eyes, give to its front a different appearance; the form of the body is likewise more slender.

Four regular series of elongated spots pass from the head to the tail of this species, and the sides are covered with regular smaller spots, decreasing in size and intensity of colour as they approach the central line of the under part of the body; these peculiarities, together with two transverse bands which pass across the anterior part of the throat, form its principal specific characters.

This species measures about one foot eleven inches in the length of the head and body, and the tail about eight inches, which is also about the average height of the animal.

The Kenouk is found in large forests in every part of Java. It forms a retreat in hollow trees, where it remains during the day; at night it ranges about in quest of food, and often visits the villages at the skirts of the forests, committing depredations among the Hen-roosts. The natives ascribe to it an uncommon sagacity, asserting, that in order to approach the fowls unsuspected, and to surprise them, it imitates their voice. It feeds chiefly on Fowls, Birds, and small Quadrupeds, but in case of necessity it also devours carrion.

This animal, says Dr. Horsfield, is perfectly untameable; its natural fierceness is never subdued by confinement. The same character is given to the Bengal Cat; but it has not the disagreeable odour ascribed to that species, nor does it frequent reeds near to the water, to feed on Fish, Snails, and Muscles.

The *Rimau Bulu* of the Malays is described and figured by Dr. Horsfield, under the epithet *Sumatrana;* he says it is one of the various species of Felis which are found on the Island of Sumatra.

It is really quite disheartening to the Tyro in zoology to be told of the various species of one genus of mammalia to be found in a single island. Either any thing like perfection in a catalogue of animals is perfectly unattainable, or the species are unduly multiplied. It cannot be doubted on the authority of such an observer and such a scholar as Dr. Horsfield, that there are strongly-marked external differences in various individuals and, perhaps, races of the genus ; but are not all these analogous to what the French call the *Chien des Rues*, the endless varieties of the Dog, and will they not all breed together ? If so, these external characters on which their distinctness is founded, will be found to be evanescent, and the zoologist be constantly doomed to the more difficult task of unlearning much of what he has acquired.

We shall merely observe that this particular Sumatra Cat, or the Rimau Bulu, is about the size of the prece ng, the colour more yellow, and the spots blacker; these are also much more irregular both in disposition and shape. Our author seems to refer it to the Bengal Cat of Pennant.

There appears, however, to be in Java another wild species of the Cat, much larger, and very remarkable for the beautiful regularity of its spots, which our author names from M. Diard, its describer, *Felis Diardi*. Major Smith has long had it in his collection; and we have also a drawing of it.

Its size is nearly that of the Ocelot. The ground colour of its fur is yellowish-gray. The throat and back are co-

1. TIGER CAT OF JAVA .   2 . CAPE CAT OF FORSTER .

FELIS DIARDI. Cuv .   *FELIS CAPENSIS.*

London Published by G.B. Whittaker. June 1825.

vered with black spots, forming longitudinal bands. Other similar spots descend down the shoulders perpendicular to the preceding. On the thighs and part of the flanks are black rings, open, with the centres gray, and upon the legs are black full spots. The yellowish-gray and the black of the tail form some dusky rings.

Mr. Burchel, in the second volume of his travels in Africa, describes, from at least fourteen skins he met with, a species of South African Cat, in size not larger than the common domestic species. The general colour is tawny, fainter on the under parts, but entirely covered with black spots, rather long than round, neither annulated nor ocellated. A few of the spots on the back of the neck are sometimes elongated into stripes, while those on the fore-part of the shoulder join and form very black transverse stripes or irregular bands, of which several surround both the fore and the hind legs. In some older individuals the upper spots seem faded nearly to a brown. All these marks on the lower part of the body are extremely black; and the under parts of the feet are the same, whence Mr. Burchel appears to have named the species, *F. Nigripes.*

The tail is of the same colour as the back, and confusedly spotted, at least, to four inches from its base; but it was in no part annulated.

The top of the head is of a darker colour than the body. The ears ovate, obtuse, and of an uniform grizzled dark brown covered with very short close hairs, the anterior edge being furnished with upright white hairs as long as the ear itself. The hair over the eyes is whiter; the cheeks are of the same colour as the sides; and the whiskers are white.

This, though not bigger than the Common Cat, may perhaps be placed in the little group of Servals; but as the length of its tail is not ascertained, its proper place in the genus must remain doubtful.

We have seen that the Baron, in a note on the text

(v. i. p. 46,) treats the Cape Cat of Dr. Forster as a Vivera. He since, however, has been inclined to consider it as a young specimen of the Serval.

Shaw has figured this Cat from one of two drawings by Dr. Forster, now in the Banksean library; and, as much uncertainty exists in regard to it, we have copied both these drawings, in order that they may be compared with the rest. The head in profile has a peculiar appearance, and if correctly drawn, stamps it with an originality that can hardly be mistaken should other specimens again occur.

The group of this numerous family of the Felinæ, which has the common Domestic Cat for its type, must receive but a brief notice, to which shall be added one or two figures, which seem to demand some attention.

D'Azara describes three species of Wild Cats, found in America: first, the *Yagouaroundi*, the colour of which is uniform, and without spots; each hair is annulated, black and white; but as the tip is always dark, this colour prevails. It is very savage, and inhabits the borders of the forests. This has been already noticed among the group of Ocelots.

The *Eira*, which is of a clear red colour, with white whiskers, and a white spot on each side of the nose.

And the *Pageros* or *Pampa Cat*, which has the upper part of the body a clear brown, and the lower parts white, with transverse stripes.

To which is added, in his Travels, the *Negre*, or *Black Cat.*

We are enabled, by the kindness of Major Smith, to present a figure which seems to refer to the second-mentioned of these. Indeed, it so nearly accords with the short description d'Azara has given of his Cat, as to leave little doubt of its being the Eira.

This is a miniature Couguar; and the drawing from which Major Smith copied it, is the original whence Margrave and

C. Hamilton Smith Esqr del.^t

1.   2

CAPE CAT OF FORSTER.   THE YAGOUAROUNDI OF D'AZARA.

London, Published by G. B. Whittaker, Decr 1.st 1826.

1

THE EIRA OF D'AZARA

2

THE WILD TORTOISESHELL CAT

OF HOFFMANSEGG.

C. Hamilton Smith Esqre del.^t

I. Landseer

London Published by G.B. Whittaker. Decr 1.st 1825.

Piso have taken their figures, and Buffon his name of Couguar. It is deposited in the very curious collection made
under the eye of the celebrated John Maurice, Prince of Nas-
sau (commonly styled Prince Maurice), who commanded the
Dutch forces in Guiana in the seventeenth century, which
collection is now in the royal library at Berlin. The figure
is as here represented, with two names, Cuguaçuarara, and
Cuguaçuguarana, above; and in the Prince's hand is
written, " *sehr furios, und nicht grosser als ein kleine katze,*"
very furious and not larger than a small Cat. This figure
is copied in oil in another book, with the same names, and
a note of Margrave, who, by some mistake, has confounded it with the South American Couguar, or Puma;
and in examining the description, he has extracted the
word Couguar out of the Brazilian denomination. D'Azara,
who describes this animal, states that there is some uncertainty as to its name; but he believes that this, as well as
his Yagouaroundi, is known by the name Eira. The name
Haira was also given to a species of Wild Cat sent to Buffon
from America. The original drawing whence the figure
was taken corresponds so exactly with D'Azara's animal, as
to leave no doubt of its identity; while the note upon it renders it at least prudent to adopt the name of Puma, and
to drop that of Couguar, for the animal vulgarly known by
the name of the American Lion.

We have also engraven the figure of a beautiful stuffed
Cat in the Museum of Erlangen, brought from South America by Count Hoffmansegg. It is almost two feet long,
and the tail ten or eleven inches. The hairs are extremely
soft, long, and silky. The ground colour is white, but the
animal is variously clouded with shades of brown and yellow. This seems to be the Spanish or Tortoiseshell Cat in a
wild state, or it may be the Pageros of d'Azara.

The *Common Cat* is said to be originally from the forests

of Europe. In the savage state it is of a brown-gray colour, with transverse deeper stripes ; the tail has two or
three dark bands, and the extremity is black. The genuine
Wild Cat is to be found in the remote parts of Great Britain, and may be called, as Mr. Pennant observes, the
English Tiger. Its manners are similar to those of the
Lynx, living in woods, and preying during the night on
every animal it can conquer.

In a domesticated state, the Cat varies greatly in colour
and the length and fineness of the hair, but much less than
the Dog ; nor is it so submissive or capable of attachment
as the latter animal, ever retaining much of its primitive
ferocity, perfidy, and cruelty, and never entirely to be
trusted. It has, however, a considerable and blind attachment to its domicile *.

However prevalent and however inexplicable the tendency
to variety, particularly in domesticated animals, many facts
seem to evince that nature seems as it were unwillingly
forced into the operation, as may be observed in the domesticated Cat; when, as is sometimes the case, it escapes
from the society of mankind and returns to its primitive
mode of life, its offspring soon return to the dark striped
character of the ordinary wild species.

It is observable also, that such varieties of the Domestic
Cat as differ most in appearance from the common wild
species are proportionally more different in manners and
habits. The former sort will eat occasionally vegetable
food, which the wild varieties are not known to do; a na-

* It must be observed, however, that the disposition of the Domestic Cat depends materially on its treatment. Cats are a perse-
cuted race, but when treated with kindness, are very nearly as
capable of personal attachment as Dogs. I am also inclined to
think that the tortoiseshell, and lighter varieties, are of a gentler
disposition than the others.—P.

tural or physical indication of which is to be found in the intestines of the domesticated, which are longer than those of the common wild species.

The varieties of the Domestic Cat are considerable in number : as the Brinded Cat, with black feet and annulated tail ; the slate-coloured or blue-gray, called the Chartreuse Cat ; the tortoiseshell or Spanish Cat ; the white or slate-coloured, with long fur, called the Persian Cat ; and a beautiful long-haired species, called the Angora Cat, which is remarkable for sometimes having one eye blue and the other yellow ; the Red Cat of Tobolsk, mentioned by Gmelin ; the Pendant-eared Cat of China ; and the Pensa Cat, described and figured by Pallas in his Travels, which, indeed, seems likely to have been hybridous, though it was prolific.

There is also, according to Sir S. Raffles, a variety of the Domestic Cat peculiar to the Malayan Archipelago, and remarkable for having a twisted or knobbed tail, in which particular it agrees with that of Madagascar. Sometimes it has no tail at all. This coincidence with the Madagascar variety, says Sir Stamford, is the more remarkable, as the similarity between the language and customs of the inhabitants of Madagascar and of the Malay Islands has frequently been a subject of observation.

There is also an hereditary variety of the Cat in this country, which is without any visible tail. It is not uncommon in Cornwall ; and Dr. Leach received one from the Isle of Wight, which, however, could not be reconciled to its new habitation.

It appears by the *Bibliothèque Universelle*, that a hybridous race has lately been propagated between the Domestic Cat and the Pine Marten, which, contrary to the more ordinary course of nature, is prolific ; and as these animals are said to breed freely, they seem likely to become a distinct hereditary species. They appear to have more of the character of the Marten than the Cat, as the snout is elongated, and the claws are not retractile ; but they are perfectly domesticated, and the fur is very fine. The teeth are not described. The account must, however, be taken with caution ; as, although the animals in question partake as much or more of the character of the Marten than of the Cat, the original intercourse which produced them is merely supposititious. May not this be the Pensa Cat mentioned in Pallas's Travels ?

The fur of the Cat, when dry, will yield electric sparks by rubbing ; and if the animal be placed on an electrical stool with glass legs, and rubbed for a short time in contact with the wire from a coated jar, the jar will be effectually charged with electric matter.

Cats dislike being wetted, and are averse to many scents ; but they are passionately fond of the smell of the valerian root.

Such as have lost their young have been known to transfer their maternal affection to Leverets, young Squirrels, and even Rats. These and similar facts, equally common and notorious, as the maternal affection of birds for the young of a different genus which they may have hatched, evince a special interposition of Providence for the propagation of animals, and show that what we call instinct is totally different from any thing like reason ; and is in fact an impulse acting on animals independent of volition, for the most important certainly of all purposes to them, their preservation.

After the Cats generally, may be placed the LYNXES, or Cats with ears terminated with a pencil of hairs ; their size is moderate, and their tail generally short.

The *Caracal, Siagous,* or *Lynx* of Barbary and the Levant, *(Felis Caracal,* L.,) is distinguished by its uniform vinous red colour ; by its ears, black without, and white within :

and by its tail, which reaches to the heel. There is some white above and below the eye, round the lips, under the jaw and throat, as well as under the body, and on the inside of the thighs. A black line passes from the eye to the nostrils, and there is a black spot about the whiskers. It is about eighteen or twenty inches high, and about two feet six inches long.

The Long-tailed Caracal of Edwards and Buffon does not differ from this, as the Baron informs us. The first Caracal of Buffon was mutilated as to the tail.

It appears to be the Caracal that the ancients have frequently named Lynx, for Pliny says, l. viii. ch. 30, that the Lynx is a native of Ethiopia; and Ovid says it comes from India:

" Victa racemifero Lyncas dedit India Baccho."

Elian, l. xiv., ch. 6, gives him pencils of hairs to the ears. Oppian, Cyneg iii., v. 84, who makes two races, the small red and the large yellow, does not mention the spots; and the Lynx of the Mosaic of Palestine is drawn with a long tail

We may nevertheless conclude that the name was applied sometimes to the Common Lynx:—

" Monstrate mearum
Vidistis si quam hic errantem forte sororum
Succinctam pharetrâ et *maculosæ* tegmine Lyncis."

Probably, however, Virgil may have supposed that the Lynx was like the Panther or Leopard, and the other animals consecrated to Bacchus

The name Caracal is from the Turkish *kara*, black, and *kulach*, ear. The Persian name, *sia-gusch*, has the same meaning, *sia*, black, *gusch*, ear.

The *Chaus, Booted Lynx,* or *Lynx* of the Marshes, (*F. Chaus,* Guld.) is intermediate in size between the Common Lynx and the Wild Cat, and in the length of its tail between the Caracal and Common Lynx. It is yellowish-brown above, with some deeper shadows, lighter on the breast and belly, and whitish on the throat; the limbs and cheeks have a yellowish tint; two black bands mark the arms and thighs. The tail reaches the calcaneum, whitish towards the point, with three black annuli; behind the paws is black, like the tips of the ears, but the rest of the convexity of the ear is yellow.

This is the Booted Lynx of Bruce, whose individual appears to have been a small one. It is found from Barbary to India.

The *Common Lynx* appears to be so much subject to variety, or to be so nearly allied to other species, that it is extremely difficult to discover its constant specific characters from those that are subject to change. It is about twice the size of the Wild Cat; the back and limbs are generally bright red, with blackish-brown dots; round the eye is white; three lines of black spots on the cheeks join a large black oblique band on each side of the neck under the ear; the fur of these parts, longer than elsewhere, forms a sort of lateral beard. The forehead and top of the head are dotted with black. On the top of the neck are four black lines, and in the middle one irregular and interrupted. The dots form two oblique bands on the shoulders, and transverse bands on the fore legs. The feet are yellow and spotless, but the tarsus of the hind feet has a brown band.

The convexity of the ears is black at the base and tip, ashy in the middle; the tail is yellow-white underneath, and dotted with black like the back.

Others have the spots only a little deeper red than the ground colour; the upper part of the tail red, the under part white, and the tip black, as that of Buffon, tom x., plate 21.

The Swedes acknowledge considerable differences in the Lynx, from which they make one race with black spots under the name of Cat Lynx, another with pale spots under the name of Wolf Lynx, and a third with bands, under that of Fox Lynx. Linnæus at first separated, but afterwards united them. Retzius considers the two first as specifically different.

The physiognomy of the Lynx is rather gentle than savage; and, indeed, it is said to be less ferocious than most of the species of this genus. It walks and leaps or bounds like a Cat, and hunts Wild Cats, Martens, Ermines, Squirrels, &c., pursuing them up into trees, where also it will lie in wait to drop on Deer, Goats, &c., that may pass beneath. It is sanguivorous; and having seized on a prey, is said frequently to suck the blood, and then leave it for another victim; whence it has been asserted, that the Lynx has the least memory of all animals. Its skin is changed by climate and season; and in high latitudes, particularly in winter, the fur is much finer and thicker, and more esteemed.

Why the treacherous Lyncus should have been transformed into a Lynx, and this animal be in consequence held up *in terrorem* to the world as an example of perfidy, is not stated by Ovid, who, while he relates the tale, " Lynca Ceres fecit," like a true chronicler, abstains from all comment. A namesake of the Scythian king, Lynceus the Argonaut, who, by-the-by, was a sheep-stealer or something worse, appears also to have been in some way allied to this animal, in the opinion of antiquity; as the powers of vision of both were considered equally extensive and surprising, and no doubt with equal truth; but if so, the eyes of the Lynx must have suffered in these degenerate days. Other marvellous stories were also told by the ancient naturalists of the Lynx, which have gained credit in later times with the vulgar, and with those who are easily credulous, and too idle to seek for truth at the expense of trouble.

The Lynx was formerly spread over the Old World, was

common in France, and has but recently disappeared from Germany. It is found in Spain and in the north of the European continent, but it is not yet certain whether it inhabits Africa.

America produces certainly two species of the Lynx, and, probably, not more. One of these is gray, with the end of the tail black. This is the Canada Lynx of Buffon, Supp. iii., pl. 44, and his Lynx du Mississippi, of Supp. vii., pl. 33.

Some individuals have the fur so thick and long, especially on the paws, that they have a very different appearance from the European Lynx. The fur is in general yellow, with the points white, which makes the general colour grayish-ash; on the back the bottom of the hairs is blackish, which gives a general brown tint. The blackish band on each side of the neck is nearly effaced, and there is no black at the base of the ears. The head and body are nearly three feet long, and the tail about four inches, and it is about two feet high.

Others have less fur, are rather smaller, and shew the darker colour more distinctly. We suspect, indeed, that in regard to the fur, the same individual varies considerably with the season.

The other, or *United States' Lynx*, is of the size, and has the form and distribution of spots, of our first European species. The ground colour is gray; its spots are more numerous, deeper on the back, and paler on the sides and limbs. They, however, vary in number and size. The tail has four black rings and four gray.

The above observations on the species, by which it will be seen their number is very much limited, are almost entirely from our author in his *Ossemens Fossiles*. The table will shew the claimants to distinct species of this group.

In the clear definition of species, the great goal of Zoology, no branch of it, perhaps, is more imperfect than that

THE LYNX OF SIBERIA.

London, Published by G. B. Whittaker Sep.r 1825.

C. Hamilton Smith Esq.r del.t

of the PHOCÆ or SEALS ; nor when we consider the existing state of ignorance in relation to so many other Mammalia, more, in fact, within our reach than these marine animals, can we be surprised that but little should be known about them. Governments, societies, or individuals of wealth and power, may send out men of science to explore the most distant countries ; and scientific zeal may stimulate others to investigate the wonders of nature, in her most sequestered recesses, but we have not the means, except by deduction and analogy, of ascertaining the habits of these half amphibious animals, while procuring their sustenance at the bottom of the sea ; nor have we often, or in an efficient manner, the opportunity of watching them in their favourite haunts, the isolated steril rock, or the most retired and deserted strand.

Many reasons seem to concur in pointing to the situation of the Phocæ, in artificial arrangement, as among the other marine Mammalia ; it seems an easy transition from them through the Dugong to the Cetacea, and, in fact, Illiger has so arranged them under the general name of Pinnepedie, but it is far from our wish to invent new systems, or even to reform the old ; we shall merely observe, that such a transposition of these animals, as that alluded to, might, perhaps, be made with advantage to the general symmetry of the Cuvierian system.

Until very lately, the Seals were not supposed to present any very decided physical grounds of diversity. The presence or absence of an external ear, no very influential character, had, indeed, been employed to separate the Seals, properly speaking, from the Otarys, the former, as was supposed, wanting the external ala, and the latter having it ; and a further sub-division of these Seals, properly speaking, has been still more recently made, by a separating those species which had an elongated snout, or a cutaneous appendage to the head, from such as had neither.

In a physiological point of view, these differences become almost unimportant ; nor are they strictly grounded, in fact, for the Common Seal has, in reality, an external conch, very small, and hidden, it is true, but very distinct and perfect.

One organic distinction, of a general and influential character, it is, however, now ascertained, does exist among these animals ; and whether this alone will be found to divide the whole of them into two sub-genera, or more equally influential, will arise to furnish the foundations for other sub-divisions, time and research must determine.

The character alluded to is that of dentition. In the Bats, we have had occasion to observe that the two leading systems of dentition prevailing in these animals, seem to separate the frugivorous from the remaining insectivorous species; and we shall have occasion to notice that the Marsupiata, which belong to the same order, are very decidedly divided by the characters of the dentition of several groups or sub-genera. So in the Phocæ before us, some have the cheek-teeth of a sharp, pointed, cutting character, while others have them conical or obtuse. True it is, that the external part of the cheek-teeth, however variously destined, and however different in appearance, present, when critically examined, rather an easy transition than a positive change; the tubercles in the one case are more or less developed, and more or less rounded or acute than they are in the other. But in the points of difference, which separate the two divisions of the Seals, now under consideration, there is another material character in the root of the cheek-teeth, inserted in the jaw. In the first division, the root, or rather roots, are several ; in the second, there is but one, as in the cetaceous teeth. Hence, they may be conveniently distinguished ; first, by having the cheek-teeth with several tubercles, more or less indented, more or less sharp or obtuse, but always with several roots: or, se-

TEETH OF SEALS.

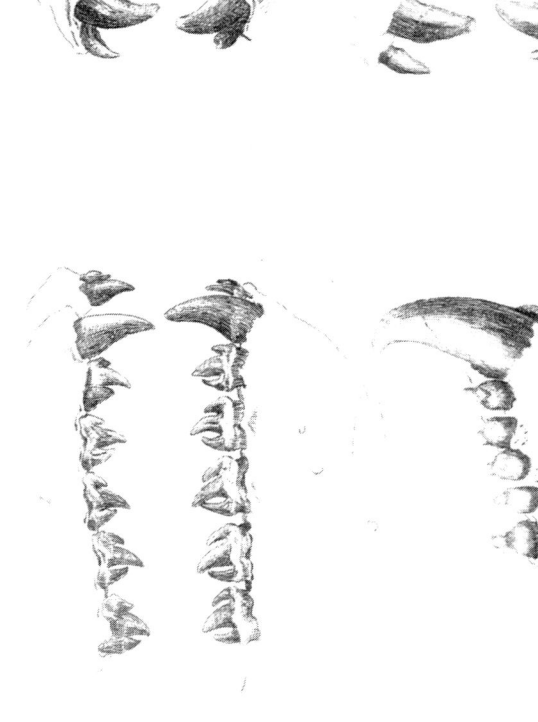

1 *P. leptonyx* ⎫
2 *P. vitulina* ⎬ with several roots.

3 *P. proboscidea* ⎫
4 *P. ursina* ⎬ with simple roots.

London, Published by G.B.Whittaker March 1827.

condly, by having the same teeth, however tuberculated, provided simply with a single root, like those of the cetacea, and, therefore, called cetaceous-teeth.

It will save a great deal of description which, after all, can never convey a perfect idea on this subject, to refer to the plates of teeth of the Phocæ ; and though we cannot very conveniently delineate that part of the teeth which is buried in the jaw, that is the root, it will be observed tha-those species only are put in one plate, which have the cheek-teeth furnished with several roots, and those, alone, in the other, which have the simple root, or cetaceous formation.   The external parts of all of them will be immediately observed to present much diversity, nor can it be supposed that such diversity can be unaccompanied with analogous differences of impulses and habits.

The harmony of nature is a constant theme of observation to the reflecting naturalist, and in no particular can that harmony be more universal, or, indeed, more essential, than in the physical adaptation of the organs appropriated to supply food to the animal, and in the impulses and passions of the same : without adverting at all, therefore, to the peculiarities of the root of the cheek-teeth of the several species which compose the Phocæ, it is impossible to contemplate the differences which distinguish their external portion, without feeling satisfied, from analogy, that the pursuits and habits of these several species must be as various as are these physical differences.   Our knowledge of the differences of the modes of life of these animals, by no means accords with what we now know of these differences of conformation ; we cannot, therefore, but conclude that we have very much, as yet, to learn in regard to the manners and habits of the Seals.

Little more is known of the Common Seal, though an inhabitant of our own seas, than of those which are met with in the most distant latitudes. The high intellectual qualities of the Seal, were, however, observed and appreciated by the ancients. Diodorus, Ælian, and Pliny, speak of them at some length; and all travellers and naturalists, who have treated of the Seals, since the resumption of Zoological studies in Europe, have related additional proofs of them. Notwithstanding the numerous facts known on this subject, the analogy existing between the intelligence and organization of these animals has not yet been established, though a point of the highest consequence, and one, without knowing something of which it is impossible thoroughly to appreciate the moral nature of any intelligent being. M. F. Cuvier, anxious to supply the desideratum on this subject, paid a more than ordinary attention to three Seals, in possession of the French Menagerie, and we shall give our readers the result of his observations in as few words as possible

These animals were very young, and differed little in magnitude. They were about three feet in length. On coming out of the water, they were not of the same colour as when dry. In the first situation, the black spots on the back were much more visible than in the second, and the ground-colour of the coat was gray in one instance, and a deep yellow in the other.

The black spots were more or less extended in the different individuals, and the under part of the body more or less pale than the upper. But, in all, the spots united along the spine, and formed a broad dorsal line, extending from the lower part of the head to the tail. One individual, of a fawn-colour, had an additional black spot upon the neck, in the form of a crescent, which was distinctly visible in every position of the animal, and its head was continually surrounded by a circle of oiled hairs, announcing in these parts the presence of some peculiar glandular organ.

THE SEAL.

PHOCA VITULINA L.

London. Published by G.& W.B.Whittaker. Sep.ʳ 1824.

These differences, probably, appertain to the distinction of sex, as this fawn-coloured Seal was a male, while the gray specimens were females.

The hairs are all silken, flat, pointed, harsh, and compact. Their length rarely surpasses six or seven inches. The skin secretes an oily matter, which contributes to secure the animal from the effects of humidity.

The Seals have five toes on the fore-feet, perfectly free, and five on the hind, united by a membrane, which constitutes them genuine oars, and both are armed with nails. The hands are the only parts of the anterior limbs which are external. The hinder limbs are parallel with the trunk, and are visible only from the calcaneum.

These animals are not less remarkable for the form of their sensitive than that of their locomotive organs. A short muzzle, orbits without brows, a broad front, and an immense and rounded cranium, give them a physiognomy not to be found in other Mammalia, &c. Their eyes, large, round, and parallel with the head, have a pupil like that of the domestic Cat. It dilates into a broad disk, in a feeble light, but contracts in the open day. The eyelids are narrow, and seldom completely close.

The animal does not appear to have occasion to clean the surface of the eye so often as the other Mammalia. When these organs move, the skin of the forehead and cheeks form wrinkles, which show that the fleshy pannicle takes a part in this motion. The third eyelid is tolerably developed, and perfectly visible, but the animal would seem but rarely to use it.

The nostrils, situated behind the end of the muzzle, have two longitudinal apertures which form nearly a right angle. They are seldom opened, except when the animal is desirous of expelling the air from its lungs, or introducing fresh. They then assume a circular form. Respiration in the Seal is very quick, and extremely unequal, and often performed after very long intervals. There is generally from eight to ten seconds between each inspiration, and this function is sometimes suspended for half a minute without apparent inconvenience. It would seem that the nostrils are habitually closed, and that the act of opening them is attended with some effort. The quantity of air, however, that enters the lungs, must be considerable, to judge from the motion of the sides, and the air expelled at each expiration. The quantity of air inspired compensates for the paucity of the inspirations, for few of the Mammalia seem to possess so great a natural heat as the Seals. These animals also have a large quantity of blood.

The external ears are but a small rudiment of a triangular form. They are situated below the eye, a little in the rear. The bony part, however, is in the same place as in other Mammalia. This rudiment is closed when the animal dives. The tongue is soft, the lips extensible, and in the mustachios apparently resides the greatest sensibility of touch. They communicate with nerves remarkable for size, and in which the slightest touch produces a sensation.

The teeth of the Phocæ are very peculiar. Six incisors in the upper jaw, and four in the lower. Canines in number and form like the rest of the Carnassiers. Five molars on each side of the two jaws, trenchant, triangular, and analogous to the false molars of the other Carnassiers, except that they are a little thicker at the base, and the edge is more sloped. The first of these teeth is smaller than the others, and placed immediately at the base of the canine.

These observations were made upon young Seals. The molars of the adult are probably more numerous, and, in fact, Lepechin enumerates four more.

The structure of this animal's limbs evidently shows, that it was intended to live in the water, and all its movements on land are slow and painful. It seldom uses its paws but for swimming, and unless it climbs, it never uses them for locomotion on land. When it wants to walk, it presses the hind and fore parts of its body alternately on the ground, bending its back upwards, something like a Snail. The paws are inactive in this operation. In climbing, however, the claws are used with good effect. The hind feet are of use only in swimming, and they are always in requisition for that purpose.

When the Seal is disposed for rest, he stretches himself on one or the other side of his body, and the head is habitually withdrawn between the shoulders.

The senses in these animals seem to possess no great degree of acuteness, a fact which we should have been led to pre-suppose from their mode of life, consisting, as it does, in a state of almost continual repose. Their sight is perhaps the best; they can see at some distance, but better in a weak than strong light, and they do not seem to be able accurately to distinguish forms. To this conclusion our naturalist was led, by their constantly approaching to partake of food, which they as constantly rejected, and the form of which was totally different from that of the only aliment they would taste. Hearing must be but a feeble sense with the Seals, as the external organ is so little developed, and as the animal passes so large a portion of its life at the bottom of the water with the ears closed against the vibrations of sound; the want of exercise must combine to render this sense defective.

Were we to judge solely by the external organ, these animals would seem to derive no greater advantage from the sense of smelling, than from the senses of which we have already spoken. The nostrils, like the ears, must remain closed when the animal is excluded from the external air, and as it pursues its prey in the midst of the waters, it certainly cannot employ its scent in the usual manner, to discriminate and select it. Notwithstanding this, if the cornets of the nose have any influence over the sense of smelling, the Seal should be able to discriminate with ease the weakest odours, as there is no animal at all comparable to it for the complexity and convolution of the cornets. But one mode of smelling can remain, and that is, to retain the odorous emanations of bodies enclosed within the mouth in contact with the pituitary membrane, thus introducing them into the nose through the medium of the palate.

This conjecture will not appear ill founded, if we consider what very little service these animals must derive from the sense of taste. Their mastication goes no farther than to reduce the fish to such dimensions as may render it barely capable of passing the larynx and the œsophagus. To produce this effect, they generally confine themselves to pressing the fishes between their teeth, not so as to divide them in pieces, but merely to contract them in size. Sometimes they will tear their prey with their claws, but they are often observed to swallow it entire, even when apparently it is too large for their mouths. Thus they are frequently compelled to raise their heads to facilitate the operation of deglutition, so that the weight of the aliments may contribute to make them slide into the œsophagus and stomach, and favour the efforts of the muscles.

There is little to add on the subject of the sense of touch. It is evident that the Seal must have very limited notions of such qualities of bodies as are transmitted to our understanding through the medium of this sense. It employs it more likely to ascertain the presence of objects, and to appreciate their form, dimensions, or solidity. This is an object which must be accomplished in the most suitable manner by the mustachios for a carnivorous animal, which most generally cannot be informed of the presence of its prey by sight, by hearing, or by smell.

We have already observed the imperfect manner in which mastication is performed by the Seals, and how they swallow bodies without chewing them. Nature has not only provided them with the means of distending excessively all the parts through which the aliments must pass, but has also supplied them in abundance with a viscous saliva, which fills the mouth to such a degree, that during deglutition it escapes in long threads, and this phenomenon is also observed to take place, even when the Phoca only perceives its prey. While mastication and deglutition are performed by these animals on land, it is not possible that they should meet with any impediment; but it frequently happens, that the Seals devour the prey which they have caught at the bottom of the waters, and it is difficult to suppose that, in this case, the operation can be performed in the same manner as in the other. In fact, when the fish is on land, the Seal seizes it with its teeth, crushes and swallows it, letting it fall, as it were, into the stomach, rather than directing it thither; but in the water it catches the prey by a sort of suction. It does not completely open the mouth. It only separates the extremities of the lips, lowering a little at the same time its under jaw. The fish is thus drawn into the vacuum left in the mouth, if it presents itself suitably for that purpose, by the head, the tail, or the point of the fins.

Should it present a large surface, surpassing the little aperture in the mouth, the Seal is obliged to adopt new measures, and to attack it again. There is another point upon which we are destitute of information. The Seal must swallow its prey under the water. If this operation were performed as on land, it is obvious, that along with its food, it must of necessity swallow much of the liquid element. How this is avoided, the observations of naturalsts have not, as yet, enabled us to ascertain.

The voice of the Seal is a sort of barking not unlike that of a Dog, but much feebler. The Seals on which these observations were made, barked usually of an evening, or on change of weather. They exhibited anger by a sort of hissing, resembling the swearing of a Cat.

What has been said concerning the organs of the Phocæ, must be sufficient to leave little doubt concerning their imperfection, and were we to judge of the intelligence of these animals from the facts now stated, we should not hesitate to regard them as the most stupid and brutal of terrestrial Mammalia. Those Seals, however, with members so imperfect, with senses so obtuse, are enabled, from their few sensations, to derive results infinitely superior to those obtained by animals, in appearance, the most felicitously organized. An additional proof of the predominant influence of the brain in all that is intellectual.

The animals on which these remarks were made, experienced no fear in the presence of men, or any other animal. Nothing ever induced them to fly, except approaching so near as to excite in them the apprehension of being trodden under foot, and even in this case they only avoided the danger by removing to a little distance.

One of them, indeed, would sometimes threaten with its voice, and strike with its paw ; but it would never bite, except in the last extremity. In taking their food, they evinced a similar gentleness of character. Though very voracious, they could behold it withdrawn from them without fear or resistance. They would suffer the fish which had been just given them to be taken away with impunity, and some young Dogs, to which one of those Seals was attached, would amuse themselves in snatching the fish from his mouth which he was just ready to swallow, without his testifying the least anger. When two Seals, however, were allowed to eat together, the usual result was a combat carried on with their paws, which ended by the weakest or most timid leaving the field in possession of his antagonist.

With the exception of some species of the Simiæ, there is scarcely any wild animal more easily tamed than the Phoca, or capable of a stronger degree of attachment. One of the individuals before-mentioned, showed, at first, some degree of shyness, and fled at the show of caresses ; but, in a few days, his fear was totally at an end. He soon discovered the nature and intent of such movements, and his confidence became unbounded. This same Phoca was shut up with two little Dogs, who used to mount upon his back, bark at, and seemed to bite him ; and although sports of this kind were at variance with his habits and nature, he soon learned to appreciate their motive, and to take pleasure in them. He never replied to them, but by gentle strokes of his paw, which seemed rather intended to excite than to repress them. If the Dogs escaped, he would follow them, though walking over ground covered with stones and mud must have been a painful effort to him ; and when cold weather came, he and the Dogs would lie closely together, to keep each other warm.

The fawn-coloured individual was peculiarly attached to the person who had the care ot him ; he soon learned to know this person at any distance within his range of vision. He would hold his eyes fixed upon him while he was present, and run forward the moment he saw him approach. Hunger, to be sure, entered for something into the affection he testified towards his keepers. The continual attention which he paid to every motion connected with the gratification of his appetite had made him remark, at the distance of sixty paces, the place which contained his food, although it was devoted to several other uses, and though it was entered but twice a day for the purpose of procuring his nutriment. If he was at liberty when his keeper approached to feed him, he would run forward, and solicit his food by lively motions of his head, and the most expressive glances of his eye. This animal exhibited many other in stances of considerable intelligence.

M. F. Cuvier has since seen an individual of this species, as well-educated as any Dog could be.

Of the common species there are many varieties, differing principally in colour, but not deviating, of course, in craniological characters. That we have engraved, under the name of the Common Seal, seems to be the spotted variety, met with most commonly on the Dutch coast. Linnæus, as the Baron observes, under his Phoca Vitulina, or Common Seal, has quoted several species.

Another species, whose skin is the most esteemed for commercial purposes, is an inhabitant of the Frozen Ocean. It has been often described, but seems so much subject to variety, as to present the appearances of many distinct species. The sexual difference, also, is very great, as our figures evince, of both the male and female. This is the *Phoca Grœnlandica* of Fabricius, the *P. Oceanica* of Lepechin, (Acta Acad. Petrop. 1777,) and the *Harp* or *Heart Seal* of English traders. It is generally of a grayish-white colour ; and the sub-contiguous blotches, represented in our figure, are generally described as more regular in their conformation, forming an arch or crescent pretty complete. It attains nine feet in length.

The females and the young are covered with unequal spots, generally angular in shape, and spread irregularly over the whole body. The specimen whence our figure was taken, was very much darker in colour than the male.

The osteological characters of this species are not known. Lepechin seems to describe it as having four incisors in each jaw, and six cheek-teeth with three points ; but the Baron has observed, that his description of the dentition

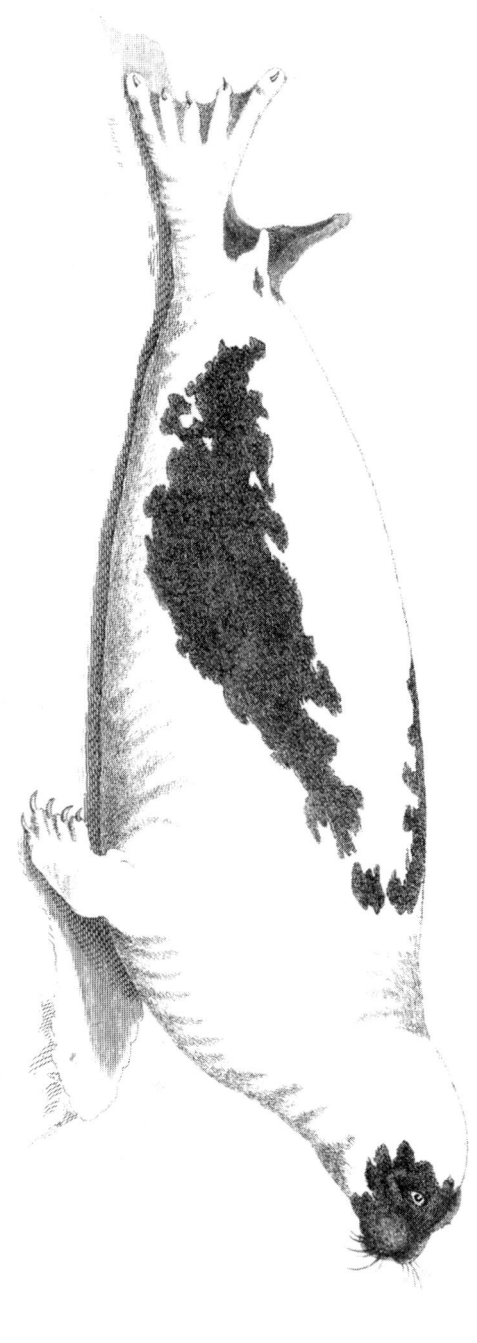

Major C. Hamilton, Smith del.

THE GREENLAND OR HARP SEAL.

P. GROENLANDICA. male.

London. Published by G. B. Whittaker Dec.r 1824.

Ma.s Prince Maximilian. of Neuwied.

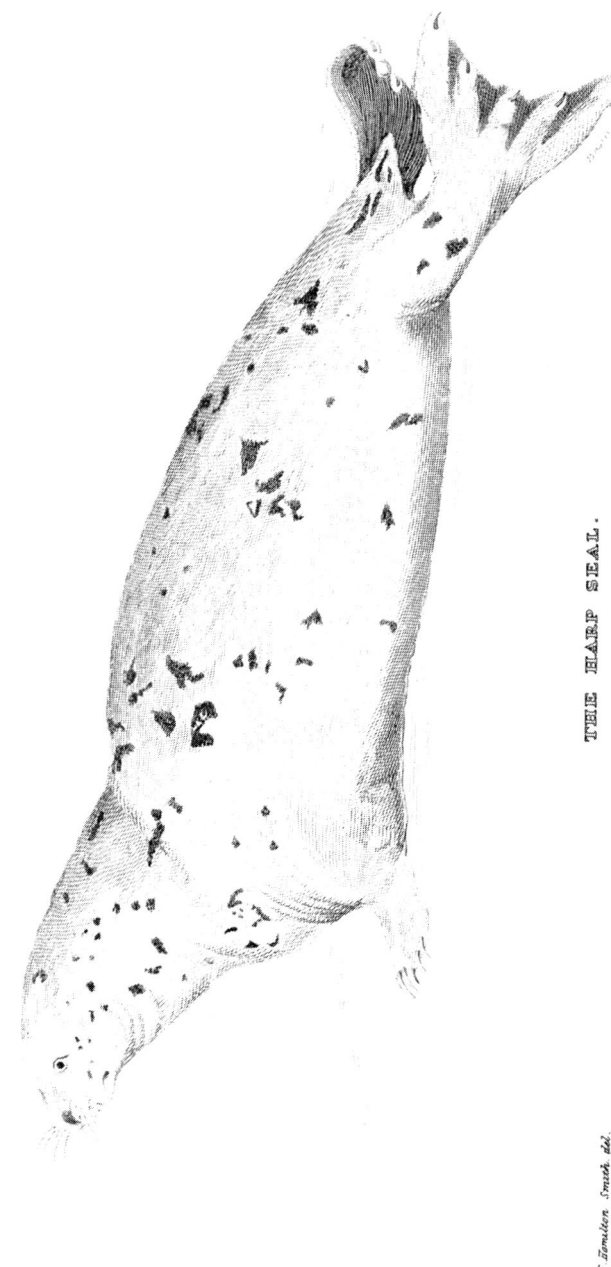

Major C Hamilton Smith. del.

Nat. Pr. Maximilian of Neuwied.

THE HARP SEAL.

PHOCA GROENLANDICA – Female.

London. Published by G. B. Whittaker Dec.r 1824.

appears to be inaccurate, from some error of the press. Fabricius gives it six incisors above, and four below, whence it has been concluded that those writers described different species.

Cuvier suspects that the head in the Surgeons' College, engraved in the Philosophical Transactions, pl. 28, and described by Sir Everard Home, is, in fact, of this species.

M. de Blainville has named *P. Leptonyx*, a new species, lately received in the French Museum, and which appears to be the same as that, the head of which has been long in the Surgeons' College here, and engraved in the Philosophical Transactions for 1822, and pl. 29, described by Sir Everard Home.

It is about eight feet long, blackish-gray, slightly tinted with yellow, becoming yellower by degrees on the sides by small yellowish spots mixed with the general colour. The flanks, under part of the body, feet, and over the eyes, are pale yellow. The mustachios short, and the nails much shorter than in other species.

The *Hooded Seal, P. Cristata*, Gm., has the power of bringing a fold of skin, placed on the forehead, forward, so as to cover the eyes, which it does when threatened, or about to be struck. This singular appendage appears to be filled with blood vessels, and to contain a vast quantity of blood ; when at rest, or drawn back, it considerably enlarges the apparent size of the neck and shoulders. For the rest of its particulars, we refer to the table.

One of the largest and most celebrated of these animals, s that described in Lord Anson's voyage, found on the Island of Juan Fernandez, and which he names the *Sea Lion*. Peron has described it since, under the name of the Elephant or Proboscis Seal.

It attains from twenty-five to thirty feet in length, and appears to be almost uniformly brown. It is the male which is distinguished by the extensible faculty of the nose.

or muzzle in the moments of anger, it then assumes nearly the shape of the short proboscis of the tapir; but according to recent observation, it is not so when the animal is perfectly at rest, the nose then being drawn back and thickened.

These animals, at certain times of the year, are said to be so excessively fat, as to resemble skins of oil; the tremulous motion of the blubber being plainly perceivable beneath the skin. A single animal has been known to yield a butt of oil, and to be so full of blood, that what has run out has filled two hogsheads. This species is an abundant source of trade in the antarctic seas.

The figure of this animal, in Lord Anson's voyage, differs very widely from that of Peron; the former has the nasal appendage or peculiarity almost in the shape of a cock's comb, and the latter represents it as a short proboscis. We have no original drawing to illustrate the subject.

The above-mentioned species may be said to be established, since their bony parts, at least, have been examined and observed upon by our author. Of the various other species, real or pretended, which have been named, we shall say nothing in this place, but must refer to the table.

The Seals distinguished by visible external ears, were noted by Buffon, but continued mixed with the earless species by subsequent zoologists, until Peron designated them by the name of Otaries. They differ from the Seals, properly speaking, in other points than that of the external ear.

Their arms are better calculated for swimming, being placed more behind the animal, which gives them the appearance of having a long neck. The fingers are more enveloped in the skin, and are destitute of nails. Their hind extremities have the membrane more divided.

Gmelin cites three species of the Otary, or Eared Seal; the *P. Ursina, P. Jubata,* and the *P. Pusilla* of Daubenton, the Little Seal of Pennant and Shaw. The resu lt of all the observations of navigators in the Pacific Ocean appears to be, that they have seen a Red Eared Seal and a Brown Eared Seal, the former of which is the Sea Bear, and the latter the Sea Lion. All the other species, or pretended species, with the exception of the *Pusilla*, are involved in doubt and uncertainty. We shall, therefore, briefly notice these three, and refer to the table for the rest, with this general observation.

The *Sea Bear, P. Ursina,* called also by M. de Blainville, *P. Byronia,* from the skull in the Surgeons' College, brought by Lord Byron, is described as growing to eight feet in length, but the female is much smaller. The greatest circumference of the body is about five feet, but near the tail, it does not exceed twenty inches. The nose projects ; the nostrils are oval. The ears are small and pointed, hairy without, but naked within. The general colour of the animal is black, but the hair of the old ones is tipped with gray, and the females are cinereous.

We are told, that these animals live in a polygamous state, and that each male has from eight to fifty females. Though they may be found by thousands on the shore, each family is perfectly distinct ; they are very jealous of their separate station, and if an individual of one family trespass on the station of another, not merely a single combat, but a general battle frequently ensues.

This species, in common with others of the genus, evinces considerable intellectual development, and exhibits a great degree of feeling and attention in the care of the young.

The *Sea Lion of Forster,* (Captain Cook's second voyage) is from ten to twelve feet long, but the females do not exceed seven or eight feet. The body is thick, cylindrical, and very fat. The head is small, and the muzzle similar to that of a great Dog, truncated at the end. The male has the head and the upper part of the body covered with thick, rough, and stiff hairs, about three inches long, of a deep yellow or tan-colour, falling over the forehead and cheeks, and forming a mane on the neck and chest, which is erected when the animal is irritated. On the rest of the body the hairs are short and soft, of a brownish-yellow, and lying close to the skin. The female has no mane, and the whole animal is covered with the soft hairs which are of a lightish colour.

The habits of this species are very similar to those of the last. Forster tells us that the voice of the male is like the roaring of a Lion or an irritated Bull, and that of the female like a Calf or a Lamb. The old males live separate from their females out of the times of sexual intercourse, but during these periods form families or societies like the Sea Bears.

As for the other species of the Seals, they are very little known, and what little is known is not particularly interesting to the general reader. Nothing can be more confused, contradictory, and unsatisfying than the reports of authors on this subject. In his Ossemens Fossiles, the Baron has done a great deal to introduce clearness, arrangement, and authenticity into the zoology of the Phocæ. To follow him, however, would suit neither the character of our present work, nor the limits to which it must necessarily be prescribed. Nor would the result be very satisfactory in regard to establishing the species. With the exception of a few, and these we have principally mentioned, his observation rather tends to throw discredit and doubt on the many species noticed than to authenticate or illustrate them.

In the table we shall simply follow M. Desmarest, as far as our plan permits. When Cuvier himself is unable to come to a determinate list of species, and of their synonyms it would be presumptuous in us to attempt it. Of M. Desmarest's Table the Baron says, that it is a very complete collection of the descriptions of Seals to be met with in

THE MORSE.

TRICHECUS ROSMARUS. Gm.

J.ᵉ Basire sculp.ᵗ

London Published by G.&W.B.Whittaker. Sept.ᵗ1824.

different authors: but from it may be judged how far these descriptions are insufficient and contradictory, and to what extent it is necessary to clear them up by actual observation.

The MORSE (TRICHEUS,) has oftèn been joined by authors to the Lamantins, Dugongs, &c., but yet it materially differs from them by one very important character. This character is the possession of posterior extremities, like the Seals, to which last animals they bear a much more general resemblance than to the first mentioned Mammalia, which the Baron, with his usual judgment and discrimination, has more appropriately classed with the Cetacea.

The teeth of the Morse are different from those of the Seals, properly so called, in form, number and position. In the upper jaw are two immense tusks or canines, arched below, longer than the head, compressed laterally, and obtuse at the extremities. Between these are two incisors, scarcely apparent, and conformed like molar teeth. Between these again in the young Morses are some still smaller and pointed. The molars are four in number on each side: their form is cylindrical, and their coronals obliquely truncated. Two between each of them fall out at a certain period. In the under jaw but four molars are observable like those in the upper on each side. There are neither incisives nor canines, and the symphysis of this jaw is prolonged like that of the Elephant, and sufficiently compressed to find room between the two tusks. The muzzle is considerably inflated, which is owing to the prodigious development of the alveoli of the tusks. The cranium is rounded. No external ears ; the body is elongated and attenuated in the hinder part. The tail is very short. The fore-feet answer the purposes of fins or oars, like those of the Phocæ. The hinder ones are in the direction of the body, and their two external toes are the longest.

The name of *Trichecus* which comes from τριξ a hair, now applied to the Morse, was given at first by Artedi to the Lamantius.

The Morses seem to live on prey like the Seals. Their stomach is exactly alike. They are found in abundance in the northern Atlantic Ocean, and also in the polar regions of the Pacific. As yet but one species is known, though it is by no means impossible, as Shaw remarks, that each of these great seas has one peculiar to itself. The difference of such species would consist in the proportional magnitude of the tusks, and their more or less convergent direction.

The only species named the *Morse* (*Tricheus Rosmarus*,) is vulgarly called the Sea-cow, Sea-horse, &c.

These are animals of a very large size, brown colour, very like the Phocæ in general form, but heavier, and closely resembling them in their way of life, They inhabit similar places, and are generally found together. Both hold equally to the land and water, mount on the icebergs, suckle and bring up their young similarly, subsist on similar aliments, and live together in numerous societies in the same manner. It would seem, however, that the Morses do not travel so far as the Seals, and are more attached to their native climate. They are never found except in the North Seas, and accordingly were unknown to the ancients, who were well acquainted with the Seal tribe.

Most travellers in those seas have spoken of the Morse, but those to whom most credit may be attached are Zorg-drager and Cook.

" The Seals and Morses," says the first of these writers, " come during the heat of summer into the seas near the Bay of Horisont and that of Klock, in troops of eighty, a hundred, and even two hundred, especially the Morses, which remain there many days until hunger forces them back into the main ocean. Many Morses are seen towards

Spitzberg. On land they are killed with lances. They are hunted for their tusks and fat. The oil is nearly as much esteemed as that of the Whale. Their tusks are also very valuable. The interior of these teeth is considered more valuable than ivory, and is of a substance harder and more compact in the larger than in the smaller teeth. A moderate-sized tusk weighs three pounds, and a common Morse will furnish half a ton of oil. When one of these animals is encountered on the ice or in the water, the hunters strike him with a strong harpoon, made expressly for the purpose, which will often glide harmlessly over his thick and hard skin. When it penetrates, the animal is drawn towards the vessel with a cable, and then killed with a lance peculiarly formed. He is then dragged to the nearest land, or flat iceberg. They then flay him, throw away the skin, separate the two tusks from the head, or simply cut the head off, cut out the fat, and carry it to the vessel."

The female brings forth in winter, but one at a birth.

Some say these animals eat the shell-fish at the bottom of the sea. Others assert that they only eat a sea-weed, with large leaves, and are not carnivorous. Buffon thinks these opinions ill-founded, especially as the animal never eats when on land, and is driven back to the sea by hunger.

The form of the molar teeth would indicate the Morse omnivorous ; but its stomach, like that of the Seals, simple and membranous, would shew that it lived in the same way as these animals.